Mining Principles
and
Practices

Mining Principles

and

Practices

Dr Nishakar Panda

KALPAZ

Published by

Kalpaz Publications
C-30, Satyawati Nagar,
Delhi – 110052
Ph.: 011-47034999, 9811692060
www.kalpazpublications.com
E-mail: kalpaz@hotmail.com

Distribution Network
gyanbooks.com
India, USA, Canada, UK, Australia, France

ISBN: 978-93-5722-695-0 HB
First Published, 2023

Printed at: Gyan Press, Delhi.

MINING PRINCIPLES And PRACTICES

Author: DR. NISHAKAR PANDA

TO

MY ESTEEMED GUIDE

Dr L.N. Dash, Ph.D., D.Litt.

Professor of Economics

Contents

Preface

Sustainability has been a catchphrase and a buzz word around the globe but emergence of Inclusiveness, Responsibilities and Ethics are recent approaches, gradually being adopted as development principles across the developed, developing and under developed countries. Mining has a significant and vital role to play in economic development of a country and constitutes the major source of revenue for government among all the development projects. Minerals are indispensable not only for the well-being of man, but also for his survival. The mining is, no doubt, basic and strategic sector for industrial and economic development but mining has its own and inherent trail of negative and adverse impacts. Besides, the mining resources are not infinite and unlimited. The mining operations, essentially deplete the underground resources besides entailing adverse impacts. The negative impacts of mining can be controlled through application of the principles and practices of inclusiveness and sustainability. Ultimately, inclusive approach and sustainable practices should be at the heart of mining operations during its entire lifecycle from exploration to closure primarily in environmental, economic and social dimensions.

Inclusive mining approach is capacious, and does not leave any part or exclude any group of people/stakeholder where as Sustainability in mining means creating a balance of minerals – 'that is meeting today's needs without compromising future generations'. Inclusiveness and sustainability can be achieved through responsible and ethical mining. A modest attempt has been made in this work to deal with the concepts, critically examine the pertinent issues and key challenges pertaining to inclusive, sustainable, responsible and ethical mining with its ramifications, highlight and address some of the unexplored areas. This work is a mix of empirical and theoretical analysis. An analytical approach is developed through extensive and systematic literature review as well as analysis of data/information. The book contains ten chapters. The first chapter is an introductory chapter and it is intended to build a brief conceptual framework of Mining and Minerals, Inclusiveness and Sustainability with flashes of the need, methodology and scope of the work. In the succeeding three chapters, it is attempted to provide a contextual background on inclusive, sustainable, responsible and ethical mining. Triple Bottom Line are thoroughly analyzed in the next chapter. Three vital pillars of mining e.g. Environment, Economy and Society with Mining Induced Displacement and Rehabilitation (MIDR), multiple other dimensions associated with these three major issues of mining with experiences of various nations, our own country India and a mining rich developing region/state are examined in subsequent three chapters. Acts, Laws, Rules and Guidelines in the context of mining are highlighted in the next chapter. Policy prescriptions, alternative strategies and suggestions towards an inclusive, sustainable, responsible and ethical mining are outlined in the last chapter.

I am grateful to the numerous readers of my earlier books for their appreciation and compliments. I owe a word of thanks to them all. Thanks must go to my friends and well wishers who have constantly inspired me to bring out this volume. I greatly valued their suggestions and untiring encouragements. The National Library, Kolkata, NIRD and PR Library, Hyderabad and Odisha Legislative Assembly Library helped me in collection of information and documents. These Libraries allowed me to go through the relevant books, journals, documents and periodicals. I am thankful to all the staff of these Libraries. Required information and secondary data were collected from concerned Central, state and quasi government organisations/agencies and institutes. I am thankful to all of them for their support and cooperation.

I am greatly indebted to my wife and children and place on record my sincere gratitude for my wife; Anusuya, son; Chandan and daughter; Ananya for their support, cooperation and contribution in my endeavour. They all tolerated my negligence. It is but for their constant and continuous encouragements, the work has been completed.

1st September, 2023

Dr Nishakar Panda

List of Tables

List of Annexures

List of Figures

MAP

Acronyms

ASM	Artisanal and Small-Scale Mining
CSR	Corporate social responsibility
EIA	Environment impact assessment
EMP	Environmental Management Plan
FFRM	Framework for Responsible Mining
GIS	Geographical Information Systems
GMI	Global Mining Initiative
ICMM	International Council on Mining and Minerals
IIED	International Institute for Environment and Development
IISD	International Institute for Sustainable Development
MIDR	Mining-induced displacement and rehabilitation
MMSD	Mining, Minerals and Sustainable Development
NGO	Non-governmental organisation
RAP	Resettlement action plan
RMDI	Responsible Mineral Development Initiative
SIA	Social impact assessment
SLA	Sustainable Livelihoods Approach
UNDP	United Nations Development Programme
UNEP	United Nations Environmental Programme

Glossary of key concepts

Accountability - The individual(s) or the organisation(s) that bears the responsibility or onus for proper implementation and face the consequences in case the objectives or the targets are not met.

Biodiversity offsetting - creating an equivalent environmental and ecological benefit towards compensation for the damage to habitats and species caused by mining.

Capacity building - activities undertaken for skill and capability development of individuals, organisations or institutions to solve the problems.

Community engagement – to involve project affected community(ies) due to mining operations in rehabilitation and development process and receive their feedbacks and expectations.

Complaint register - maintaining a record containing the public grievances or matters of dissatisfaction.

Displacement - loss of house, property and other assets due to land acquisition for mining or any other development project.

Drainage - removal of water, surface or ground water from a particular area.

Environmental Impact Assessment – survey and analysis of the impacts of a proposed project on environment and ecology.

Ecosystem - a healthy system maintained by plants, flora and fauna, animals and microorganism communities and interacting as a functional unit.

Environmental management system - a company's system for preservation, protection and maintenance of environment.

Environmental protection plan - a plan of action to prevent environmental degradation and its adverse impacts.

Extractive Industries – it refers to the mining, oil, gas industries.

Local community - the settlers or inhabitants in the vicinity and surrounding areas of a project.

Mitigation - the minimisation, reduction or abatement of an adverse impact.

Mine closure - a process of ceasing the mining operations, relinquishment and decommissioning activities.

Polluter-pays principle - the company which causes the environmental degradation should bear the cost of measures to reduce adverse impact or pollution.

Primary stakeholders - the individuals, communities or parties directly affected by a project.

Project-affected Families - households whose means of livelihood and assets are directly or indirectly affected by a project.

Project-displaced Families - households who lose their house and properties and are to be physically shifted due to a development project.

Resettlement - a process when people or families are shifted from their original habitation due to a development project and rehabilitate in another place with compensation or facilities.

Resettlement Action Plan - a detailed plan about how a specific resettlement process will actually be conducted.

Resource curse - refers to a situation when revenue from natural resources causes corruption, conflict and bad governance.

Secondary stakeholders - the individuals, communities or parties indirectly affected by a project.

Social Impact Assessment – a way of analysing, monitoring and managing the social consequences of resource development.

Stakeholders - the individuals, communities or parties those who have an interest in or directly or indirectly affected by a project.

Tailings - the residue of ore which are poor in quality. They are the waste material after the commodity of value is extracted from the ore.

Chapter - 1

Concepts and Scope

Mining and Minerals

Mining is the extraction of valuable minerals or other geological materials from the earth, usually from an ore body, vein or (coal) seam. The term also includes the removal of soil.[1] Materials recovered by mining include base metals, precious metals, iron, uranium, coal, diamonds, limestone, oil shale, rock salt and potash. Any material that cannot be grown through agricultural processes, or created artificially in a laboratory or factory, is usually mined. Mining of stone and metal has been done since prehistoric times. Modern mining processes involve prospecting for ore bodies, analysis of the profit potential of a proposed mine, extraction of the desired materials and finally reclamation of the land to prepare it for other uses once the mine is closed.

Minerals, Metals and Mining Industry

Mining, nowadays, has become the material basis for life. According to George Orwell, "In the metabolism of the Western world, the coal miner is second in importance only to the man who ploughs the soil". Global economy cannot sustain without metals and mineral commodities. The dependence of mankind on minerals and metals is well illustrated in the research by the 'Minerals Education Coalition' in 2019. It is assumed that every American born will need 3.19 million pounds of minerals, metals and fuels in their lifetime. An American's needs consist of 7.70 million cu.ft. of natural gas, 1.36 millions lbs. of stones, sand and gravel, 330,573 lbs. of coal, 75,327 gallons of petroleum, 62,648 lbs. of other minerals and metals, 53,071 lbs. of cement, 30,190 lbs. of salt, 20,127 lbs. of iron ore, 14,337 lbs. of phosphate rock, 12,182 lbs. of clay, 2,066 lbs. of bauxite (aluminium), 980 lbs. of copper, 953 lbs. of lead, 466 lbs. of zinc, 1.54 Troy oz. of gold. Besides, every person every year in the US needs 40,633 pounds of new minerals that has to be provided to make things for daily use. These new minerals include: 9,924 lbs. of stone, 7,345 lbs. of sand and gravel, 675 lbs. of cement, 624 lbs. of other non-metals, 384 lbs. of salt, 256 lbs. of iron ore, 182 lbs. of phosphate rock, 155 lbs. of clay, 34 lbs. of soda ash, 26 lbs. of aluminium (bauxite), 23 lbs. of other metals, 12 lbs. of copper, 12 lbs. of lead, 6 lbs. of manganese and 6 lbs. of zinc. That apart, the energy fuels that is being used by each person each year to generate the energy include 97,988 cu. ft. of natural gas, 4,026 lbs. of coal, 958 gallons of petroleum and 0.13 lb. of uranium. It is pertinent to indicate that the new minerals are being used for multifarious purposes and uses to satisfy human needs. To cite some illustrations; stone is used to make roads, bridges, buildings and various other construction purposes, sand and gravel to make concrete, asphalt, roads, blocks bricks etc., cement for roads, bridges, buildings, schools and houses, other non-metals used in glass, chemicals, soups, paper, cell phones, computers, salt used in various chemicals, food and agriculture, iron ore to make steel for its multiferous uses like building, cars, trains, planes, constructions so on and so forth, phosphate rock fertilisers essential for growing food, animal feed etc., clay for making lunch and dinner wares, wall and floor tiles, paper, bricks, cement, soda ash used to make all kinds of glass, powdered detergents, medicines, food additive, treatment of water, photography etc., aluminium (bauxite) for making autos, airplanes, buildings and containers of all types, other metals used in electronic, TV and video equipments etc., copper used in electrical and electronic parts, buildings etc., lead used in batteries, electricals, communication, manganese for making all steel products for machinery, transportation, construction etc. and zinc useful in making healthcare and nutrition products, skin creams, numerous metals and alloys, rubber, paint, to make metals rust resistance etc.

The mining and metals industry spans a complex interdependent web that includes a formal component, an informal component and many affiliated interests-support services (such as investors, contractors and suppliers), government, Indigenous Peoples and their organisations, mining affected communities, civil society organisations, organised labour, academia and research institutions and downstream users.[2]

In India, minerals are typically classified into major and minor on the basis of their end use. Minerals, which are mostly used locally for local purposes are categorised as minor while the others are major. Most minor minerals are used in construction. For instance, boulders, gravel, ordinary sand, limeshell, kankar etc. are minor minerals. Minerals, like coal, lignite, iron, bauxite, lead copper etc are categorised as major minerals. However, certain minor minerals could be treated as major minerals for regulatory purposes. For example, ordinary sand is treated as a major mineral if it is used in coal mining, manufacturing of ceramics, optical industry and so on. On the other hand, dolomite (a carbonate mineral composed of calcium magnesium carbonate), typically regarded as a minor mineral if it is used in buildings or for road construction.

Broad categories of the minerals and metals produced by the industry include the following: (i) Ferrous and non-ferrous metals: Iron and steel, nickel, lead, zinc, copper, aluminium, tin, tungsten, molybdenum, tentalum, cobalt, bismuth, cadmium, titanium, zirconium, antimony, manganese, beryllium, chromium, germanium, vanadium, gallium, hafnium, indium, niobium, rhenium, thallium and rare earth metals. (ii) Precious minerals and metals: diamonds, gold, silver, platinum group metals (platinum, palladium, iridium, ruthenium, rhodium, osmium). (iii) Energy minerals: coal, uranium etc. (iv) Industrial minerals: clays, silica sand, talc, salt, limestone, gypsum, pumice, potash, refractory bauxite, chromite, rutile and limonite (titanium), zircon, magnetite, borates, fluorspar, sulphur and phosphate rock. (v) Building materials: stone, sand and gravel.

The mining consist of a number of sub-activities like servicing, construction, exploration, processing, extraction, refining etc. and there are a number of mining industries comprising of a mix of large/major and small/minor companies internationally and nationally. Some of the companies having access to large amounts of capital develop a major mine on their own. Minor companies with less access to capital tend to focus on exploration and other operations and some companies produce minerals in joint collaboration. There are also various small mining like artisanal and small-scale mining operators with informal enterprises. According to the World Bank estimates, around 15-20 million artisanal miners currently operate in nearly 30 countries.[3]

Lifecycle and stages of a full mine project

The mine projects to produce this broad range of minerals and metals as well as the resulting products follow a lifecycle as presented in a hypothetical example at Table 1.1 which shows the full mine project lifecycle and is roughly indicative of activity and mine-related employment.

Table 1.1: The mine project lifecycle

Sl. No.	Activity	Lifecycle (Years)
1	Exploration	1-10 Years or more
2	Site Design and construction	1-5 years
3	Operation	2-100 years
4	Final closure and decommissioning	1-5 years
5	Post-closure	10 Years to perpetuity

Source: IMDR Report.

Allowing for the time required for exploration and to find an ore body to develop a mine, as well as the decades or more for construction, operation, closure and post-closure management, a 20–30 year mine can often involve five to seven generations of relationships between an industry player (more than one company will likely be involved) and the host community. Many mining operations have an even longer lifespan which further extends the multi-generational relationship with communities. A critical feature of metals is their capacity to be recycled. While metals have been recycled for many centuries, the realisation in the last few decades of the potential cost savings and reduction of environmental and social impacts achieved by recycling.[4] Thus, the life of a mine comprises of a series of steps. However, this lifecycle varies in length from a few to many decades, sometimes stretching to centuries. Typically, there is an exploration and evaluation phase (1- 10+ years), followed by a site design and construction phase, also popularly known as the mine development phase (2-5 years), followed by the mining extraction phase (2-100 years), and then closure, decommissioning and reclamation (5-30+ years).

Each phase of the full mine project lifecycle entails different environmental, economic and social challenges. Moreover, within a mine's lifecycle there are periods of 'booms' and 'busts', which are driven by volatile commodity prices and other socio-political factors.[5] 'Booms' are fuelled by high mineral demand, which drive high commodity prices and incentivize higher production. 'Booms' are inevitably followed by 'busts', which are characterised by supply surpluses, falling commodity prices, reductions in company capital costs and spending, investment postponements and mine closures.[6] Exploration typically proceeds in a succession of stages, wherein each subsequent stage 'zooms in' to an area to get more detail about the presence of the deposit and economic viability of its extraction. The main objective of mineral exploration is to locate mineral deposits. An additional objective of mineral exploration today is to find significant extensions of pre-existing deposits associated with operating or abandoned mines. There are two types of exploration viz. Greenfield exploration: new, remote territories and Brownfield exploration: close to existing mines. Greenfield exploration projects can be subdivided into grassroots and advanced projects. Geological surveys are fundamental to exploration. Geological surveys are normally used on large tracts of land and mapped to identify areas of potential mineral deposits. At times, airborne magnetic surveys are used to cover large tracts of land quickly to identify anomalies that may indicate a mining zone. After the target area is identified, a more detailed ground survey is undertaken using techniques such as seismic surveys, direct sampling and drilling.[7] Immediately following the geological exploration, mining companies conduct feasibility studies combining geological information with economic analysis. Before proceeding to the planning stage, mining companies also study various risks, uncertainties, other forces and factors like the current demand for the mineral resource, projected future demand, market situations, locational advantages/disadvantages of the mine, availability of the workforce, political situations, stability of the government, socio-economic viability of the project etc.[8] The feasibility and construction stages of a mining project follows the Planning stage. The feasibility stage of a mining project usually includes Scoping Studies, Prefeasibility Studies, and final Feasibility Studies prepared efficiently by the professional experts. This is the critical stage to ascertain that the project is technically feasible, socially sound and economically viable and the project can go ahead with construction and operation with preparation of a detailed design. The next logical step is mine planning stage which addresses the issues of technical considerations such as mining methods, processing methods, as well as inputs required to conduct mining, including infrastructure and workforce options after which the mine construction begins with preparing the designated land area for the construction site. Preparation process includes land shaping with the removal or stripping of vegetation, rocks and mountainous terrains, and any previous buildings or structures. The most important process is putting in place the plans for managing social, economic and environmental impacts. After clearance of the land, boundary wall or fencing is erected and ancillary infrastructures such as haulage roads, railway lines, and bridges need to be built. Workers camps or “mining towns” need to be built to house the workers. The type of mining activity will determine the type of construction needed. Whereas underground mining activities require various tunnels to be built, surface mining need the various structures surrounding it. Processing plants are normally built on the site of extraction. Tailings dams and waste rock dumps need to be constructed to encompass the wastes that will be generated from the mine. At the mine development phase, companies or the mining authorities should systematically try to address non-technical matters such as sustainability and community expectations. A recent study, published by the CSR Initiative at Harvard Kennedy School[9] together with Shift and the University of Queensland, found that the feasibility and construction stages of projects are highly prone to serious conflicts that can lead to the suspension and abandonment of projects. The construction stage can represent dramatic transitions for local communities with major project impacts experienced for the first time, including impacts arising from a large influx of workers.[10] Mineral extraction phase begins with extraction of minerals. Mineral extraction methods can be divided into two types: surface and underground. The method used depends, primarily, on the depth at which the ore is found and the cost of the methods of extraction. Surface mining methods, such as open pit mining, strip mining, quarry mining, leach mining, dredge mining etc., extract minerals and metals relatively close to the surface. Surface mining techniques are used when there is large tonnage, high rates of production and when the overburden i.e. soil and rock overlying the mineral deposit, is thin. Examples of surface mining include coal, brown coal and bauxite mines. Underground mining methods, such as room and pillar, open stoping, longwall mining, sub-level open stoping, block caving etc. are used when the ore body is deep beneath the surface, and the ore grade is sufficient to cover the cost of mining at depth. Underground mining is more expensive and more dangerous as compared to surface mining. Separating the valuable metals and minerals from the rock and waste material that surrounds them is known as 'processing' or 'beneficiation'. Different minerals and metals require different types of processing. Coal and iron ore often require very little processing, as they can simply be crushed and/or washed before transport. In contrast, gold, copper and aluminium require several stages of complex processing to separate them into their pure state. Gold ore is treated with chemicals while aluminium is separated out by exposing the processed ore to an extremely powerful electric current. Mines generate enormous quantities of waste. The waste /residue or tailings produced by the refining process often needs to be stored in tailings dams. Much of the waste generated from the production of just iron ore, copper and gold. It is estimated that for every usable ton of copper, 110 tons of waste rock and ore are discarded, and another 200 tons of overburden earth moved. The ratio is more staggering for gold i.e. about 300,000 tons of wastes are generated for every ton of marketable gold. Much of this waste is contaminated with cyanide and other chemicals used to separate the metal from ore.[11] The last phase is Closure of mines. All mines eventually close. Some mines close for economic or logistical reasons, while others simply run out of viable ore to extract. In some cases mines may close on a temporary basis until markets improve. The closure of any mining operation will have an impact on the local society, economy and the environment. These impacts are managed through a mine closure plan, which addresses key issues such as environmental rehabilitation, socioeconomic impacts, removal of infrastructure and plant. Mine closure further ensures that the site is left in a safe, stable condition while the community is able to generate livelihood and thus thrive in the absence of the mining industry. The productive use of the landscape in the future is essential in planning for restoration and closure. The inputs and knowledge of the local people in the closure plan should be taken in to account. Closure and Land rehabilitation has to be based on mutual understanding as well as consultation with local community.[12]

Thus, there are various phases in a mining project, beginning with mineral ore exploration and ending with closure. Each phase of mining has different environmental impacts and therefore, requires different strategies, tools and technological interventions to address these impacts. In addition, the environmental impacts of mining projects differ considerably depending on the particular kind of mining methods they employ. The table below provides a useful framework for understanding the environmental impacts of mining and associated regulatory and management requirements.

Inclusiveness

Inclusiveness is broad based and etymologically means including everything or not excluding, not leaving anyone or anything whatsoever. It signifies comprehensiveness, including a great deal, or encompassing everything concerned. The process of inclusion engages each individual and the entire community concerned. Hence, inclusion, in a broader sense means togetherness, and encompassing interrelationship among people and inclusiveness is a holistic approach based on the premise of equity or equitable distribution of wealth and power aiming at substantial reduction of disparities and inequalities through mass participation of every sections in the society and empowerment of the poor, marginalised and excluded persons. However, the term 'Inclusiveness' mostly has a reference to development or development sectors. Inclusive development is a process that includes all people and all regions including marginalised, vulnerable people and backward, underdeveloped regions for greater human welfare. There are many cases where countries have registered high economic growth but a significantly large number of people have remained poor and vulnerable due to their exclusion from the benefits of this growth and progress. It is because economic growth in terms of income and production is a necessary but not essential condition for peoples' welfare or nation's welfare for that matter. A country's sustainable and transformative socio-economic development will be at stake without inclusiveness. Hence, the role of inclusiveness is important because growth and development are not evenly distributed across sectors and regions.[13] The benefits of this growth, therefore, should be encompassing, inclusive and shared equitably across all sections of society. Inclusive development, thus "hinges on the inclusion of excluded people and the utilisation of their capabilities[14]. On the other hand, the exclusion of individuals and groups can become a major threat to social cohesion and dangerously affect the humanity. Exclusion may sometimes lead to marginalisation. The UNDP's inclusive development approach integrates the standards and principles of human rights.[15] The World Bank's report – 2016 discussed the notion of shared prosperity[16] i.e. Growth is inclusive when the poorer segments do at least as well as the overall population. The objective of development should be "growth with as much inclusiveness as possible, and with as much inclusiveness of the poorest as possible"[17] with an "integral focus on the achievement of equity and the rights of citizenship".[18] The focus of development should be on human development, reducing poverty and increasing entitlements[19] and also an increasing focus on capabilities and freedom.[20] It is necessary that "the capacity of economic, political and social systems should provide the circumstances for the well-being on a sustainable and long-term basis".[21]

Thus, 'inclusiveness' is a nebulous term in the lexicon of development. It has vast and varied dimensions with wide ramifications. There is no single or short-cut definition of the concept. The economists and development specialists have defined the term differently, which are, of course, interrelated and mutually inclusive. It 'brings into play all dimensions of wellbeing and welfare, it focuses on the equitable sharing of the benefits of growth and related distribution of well-being in society across divides within societies, across income groups, genders, ethnicities, religions, religious groups or others'. The main focus is on the distributional aspects besides poverty or in other words it looks in reducing inequalities and decreasing poverty. Inclusiveness refers in some sense to the distribution of well-being, however measured. A given average for a population can be distributed in an infinite number of ways, ranging from perfect equality to extreme equality. It is the improvement of the distribution of well-being along these dimensions at the same time that the average achievement improves.[22] It is an "improvement in distribution of well-being along the dimensions beyond growth, at the same time, average improvement in achievements is realised".[23]

Inclusiveness in mining or for that matter inclusive mining is a new dimension of mining development that is based on the principles of equity and participatory approach with a focus on the poor and marginalised sections considering political forces and socio-economic and ecological factors.

Sustainability

The concept of 'Sustainability' is a buzzword in recent policy and political domains across the globe. In simple words, Sustainability means leaving enough for future while meeting the present needs. It has always a reference to both present and the future in terms of supply, distribution, consumption etc. Many authors and development practitioners have tried to define this concept. Sustainability 'literally is a capacity to maintain some entity, outcome or process over time'.[24] Sustainability is the 'efficient and equitable distribution of resources intra-generationally and inter-generationally with the operation of socio-economic activities within the confines of a finite ecosystem'.[25] Sustainability is a 'dynamic equilibrium in the process of interaction between the population and the carrying capacity of its environment such that the population develops to express its full potential without producing irreversible adverse effects on the carrying capacity of the environment upon which it depends'.[26] Sustainability 'brings into focus human activities and their ability to satisfy human needs and wants without depleting or exhausting the productive resources at their disposal. This, therefore, provokes thoughts on the manner in which people should lead their economic and social lives drawing on the available ecological resources for human development from this standpoint'.[27] However, 'transforming global society, environment and economy to a sustainable one is one of the most uphill tasks confronting man today since it is to be done within the context of the planet's carrying capacity'.[28] The 'ultimate objective of the concept of sustainability, in essence, is to ensure appropriate alignment and equilibrium among society, economy and the environment in terms of the regenerative capacity of the planet's life-supporting ecosystems'[29] and this calls for innovative approaches to managing realities'.[30] Therefore, 'it is this dynamic alignment and equilibrium that must be the focus of a meaningful definition of sustainability'.[31] The 'definition must emphasise the notion of cross-generational equity, which is clearly an important idea but poses difficulties, since future generations' needs are neither easy to define nor determine.'[32] However, 'sustainability seek to prioritise and integrate social, environmental and economic models in addressing human challenges in a manner that will continually be beneficial to human.[33] The 'economic models seek to accumulate and use natural and financial capital sustainably; environmental models basically dwell on biodiversity and ecological integrity while social models seek to improve political, cultural, religious, health and educational systems, among others, to continually ensure human dignity and wellbeing'[34].

In development literature, 'Sustainable Development' (SD) is developing in such a way that future people will have "enough and as good".[35] The term sustainable development was first coined and defined by Dr Gro Harlem Brundtland in a report entitled Our Common Future in 1987 as "Development that meets the needs of the present without compromising the ability of future generations to meet their own needs".[36] Since then, this has become the catchphrase in development discourse, having been associated with plethora of definitions, meanings and interpretations. This consists of two different concepts 'Sustainable' and 'Development' and, therefore, has been defined from different angles and perspectives, literally as well as structurally. Various authors like Hunter, Mowforth and Munt, Bramwell and Alletorp, Forsyth Berry and Ladkin, Dewhurst and Thomas, Horobin and Long, Landrum and Edwards, Bohdanowicz, Hitchcock and Willard have discussed about the concept. Many academics, researchers and development practitioners[37] apply the concept to connote improving and sustaining a healthy economic, ecological and social system for human development. Etymologically, SD would mean "development that can be continued either indefinitely or for the given time period".[38] 'SD is a core concept within global development policy and agenda. It provides a mechanism through which society can interact with the environment while not risking damaging the resource for the future'.[39] It is a 'development paradigm that calls for improving living standards without jeopardising the earth's ecosystems or causing environmental challenges such as deforestation and water and air pollution that can result in problems such as climate change and extinction of species'.[40] This is an 'approach to development which uses resources in a way that allows them (the resources) to continue to exist for others'.[41] 'SD is an 'organising principle for meeting human development goals while at the same time sustaining the ability of natural systems to provide the natural resources and ecosystem services upon which the economy and society depend'.[42] 'SD aims at achieving social progress, environmental equilibrium and economic growth'[43] and hence, 'there is the need to move away from harmful socio-economic activities and rather engage in activities with positive environmental, economic and social impacts'.[44]

The necessity for SD is based on simple economic principles. There is an exponential growth of population, it keeps increasing sometimes in galloping speed associated with an increased wants, needs and consumption but the available resources to satisfy these human wants are limited and finite. Demands are unlimited but the supply is limited and there is, thus an yawning gap and massive mismatch between demand and supply. Finite natural resources can not cope with infinite wants and this calls for an urgent need to adopt SD approach. Hence, 'conscious of this phenomenon, global concerns have always been expressed for judicious use of the available resources so that it will always be possible to satisfy the needs of the present generation without undermining the ability of future generations to satisfy theirs'.[45] SD is an 'effort at guaranteeing a balance among economic growth, environmental integrity and social well-being. This reinforces the argument that, implicit in the concept of SD is intergenerational equity, which recognises both short and the long-term implications of sustainability and SD'.[46] It is pertinent to indicate that sustainability and SD are not analogues and synonyms but the two concepts are different and distinguishable. 'Sustainability is the goal or endpoint of a process called sustainable development'.[47] While sustainability refers to a state, SD refers to the process for achieving this state.[48]

Need, Methodology and Scope

Mining should be ***inclusive***, based on ***sustainability*** principles and adopt ***responsible*** and ***ethical*** practices. Many obvious mind bugling questions like what exactly constitute the inclusive mining, what are its components, what are the principles of sustainability, are they relative or permanent, is it really possible to earnestly implement these principles, can mining be responsible, is there any stringent regulation by various national governments for miners to implement the sustainable and responsible practices? Such issues and various other challenges are still vexing the minds of miners, economists, policy planners and development specialists and have not yet been explored in its entirety nor examined critically. Not much have been discussed, deliberated and analysed to highlight and address these aspects. There is a lack of macro studies on these aspects. Adequate literature on the subject is not also available Stray studies have been conducted basing on mining specific projects. A modest attempt has been made in this work to deal with the concepts of inclusive and sustainable mining with its ramifications as well as the principles, policies and practices of responsible and ethical mining. It is essential to discuss and dissect the pertinent issues of inclusive, sustainable and responsible mining viz. Economy, Environment and Social aspects and multiple factors associated with it.

This study is based both on primary and secondary data. Relevant information were collected through case studies in the field as well as from secondary sources, such as; published materials/ documents from state and national libraries. This work is, therefore, a combination of empirical and theoretical analysis. An analytical approach is developed through extensive and systematic literature review as well as analysis of data/information.

This is an introductory chapter. It is intended to build a brief conceptual framework on Mining and Minerals, Inclusiveness and Sustainability. There are flashes of the need, methodology and scope of the study. In the succeeding three chapters, it is attempted to provide a contextual background on Inclusive, Sustainable, responsible and ethical mining. Triple Bottom Line i.e. Economy, Environment and society or in other words profit, planet and people are three vital pillars of Inclusive and Sustainable Mining. There are multiple other dimensions associated with these three major socio-economic and ecological planks of mining. These three outstanding issues with experiences of various nations, our own country and a developing region are explored in subsequent three chapters. The flashes of Triple Bottom Line with footprints of mining on Economy and Economic Development are analysed, the impact of mining on planet or environmental impacts of mining are mapped out, social issues with particular reference to peoples' dispossession, dislocation and displacement induced by mining projects as well as their resettlement/rehabilitation and other corollary factors have been dealt in the next chapter. Besides identifying some of the nagging and outstanding issues with highlights of some negative and adverse impact of mining, the experiences of some of the mineral rich countries of the world, scenario of the nation as a whole and experiences of a mining region/state has also been examined and analysed in this section. Odisha is a major mineral State of the country. It has prominence in the national front in terms of area and activities under mining. Hence, the case of Odisha region/state has been taken as a sample for regional study. It is worthwhile, after analysing the global and national scenario, to examine the mining and mineral sector of the state and it's experiences on Mining Induced Displacement and Rehabilitation (MIDR). Before delving in to MIDR, there are flashes of status of mineral sector of the State vis-à-vis all India and major mineral states, and land acquisition for mining activities of the state. The relevant international and national policies are briefly discussed in the next chapter. The last chapter is 'Towards Inclusive, Sustainable and Responsible Mining' and

'The Way Forward'. Having focused on various aspects and several issues on Inclusive, Sustainable and Responsible Mining, it is imperative to give alternative strategies, suggestions and policy prescriptions to ensure an inclusive and sustainable mining. These are outlined in the last chapter.

End Notes & References

1. Accessed from Mining.com
2. Extracts from Webpage
3. Accessed from www.mining-atlas.com
4. International Council of Mining and Metals (ICMM), *Mining's contribution to sustainable development; an overview*, June, 2012.
5. Research Report No RR5, Department of Civil Engineering, Monash University and Mineral Policy Institute, 2007.
6. Gavin M Mudd, *'The Sustainability of Mining in Australia: Key Production Trends and Their Environmental Implications for the Future'.*
7. Byambajav Dalaibuyan, *'The Story of the Discovery of the Oyu Tolgoi'*, 2016.
8. Jamul Jadamba, *'Main principles of financing mining projects'*, 2012.
9. Corporate Social Responsibility Initiative Report No. 66. Cambridge, MA: Harvard Kennedy School, 2014.
10. Davis, Rachel and Daniel M. Franks, *"Costs of Company-Community Conflict in the Extractive Sector"*, 2014.
11. Payal Sampat, *"Scrapping Mining Dependence,"* in State of the World, 2003.
12. http://www.minerals.org.au, *"Mine rehabilitation in the Australian minerals industry."*
13. Johnson, B., & Andersen, A. D., *"Learning, Innovation and Inclusive Development: New perspectives on economic development strategy and development aid"*, 2012.
14. *Ibid.*
15. Gupta, Joyeeta & Pouw, Nicky & Ros-Tonen, Mirjam, *"Towards an Elaborated Theory of Inclusive Development"*. European Journal of Development Research-27, pp. 541-559. 10.1057/ejdr.2015.30. (2015).
16. The World Bank's report *"Poverty and Shared Prosperity 2016: Taking on Inequality"*, 2016.
17. Rauniyar, Ganesh; Kanbur, Ravi; *"Inclusive Growth and Inclusive Development: A Review and Synthesis of Asian Development Bank Literature"*. Asian Development Bank.
18. Hickey, Sam; *"Thinking About the Politics of Inclusive Development: Towards a Relational Approach"*, ESID Working Paper No 1, October 15, 2013.
19. Gupta, Joyeeta & Pouw, Nicky & Ros-Tonen, Mirjam, *Op.Cit.*
20. Sen, A. K. *"Development as freedom"*, Oxford, Oxford University Press, 1999.
21. Rauniyar, Ganesh; Kanbur, Ravi, *Op.Cit.*
22. Vinaya Swaroop, *"Making growth inclusive: Challenges and opportunities"*

23. Musahara, H; *"Inclusive growth and development: Issues in Eastern and Southern Africa"*, Organisations for Social Science Research in Eastern and Southern Africa (OSSREA), 2016.
24. Basiago, A. D. *"Economic, social, and environmental sustainability in development theory and urban planning practice*", The environmentalist, Boston: Kluwer Academic Publishers. 1999.
25. Stoddart, H., Schneeberger, K., Dodds, F., Shaw, A., Bottero, M., Cornforth, J. & White, R. *"A pocket guide to sustainable development governance*", Stakeholder Forum, 2011.
26. Ben-Eli, M. *"Sustainability: Definition and five core principles, a new framework"*, the sustainability laboratory, New York, 2015.
27. Thomas, C. F. *"Naturalizing Sustainability Discourse: Paradigm, Practices and Pedagogy of Thoreau, Leopold, Carson and Wilson*": Ph.D Thesis: Arizona State University, 2015.
28. Hák, T., Janoušková, S., & Moldan, B. *"Sustainable development goals: A need for relevant indicators. Ecological Indicators"*, 60(1), pp. 565–573, 2016.
29. DESA-UN. *'The Sustainable Development Goals Report'* 2017, (2018, April 4).
30. World Bank, *'Atlas of sustainable development goals 2017'*. World Development Indicators,doi:10.1596/978-1-4648-10. (2017).
31. Gossling-Goidsmiths, J. *"Sustainable development goals and uncertainty visualisation"*. Thesis submitted to the Faculty of Geo-Information Science and Earth Observation of the University of Twente in partial fulfilment of the requirements for the degree of Master of Science in Cartography. 2018.
32. Mensah, J., & Enu-Kwesi, F. *"Implication of environmental sanitation management in the catchment area of Benya Lagoon, Ghana"*. Journal of Integrative Environmental Sciences. doi:10.1080/1943815x.2018.1554591, 2018.
33. (i) Hussain, F., Chaudhry, M. N. & Batool, S. A. *"Assessment of key parameters in municipal solid waste management: a prerequisite for sustainability"*. International Journal of Sustainable Development & World Ecology, 21(6), 519–525. doi:10.1080/13504509.2014.971452, 2014. (ii) UNSD. *"SDG indicators global database"*.
34. Acemoglu, D., & Robinson, J. *"Why nations fail: The origins of power, prosperity, and poverty"*, New York: Crown. 2012.
35. John Locke, A Phrase, 1690.
36. Dr. Gro Harlem Brundtland, a medical doctor and the former Prime Minister of Norway who, at the request of the then Secretary General of the United Nations, established and chaired the World Commission on Environment and Development. Her Committee produced a report entitled *Our Common Future* in 1987 in which the definition first appeared.
37. (i) Milne, M. J., & Gray, R. *"W(h)ither ecology? The triple bottom line, the global reporting initiative, and corporate sustainability reporting"*. Journal of Business Ethics, 118(1), 13–29. 2013. (ii) Tjarve, B., & Zemīte, I. *"The Role of Cultural Activities in Community Development"*. Acta Universitatis Agriculturae et Silviculturae Mendelianae Brunensis, 64 (6), 2151–2160, 2016.

38. (i) Dernbach, J. C. *"Achieving sustainable development: The Centrality and multiple facets of integrated decision making"*. Indiana Journal of Global Legal Studies, 10, 247–285.doi:10.2979/gls.2003.10.1.247, 2003. (ii) Lele, S. M. *"Sustainable development: A critical review"*. World Development, 19(6), 607–662. doi:10.1016/0305-750X(91)90197-P, June, 1991.
39. (i) Cerin, P. *"Bringing economic opportunity into line with environmental influence: A discussion on the coase theorem and the Porter and van der Linde hypothesis"*. Ecological Economics, 56, 209–225. doi:10.1016/j.ecolecon.2005.01.016, 2006. (ii) Abubakar, I. R. *"Access to sanitation facilities among Nigerian households: Determinants and sustainability implications"*. College of Architecture and Planning, University of Dammam, Saudi Arabia; Sustainability, 9(4), 547. doi:10.3390/su9040547,2017.
40. (i) Benaim, C. A. & Raftis, L. *"The Social Dimension of Sustainable Development: Guidance and Application"*: Thesis submitted for completion of Master of Strategic Leadership towards Sustainability. Blekinge Institute of Technology, Karlskrona, Sweden, 2008. (ii) Browning, M. & Rigolon, A. *"School green space and its impact on academic performance: A systematic literature review"*. International Journal of Environmental Research and Public Health, 16(3), 429. doi:10.3390/ijerph16030429, 2019.
41. Mohieldin, M. *"The sustainable development goals and private sector opportunities"*. EAFIT University of Medellín. 2017.
42. Evers, B. A. *"Why adopt the Sustainable Development Goals? The case of multinationals in the Colombian coffee and extractive sector"*: Master Thesis Erasmus University Rotterdam, 2018.
43. Zhai, T. T., & Chang, Y. C. *"Standing of environmental public-interest litigants in China: Evolution, obstacles and solutions"*. Journal of Environmental Law, 30, 369–397. doi:10.1093/jel/eqy011, 2019.
44. Ukaga, U., Maser C., & Reichenbach M. *"Sustainable development: principles, frameworks, and case studies"*. International Journal of Sustainability in Higher Education, 12(2), Emerald Group Publishing Limited. doi:10.1108/ijshe.2011.24912bae.005. 2011.
45. Hák, T., Janoušková, S., & Moldan, B., 2016, *Op.Cit.*
46. Dernbach, J. C. *"Sustainable development as a framework for national governance"*. Case Western Reserve Law Review, 49(1), 1–103. 1998.
47. Diesendorf, M. *"Sustainability and sustainable development"*. In D.Dunphy, J.Benveniste, A.Griffiths, & P.Sutton (Eds.), *"Sustainability: The corporate challenge of the 21st century"*, pp. 2, 19–37, 2000. Sydney: Allen & Unwin.
48. Gray, R. *"Is accounting for sustainability actually accounting for sustainability ... and how would we know? An exploration of narratives of organisations and the planet"*. Accounting, Organisations and Society, 35(1), 47–62. doi:10.1016/j.aos.2009.04.006, 2010.

Chapter - 2

Inclusive Mining: Approaches and Experiences

Inclusive Mining is comprehensive, extensive, wide-ranging and all embracing aiming to include everything in the mining process and integrate all people, types and groups especially those who are disadvantaged, vulnerable or have suffered discrimination or are living with deplorable conditions in activities of the mining from beginning to end. It encompasses all aspects of and includes everything concerned to mining.

Inclusive mining approach and practice is broad based, capacious, reckoning everything and does not exclude any particular group of people/stakeholder. It avoids bias, stereotypes, does not leave any part or group out of its ambit and takes in to account almost everything relating to mining within its scope. Inclusive mining supports multi-stakeholder dialogue, promotes local voices and explores innovations that help deliver more inclusive and accountable mining sector – with empowered local communities, enlightened decision-makers and sensitivity to the rights and responsibilities of all stakeholders.

Mining projects should address serious environmental and social problems, reduce, avoid and mitigate negative consequences for the underprivileged and affected groups as well as natural environment and enhance its contribution to sustainable and inclusive development for the country at large. Mining companies, therefore, have to take an inclusive approach to development that redresses the needs of not only the companies themselves, but also the relevant governments and communities in which they operate by identifying, developing and launching inclusive solutions to benefit all stakeholders and help reduce the inherent risks.[1]

Some of the issues and approaches of Inclusive Mining are analysed as follows.

Approaches

Inclusive Socio-Economic and Environmental Costs

Inclusive socio-economic and environmental costs of all mining related activities in the area of their operation must be ensured both by Government who earns revenue and royalties through mining leases and mining companies which maintains a profitable business. This fosters a balanced and inclusive growth in mining sector. All socio-economic and environmental costs like displacement, dislocation, dispossession, exclusion, loss of employment, income, savings, in fact all impoverishment risks, water, air, noise pollutions etc. should be taken in to account, appropriately included in the action plans and mitigated to the best possible extent. Inclusive growth in mining sector should be ensured by the government through reassessing tax rate for the mining industry to incorporate the true economic and social costs associated with mining. There is a need to broaden the scope of mitigation and compensation policies to take into account the greater cost of mining, redistribute mining related government earnings so that it reaches the affected population, incorporate mining related social costs into environmental policies to reduce pollution, encourage mining companies to address some of these negative costs through corporate social responsibility (CSR)[2] or Enterprise Social Commitment (ESC).

Inclusive Development of Community Assets and Wealth

Mining companies should adopt an inclusive approach in wealth and assets development rather than concentrating in the area of their operation. Controlling authorities must see that the benefits of mining go beyond the operational or adjoining areas and not limited to a particular group or habitation. Infrastructures and development projects should cover larger segments of population. Frederick Bird, in his paper based upon a review of studies of mining companies, most of them being Canadian, in Chile, northern Canada, Tanzania, Guatemala, Ghana, Sierra Leone and the Democratic Republic of Congo has observed that in spite of often well-meaning efforts, the wealth produced by most mining firms in developing areas largely benefits those immediately involved, sometimes neighbouring communities and often those in the governing strata. Typically, mining takes place in enclaves and fosters enclave development rather than the kind of inclusive wealth development. One way mining operations can foster more inclusive wealth development is to adopt much broader approaches towards the construction of the various physical and social infrastructures required for their own operations. By collaborating with diverse regional, provincial and/or national agencies and businesses, these infrastructures—roads, water, electricity, health services, schools and security—may be developed to serve larger populations and areas beyond their immediate operating sites. Established mining firms can also foster more inclusive wealth development by establishing more formal, collaborative relationships with artisanal miners.[3]

Inclusive Community Participation

Any development to be truly and genuinely 'inclusive', planning and policies, objectives and strategies must be discussed and negotiated with all the stakeholders, affected directly or indirectly, positively or negatively. Each and everyone affected must have a say in the negotiation process and accord their consent for the project. Peoples' aspirations, citizen's voice and community decision should be reflected in the action plans and policy papers. Without their effective participation, no implementation can be result oriented.

Participation of affected Communities in all activities of mining from grounding to operation then closure is a must for its success and uninterrupted progress. People's involvement in decision-making process, in a way, serves a link between government and company. 'Decisions about mining clearly need to balance economic, technical, environmental and social factors in participatory and transparent ways. Mining issues are complex and responsibility boundaries between the company and local, regional, and national governments are often fuzzy. Obtaining a "social license to mine" requires a mining company to gain the respect, trust and collaboration of governments and local populations. Without proper dialogue between these parties, without communication on all aspects of the proposed project—technical, economic, environmental, and socio-political—the project is likely to be rejected locally and eventually, nationally. Mining corporations that adhere to responsible mining principles can reduce local conflicts, prevent environmental impacts, and even improve their profits'.[4]

Including all stakeholders in the mining process, evoking their participation through involvement, inclusions wherever possible, interactions, intimate exchange of ideas, meaningful deliberations, discussions and dialogues, seeking their suggestions and implementing them as far as practicable and feasible ensures not only uninterrupted operation but also people centric progress. Inclusive Mining proponents suggest that mining companies should pay for people participation efforts. The citizens and communities who are affected may, sometimes, need access to independent technical and legal assistance. This is in Canada, called 'intervener financing' which Canadian government routinely finances as assistance.

Indigenous Peoples' Inclusive Approach

Indigenous Peoples reside in prospective mining areas since generations and maintain the homeland as their mother earth even though they don't have any legal proof or valid documents of land in their favour. They should be taken in to confidence, properly consulted, negotiated and should not be uprooted or evicted, displaced or dislocated without their consent. Mining companies should respect the rights of affected indigenous communities and scheduled tribes and display responsible actions towards proper rehabilitation and resettlements of the displaced communities. It must be ensured that indigenous communities and scheduled tribes are fully compensated and completely engaged through prior, informed and free consent which promotes their rights to involve and participate in any decision-making process as well as developmental initiatives in their area. In Canada, the Supreme Court of Canada gives a ruling to this effect during 2004 and made it mandatory for Mining companies to comply. The Supreme Court of Canada gives the country's Indigenous Peoples, aboriginal peoples or vulnerable ethnic minorities the right to review land-use decisions by mining companies that might affect their legal right to harvest, hunt and fish on lands for which they do not possess paper titles. In such situations, the mining companies should persue the indigenous Peoples and negotiate with them before any drilling operation. Hence, deliberations and consent through negotiations is the best possible strategy to avoid legal battles. When a proposed mine is on land directly owned by an aboriginal group, the mining company should negotiate an Impact/Compensation Contract in the public domain, and pay a royalty. In Panama, Canadian mining companies are involved in a situation in which fights between the Ngöbe-Buglé Indigenous People and Panama's military police led to bloodshed and deaths in early 2012. The protest concerned reform of the mining code, which would allow mining in indigenous territories, one of the strictest No Go Zones to mining. However, the United Nations Special Rapporteur on indigenous rights and the Inter-American Commission on Human Rights have both urged the government to halt the police violence and negotiate.[5]

Gender Inclusive Mining

In order to make mining more inclusive, gender inequality in rights and opportunities must be avoided. More and more women should be included in decision-making process. Mining must benefit all citizens equally. Both women and men need to be included in the governance and employment opportunities. Besides, women bear the brunt of the negative social, economic and environmental impacts of the mining sector more than the men and hence, their concerns should be addressed on priority basis. Mining has prominent role to play in women empowerment by ensuring their participation in decision-making process.

Inclusive workforce

Mining sector have diverse work force from highly skilled to unskilled, workers with different fields and levels of skill, women workforce and daily labourers. Fostering an inclusive workforce in this sector, therefore, poses a great challenge. "Mining offers a large range of employment opportunities from early forms of prospecting all the way to closure and reclamation. Future opportunities in mining are quite exciting with advanced technologies such as robotics and artificial intelligence to geological surveying. The labour force in mining is changing and some job prospects will shrink while others grow. The general public should be aware of what the future mining workforce will look like, it's vast and diverse opportunities and future openings in exciting fields. While mining is struggling to attract and retain a more diversified workforce, in the last few years a number of measures have been taken to rectify the situation, while only few, there are success stories related to attracting women, immigrants, veterans and millennials to mining".[6] Many people do not consider a career in mining due to sheer ignorance or misgivings about the mining and its negative impacts. These misgivings need to be dispelled to ignite curiosity and attract more people. Best practices and success stories of mining, employment potential and opportunities available in the sector should be brought to the knowledge and notice of general public in the sector. This will go a longway in ensuring an inclusive diversified workforce in mining.

Inclusive workplace and Culture

Mining have a dubious distinction of unhealthy workplaces. Sometimes, mining workers bear the brunt of unhygienic atmosphere. An inclusive and inspiring workplace should be created and maintained that enables for high performance. The environment should provide employees the freedom to actively explore new ideas and opportunities. It is an environment that fosters high levels of involvement and participation those motivates for performance; where people are empowered through engagement, inclusion and respect of their experience. To ensure Inclusive and Sustainable mining, a great Australian company and a leading provider of innovative and sustainable mining services promoted the idea of inclusive work place and culture. According to the company's mandate, "We know we need to work hard to attract, employ and develop people with exceptional skills who strive to make a difference. We know we are as only as good as the people we hire, and it is our people who shape our culture. We are well known for our progressive and forward-looking approach to our business and we are applying that same thinking to how we approach our culture. The greatest challenges are not overcome by a single person but, rather, by a powerful and united team and all of our employees play a part in maintaining the aspects of our culture that have led to our success. We've achieved what we have as we are a company that works together as one team to get things done. This is at the heart of what we do and is demonstrated through the way we work, the stories we tell, our management systems, our organisational structures and the behaviours of our people". The practice of this Australian company is worth emulating. Mining companies should ensure that their work places should be healthy, hygienic and safe.

Data Optimisation and Digital Transformation

Along with work place and work culture, data optimisation and digital transformation are essential for inclusive mining. Work force's future is greatly determined by digital transformation. Paul Mitchell has discussed on the work force's future, data optimisation and digital transformation for inclusive mining growth.[7] He emphasised that "The sector needs to be able to show how it's prioritising sustainable and inclusive growth to redefine its image as a responsible source of the world's minerals. Minerals those are critical not just to the digital age, but also the transformation under way to reduce carbon emissions and achieve the goals of the Paris Agreement. Creating long-term value through licence to operate will also set companies up for success in addressing the second greatest risk they face: the future of the work force. The work force's future moved up from seventh position in the risks-and-opportunities ranking last year, and executives are grappling with what their future workers will look like and how they can attract the right skills needed to support their businesses. An European Union (EU) study commissioned by the Minerals Council of Australia found that 77 percent of occupations in the sector will be enhanced or redesigned by 2030 because of the introduction of technological advancements. That's an immense work-force shift. - - - - - - - Addressing the future of work-force concerns is also a necessary step to tackling much-needed digital and data optimisation – the third most pressing risk on mining executives' agendas. The data landscape is complex to navigate and prioritise. Companies have yet to understand fully what data are important and how to extract value from them. These are the basics the sector needs to focus on. Digital transformation isn't about pushing the boundaries of innovation. It's about raising the visibility and performance of the business by embracing available technologies and maximizing the value they can generate. That's where data become crucial". Big data and digitisation is a buzz word in these days of innovation and technological revolution. Mining companies should sincerely work on its optimisation for transformation of their work culture.

Inclusive Financing

Financial inclusion has paramount importance in mines and mineral development. Inclusive mining will be a far cry without inclusive finance. The financial inclusion should be of both accessibility and affordability. Easy access to affordable financing to mining companies, corporations, small scale mining entrepreneurs and artisans should be guaranteed by government. This has been followed in South Africa and European Union (EU) countries. The African Guarantee Fund in partnership with the African, Caribbean Pacific (ACP) - European Union (EU) Development Mineral Programme has highlighted the concept of financial inclusion for Development of Minerals and small-scale mining enterprises. The programme seeks to improve the management of Development Minerals in Zambia and was officially launched by Ministry of Mines and Mineral Development. UNDP Country Director, Mandisa Mashologu emphasised the importance of financial inclusion of small and medium scale domestic businesses in the Development of Minerals sector. He aptly pointed out, "access to affordable finance will provide entrepreneurs in the Development of Minerals sector with economic security, resulting in a positive multiplier impact on health, education, poverty, hunger, inequality, jobs and growth". The programme is supporting sustainable and inclusive development in the industrial minerals, construction materials, dimension stones and semi-precious stones sectors through capacity development of key stakeholders such as regulatory agencies and local governments; private stakeholders including small-scale mining enterprises, construction companies, mining and quarrying associations; as well as training centres, universities, civil society organisations and community groups. The Credit Guarantee Facilities are provided to financial institutions (FIs) in Zambia, Uganda, Cameroon, Guinea (Conakry) and Nigeria for boosting mineral development. Other mineral rich countries should also adopt such financial inclusion programmes.

Inclusive Investment

Inclusive Investment in Minerals sector plays a crucial role in ensuring inclusive mining. Mining constitutes a vast arena of investment and investor's goal is not always to haunt for profit rather a responsible investor should earmark a portion towards addressing socio-economic hazards and adverse environment impacts. Corporate Social Responsibility (CSR) and Entrepreneur Social Commitment (ESC) should form an important and integral part of institutional investment in mineral extraction. According to Joseph Kirschke, CSR in mineral extraction itself is relatively new – even newer is the risk mitigation that drives institutional investment decisions, particularly among larger European banks, insurers and pension funds. Yet despite its dire need, systematic, proactive financial engagement remains elusive – even as successful examples abound worldwide. Participation by non-traditional investors in such endeavours is emblematic of a true triple bottom line: ensuring lasting positive impacts and maintenance of a social licence to operate – ahead of a resurgence in commodities growth. Social impact entrepreneurs, whose cross-industry activities advance social performance before financial returns, can make a decisive difference[8]. Some reputed companies have taken exemplary initiatives in this regard. A fixed portion of company's profit need to be earmarked towards Corporate Social Responsibilities (CSR) or Enterprise Social Commitment (ESC) activities. Inclusive investment should be the moral obligation of the company. Peripheral infrastructural development, socio-environmental upgradation etc. should be made mandatory provision by the government before granting licence to operate.

International and National Experiences

Programmes for, Policies, Principles of and Practices on Inclusive Mining differ from country to country so. Nor do the Acts, Rules and Laws are uniform across the countries. Different countries have, therefore, different experiences. Experiences of some mineral rich countries are cited below.

The government of South Africa has formulated a new Mining Charter for a modern and investor-friendly mining sector by providing regulatory certainty to investors with a reasonable assurance. The government adopted a favourable tax regime. Mine owners are assisted through favourable tax schemes and that indirectly helped to increase employment opportunities and creation of mining jobs. Appropriate safety regulations were promoted to correct potential dangers and take suitable actions against unsafe mines. In order to make mining more inclusive, South African government helped new entrants to gain a foothold by financially assisting these new entrants. Mmusi Maimane, Democratic Alliance (DA) Leader had the opinion that, "If we're talking about building a fair and inclusive mining sector in the broadest possible way, then surely helping young black South Africans to enter the industry is a great way for mining companies to make a difference. We need better partnerships with communities in order to establish a better developmental path towards prosperity for all. We will not create opportunities for black South Africans in mining by handing over bigger and bigger ownership shares to politically-connected individuals. All that will do is to sign mining's death warrant. The DA is not opposed to the handing over of shares in mining companies, but this has to be to ordinary workers and their communities, and not to a wealthy elite. And even this is only a small part of building an inclusive mining industry".[9]

Ghana has adopted some inclusive policies to boost mining sector of the country. A study conducted by The Ghana Chamber of Mines in 2013 indicates that "the mining industry is one of the most lucrative industries in Ghana. Ghana benefits from the Industry because many minerals abound within the country; chiefly among them are Gold, Manganese, Bauxite and Diamond. In the past few years, there has been an increasing demand for these minerals by developing countries. Non-renewable natural resource wealth is seen as an opportunity to obtain economic benefits that can be transformed into increased wellbeing for citizens. However, mining often involves a high environmental impact and the risk of causing, or exacerbating social problems and conflicts in mining areas is significant. According to Liebenthal et. al., mining could be a 'resource curse' bringing negative impacts on growth and development of resource rich countries with substantial negative environmental and social impacts. It is only through sustainable development initiatives these challenges and risks can be curtailed whilst ensuring that wealth creation can be long lasting for the benefit of future generations too. The mining industry can practice sustainability by promoting sustainable procurement practices whilst ensuring that all stakeholders in the mining supply chain play their roles to enhance monitoring and control of mining operations in the country.[10]

An overview of mining in Ghana reflects that the mining industry of Ghana accounts for five percent of the country's GDP and minerals make up 37 percent of total exports, of which gold contributes over 90 percent of the total mineral exports. Thus, the main focus of Ghana's mining and minerals development industry remains focused on gold. Most of the miners in Ghana are, however, small-scale miners. According to the Mineral and Mining Act, 2006 (Act 703) of Ghana, small-scale mining is defined to include entities that are owned by co-operatives and individuals. According to Chatterjee, 1993, it is also considered as a low capital-intensive form of mining in the country. It is a form of mining with little or no use of mechanisation and, therefore, makes it labour intensive. According to Barber, most small- scale mines are described as surface and or shallow underground mines. The depth of small-scale mining largely depends on the depth of the deposit, which is a natural attribute not subject to human manipulation. As a result of the large numbers of small-scale mining companies operating in the country, supervision of their activities are somehow difficult by the EPA and the Mineral's Commission which accounts for their ability to veer from the laws on mining and environmental protection and human rights". However, inclusive regulations and laws were promulgated by the Government of Ghana to address the outstanding issues of the mining industries and other stake holders.

In a sharp departure, Inclusive and Sustainable Growth in mining is still a distant dream in Mongolia. Key impediments to inclusive mining growth are; weak economic governance, macroeconomic instability, slow progress on key structural reforms, uneven progress on social/human development, narrow economic base, environmental degradation and climate change. There is a large unfinished structural reform agenda in the sector. These factors are widespread and constrain Mongolia's efforts to foster inclusive and sustainable growth in mining.[11]

The case of Philippines is slightly different. The central government of Philippines has initiated some reforms in the mining tax structure but their mining policy did not adequately address the issues of inclusive growth of mining sector. Ronald U. Mendoza pointed out that mining has become a divisive issue, yet it need not be. Efforts to clarify and sharpen the Philippines' mining policy have generated intense debate among stakeholders, spanning the business community, civil society and the central and local governments. The debate largely centers on increasing the government's share of the total mining revenues. The narrow focus on revenues runs the risk of keeping other important aspects of mining policy unaddressed, resulting in failure to generate cohesive support from all sectors. --- Very little evidence has been shared by the government on the possible impact of the new tax regime on existing and future mining investments (which would, in turn, impact on future revenues). He suggested that for the Philippines, the creation of an "inclusive growth trust fund" can be critically important, ensuring that mining wealth is invested transparently and consistently over time, and in areas that boost inclusive and sustained growth and development.[12]

However, mining in Philippines is not very much reflective of Inclusive Mining. Benefits of mining are a farfetched dream for indigenous people. Media has shown the 'dreadful conditions of the Philippine forests, floods caused by un-rehabilitated mining sites, and communities being displaced without any regard for their properties and livelihood. A lot of Filipinos have strong standpoints about mining; most of these are disapproving and negative. For most Filipinos, this is the side of mining they know of – destructive and harmful. Even the President Rodrigo Duterte stood firmly against mining, threatening mining companies that unless they will restore and rehabilitate exploited areas, he will "tax them to death" and use these taxes as funds to help affected communities. On the same hand, Former Environment Secretary Gina Lopez also showed strong disapproval of mining even at the onset of her short term, telling potential mining investors to go away if they were only to abuse Philippines's pristine forests and natural resources. All this, rallies of environmentalists, and pleas of abandoned indigenous people are what we make of mining in the Philippines'.[13]

India, which is endowed with various minerals, there are high possibilities of adverse impacts on socio-environmental spheres. The negative impact of mining can be checked by adopting the inclusive approaches in mining operations paving the way for inclusive economic growth. "The Indian mineral industry comprises of large and small mines that function under public, private sector and informal sectors covering most minor minerals being extracted. The public sector continues to play a dominant role in production of various minerals (coal, lignite, petroleum, iron and steel, bauxite and aluminum). As per the Constitution, development and regulation of mines and minerals is controlled by the Union government. The Mine and Minerals (Development and Regulation) Act 1957 was amended in 2015, to add the provision for a simple and transparent mechanism for granting mining lease or prospecting licence through competitive bidding besides assured tenure and easy transferability of mineral concession granted through auction, strict penalty provision to deter illegal mining. Although mining enterprises tend to meet the legal requirement of preparing mine closure plans, implementation falls short. Local communities are not consulted in the preparation and implementation of mine closure plans.

A large number of small mines (including quarries for extracting minor minerals) operate in most states. These present difficult challenges for sustainable development as their financial, technical and managerial limitations restrict their ability to take effective corrective measures against the negative impacts of mining. Major mining companies use advanced technologies, adopt comprehensive environment protection measures, sensitise their personnel on sustainability issues and progressively try to improve their environmental performance. There are other large, medium and even small enterprises whose environment obligation consist in strictly conforming to the prescribed legal provisions. In case of illegal mining, these niceties are not observed. It is reported that the illegal sand mining is largely responsible for the environmental degradation especially the river beds of Ganga and Yamuna rivers. Lack of checks and political interference has turned the situation grim. Even legal mines flout norms and damage the environment. A Planning Commission report titled "Sustainable Development, Emerging issues in India's mineral sector" observed that in the mineral-rich states of Odisha, Goa, Karnataka and Jharkhand, mining has brought about economic development. At the same time, it has caused significant environmental damages and negatively impacted communities in project areas. To that extent, the mining and environmental laws and regulations have not been very effective. In some cases, mining operations have been carried out without concerning for the 'carrying capacity' of the environment and other infrastructural limitations. This has put avoidable pressure on the environment and caused inconveniences to the people living in the mining areas. Illegal mining in many cases has similar effect while additionally causing loss of public revenues.

Inclusiveness and Sustainability have assumed considerable importance in major mining nations like Canada, Australia, South Africa and Papua New Guinea. These countries take a comprehensive view of sustainable development in mining that includes apart from environment, other important dimensions such as local stakeholder engagement, socio-economic development in mining project areas and transparency in communication with stakeholders. Major mining companies have set up foundations to take up socio-economic development projects in their mining areas. In developed mining nations all regulatory mining and environmental laws meant to lessen the impact of mining are strictly implemented. These include provisions for mine closure and associated reclamation and rehabilitation of mined land".[14]

Conclusion

The lifecycle of a successful mine project is more than hundred years. Operation itself continues for a century in resource abundant areas, let alone the other activities like Exploration, Site Design and construction, decommissioning etc. The post-closure and winding-up activities take around a decade. Therefore, it is quite natural that a lot of issues involved in a mining project warranting an inclusive multi-pronged approach to deal with. Inclusive model in mining sector has wider ramifications.

However, there are mostly, as identified by some authors, three key areas for action in the Framework of Mining as a partner in supporting more inclusive and resilient societies viz. lives (health and safety), livelihoods (jobs and income) and learning (education and skills). Through these lenses, companies can ensure inclusive mining by identifying 'practical actions to take across their core business activities, community investment programmes and policy engagement that support community rebuilding and resilience'. The coveted objective of Sustainability and for that matter Sustainable development in mining sector can be ensured through Inclusive initiatives and approaches.

End Notes & References

1. Andrew Lane and Riccardo Reggio. *'Mining in Africa: How Inclusive Solutions Can Mitigate Risk'*.
2. Fahmina Rahman. *'The mining debate: let's talk about inclusive growth'*.
3. Frederick Bird. *'The Practice of Mining and Inclusive Wealth Development in Developing Countries'*, Journal of Business Ethics, volume 135, pp. 631–643, 2016.
4. Robert Goodland. *'Responsible Mining: The Key to Profitable Resource Development'*, 2012.
5. The Wall Street Journal Online *"Indigenous Peoples Get Last Word on Mines"*. 26th March, 2012.
6. http://www.imii.ca/fileadmin/user_upload/IMII%20News,%20Articles/ 2018/ Inn.
7. Paul Mitchell. *'Mining companies must show how they are prioritizing sustainable and inclusive growth to redefine their image.'*
8. Joseph Kirschke. *'Mining: Inclusive Investment for a New Triple Bottom Line'*, Editor's Note, PR 13, 2015.
9. Issued by Mmusi Maimane – Leader of the Democratic Alliance, 04 Oct 2017 in News.
10. (1) Meshach Awuah Gyawu, (2) Samuel Brako and (3) Emelia Darko Adzimah; *"Assessing the Challenges to Sustainable Mining: A Supply Chain Perspective,(A Case of Li Tong Mining Company in Ghana)"*, European Journal of Business and Management.
11. Inclusive and Sustainable Growth Assessment, Mongolia, 2017–2020.
12. Ronald, U. Mendoza, *'Mining for inclusive growth'*, Philippine Daily Inquirer / October 16, 2014.
13. Aliyah Jeande Muta, *"It should n't just be mining"*, May, 2018.
14. (1) Meshach Awuah Gyawu, (2) Samuel Brako and (3) Emelia Darko Adzimah, *Op.Cit.*

Chapter - 3

Sustainable Mining: Issues and Challenges

The minerals are, no doubt, basic and strategic materials for industrial and economic development. But mining has its own inherent negative and adverse impacts. The negative impact of mining, however, can be controlled through application of the concept and principles of sustainable development to mining operations. Sustainable Mining is crucial for promotion of inclusive growth.

Sustainable Mining

Sustainability in mining means creating a balance of minerals – that is meeting today's needs without compromising future generations. Sustainable Mining is that the resources used currently should ensure their availability to future generations. It is a trade off between present wants/necessities and requirements for future. Sustainable mining refers to mining materials and energy resources, onshore and offshore, in a way that maximises the economic and social benefits while minimising the impact on the environment. Sustainable Mining Development is developing mining and minerals in such a way that future generations should have as much or "enough and as good".[1] In other words, fulfilling the current needs taking into account the 'ability of future generations to meet their own needs'.[2]

Sustainability is a management process used to secure the acquisition of goods (products), services and creation of wealth in a way that ensures that there is the least impact on society and the environment.[3]

However, as a concept, 'Sustainable mining' delineates the social, economic and environmental responsibility for the present and future generations. In recent years, it has been a catchphrase and oft discussed term. There have been growing concerns by various governments, organisations and pressure groups to promote sustainability as a result of the benefits that can be accrued by the economy, the environment and the society at large.[4] Major objectives of sustainable procurement is to protect human health; promote fair working conditions; promote social enterprises and improve local skills; reduce soil, water and air pollution, reduce resource consumption; reduce material packaging and waste; and to protect habitats and promote biodiversity.[5]

The underground resources is not always infinite and unlimited. The mining operations essentially depletes the resource besides entailing adverse and negative impacts though mining has a significant dent on economic development. Ultimately, for the mining companies to survive, sustainable practices should be at the heart of operations during mining lifecycle from exploration to exploitation to closure in three areas i.e., environmental, economic and social sustainability.

An elaborate definition for 'Sustainable Development' based in the context of Indian mining sector developed for analyzing and prioritising issues to outline the Sustainable Development Framework as "Mining that is financially viable; socially responsible; environmentally, technically and scientifically sound; with a long term view of development; uses mineral resources optimally; and, ensures sustainable post-closure land uses. Also one based on creating long-term, genuine, mutually beneficial partnerships between government, communities and miners, based on integrity, cooperation and transparency".[6] The aim of sustainable mineral development is to ensure mineral use is kept to a minimum amount without having a negative impact on economic growth. This is achieved by efficient use of minerals as well as recycling and the use of alternative materials.[7] Using recycled materials or renewable resources when building is an example of sustainable development. ... Building a new community in a previously undeveloped area without destroying the ecosystem or harming the environment is an example of sustainable development.[8] Therefore, sustainability is made up of three pillars: economy, society, and the environment. These principles are also informally used as profit, people and planet.[9] Some authors have discussed about the four types of sustainability. The term sustainability is broadly used to indicate programmes, initiatives and actions aimed at the preservation of a particular resource. However, it actually refers to four distinct areas: human, social, economic and environmental, known as the four pillars of sustainability.[10] However, there is no single universally accepted and adopted definition of sustainable mining. According to James (1999), miners can achieve sustainable development by embracing the social, environment and economic pillars. Eggert (2006) was of the view that sustainable mining can be achieved through the simultaneous pursuit of sustained or enhanced environmental quality, economic growth and social justice. Sustainable mining should be "Economically feasible, Ecologically viable, Socially desirable, Provide both profit and benefit, Use resource efficiently, Internalise environmental and social cost, Restore ecosystems and services it provides, Engage neighbours, Respect property, culture, and safety, Follow spirit and letter of law and Contribute to communities".[11] Although a number of variations exist, sustainable development is most commonly defined as development that meets the needs of the present without compromising the ability of future generations to meet their own needs. The principles of sustainable development involve integrating economic activity with environmental integrity, social concerns, and effective government systems. These principles have had a growing influence on the development of environmental and social policy in recent decades, and have been adopted and promoted by a number of international organisations, including the United Nations and World Bank. A number of industry associations, including the International Council on Mining and Metals, and the Mining Association of Canada, have endorsed the principles of sustainable development and have developed their own guidelines to promote sustainable practices among their member organisations. "Mining industry is closely linked with sustainable development, which leads to positive and negative impacts of this industry on three key principles of sustainable development, including society, economy, and environment. For ideal implementation and integration of sustainable development in the industry, the three principles of sustainability should be simultaneously applied and not sequentially at different stages of the mining lifecycle to maximise the opportunities in the mining industry to contribute to comprehensive development. A suitable framework for sustainable development will be created through the interaction of these principles with activities carried out in different stages of mining lifecycle and their implementation in the mining projects can be economically profitable, environmentally clean, and socially responsible".[12] Sustainability, therefore, means many things. It is the triple bottom line, along with people and planet, it can be profit or performance or policy. It is the shared value, informed consent, community engagement, governance, transparency, social lifecycle assessment, lifecycle costing, social impact assessment, environmental impact assessment, social risk mitigation, corporate social responsibility, sustainability reporting and sustainable development. Though there is a wide variation in circumstances and approaches, the principles of ecological sustainability, economic vitality and social equity are the grounding principles of a sustainable mining. The obvious logical questions next would be; what is the necessity or need of sustainability; is mining sustainable; if yes, what are the principles and major challenges of ssustainability? Let us discuss these in the sections that follows.

Why Sustainability?

The minerals are basic and strategic materials for industrial and economic development at large. The extractive industry of mining is one of the most sustainable industries in existence. Unlike "renewable" agriculture and logging, metals once mined remain in human use forever as they are recycled again and again to be made available to all future generations.[13] The negative impact of mining could be controlled through the application of the concept and principles of sustainable development to mining operations. Sustainable mining is crucial for the promotion of inclusive growth. In addition to a reduced environmental impact, mining firms can find the unexpected benefit of financial gains from improved sustainability. ... So there are clear financial gains to be had from sustainability". Ultimately, for mining companies to survive, sustainable practices should be at the heart of operations.[14] Since mining holds the promise of economic development, sustainability in mining is utmost important for sustainable development. Hence, 'the world needs mining to change toward sustainable mining practices to benefit communities and protect the environment. There is a need to think how we can re-build and re-brand mining, together. Mining is at a critical juncture. There is talk about "the mine of the future" and the introduction of crucial innovations such as automation and mechanisation (robots) and the extensive use of data analytics to optimise mining performance. Sustainability can be a catalyst to these changes as it addresses all facets of the mining value chain, from input sourcing and development of local supply and renewable energy to delivering a sustainable product to the market'.[15] Leading mining companies around the globe are increasingly adopting the cost-reduction and site safety measures to ensure implementation of sustainability practices. Shane Mc Lendon observes that the mining industry holds a number of strengths that mean it can not only follow sustainability guidelines, but play an active part in leading forces for change. With the environment increasingly in the spotlight, the growth of mining firms will be largely decided by how they adapt to the pressures of sustainability. While it may seem contradictory, the mining industry actually holds the power to take a key role in creating a more sustainable planet. According to Tom Melbye, a Normet senior advisor and member of the board of directors; “There is a big push in the mining industry to find more sustainable solutions, particularly in terms of equipment or methods. But in every company, you must start from the top. If senior management are really tough and focused on sustainability, it is easier to pass the message down to operator level and ensure everyone understands what is required and why it is important”. Essentially, for modern mining organisations to continue operating, it is necessary to listen to social, environmental and economic sustainability guidance while viewing environmental practices as a central component of all planning and development. Environments, Ecology, Climate change, Greenhouse effects etc. have been a guiding factor for any industrial operation now-a-days. Protective measures on these fronts are standard norms for any unit before being critical. The mining companies as well as operators are well versed to environmental and energy-related challenges both in short term and long term perspectives. Besides, reusing and recycling of materials as much as possible and reduction of waste has been a practice with most of the industries. In a sense, a metal once mined can be recycled again and again. Moreover, as a part of sustainable solutions, mine operators are using modern equipments and adopting upto date, cost effective and innovative methods of mining and safety measures. As a matter of principle, major mining industries, in recent years are normally following these sustainability guidelines.

According to Melbye, equipment selection should play a key role in these sustainable practices: “The world is committed to a CO_2 reduction. Here at Normet, we are very focused on that too, whether that’s with support systems or site equipment. For example, we are offering a range of battery driven equipment with no diesel fumes. Another issue that is particularly important for sustainability is to reduce the rebound in the spraying of concrete. With ordinary equipment, rebound this can be up to 30-40 percent, but with Normet equipment and the right operators, you can get this level below 10 percent. This leads to a big improvement in the working environment and the amount of dust creation as well as overall productivity. Plus, by reducing rebound, operators are able to use a lot less material, or using the same amount of material and covering a bigger area. All of this makes the process more sustainable”. Those apart, mining companies can make financial gains by adopting sustainable practices. ‘In addition to a reduced environmental impact, mining firms can find the unexpected benefit of financial gains from improved sustainability. Though green policies can seem costly, they actually help to give a company a competitive edge. Plus the ability to recycle and reduce waste has clear links to an improved bottom line'.

"The mining industry is starting to take sustainability seriously, realising that if they don't act on it, they won't survive," says Melbye. "After all, sustainability is increasingly connected to productivity, which is in turn linked to economics and profitability. So there are clear financial gains to be had from sustainability". Ultimately, for mining companies to survive, sustainable practices should be at the heart of operations. Many firms are beginning to understand that in order to maintain economic viability, it must take an environmentally and socially responsible approach at all levels of the operation.[16]

According to Prabhash Gokarn, 'Sustainability was a prerequisite for a license to operate and to quell threats from NGOs and progressive governments that were perceived to be only remotely linked to operational performance. The world has changed since then. Today's executives understand that sustainability can improve operational efficiency and lead to both short- and long-term benefits. Sustainable mining becomes an essential element—not only in recovering from the latest mining slump but also for the industry's long-term survival. Sustainability also drives long-term cash benefits for new mines and those that are already operational, including providing immediate cash improvements.

Mining executives know there is a significant risk to the business if sustainability does not become embedded in the culture and in day-to-day work processes. Already, activist shareholders are gaining strength and becoming more vocal, expecting sustainability to be front and center in a mine's corporate strategy. Leading mine owners and operators are increasingly developing plans to take into account the interests of the people and the environment around them, bearing in mind local circumstances and a mine's lifecycle. Their focus is on the economic, social, and environmental effects of their operations, providing a safe, healthy work environment, and raising the socioeconomic conditions of the local population by providing grants, housing, and an improved infrastructure. Their sustainability efforts reflect the value generated for the company and its stakeholders, while reducing the negative effects on communities, vegetation, wildlife, water, land, and air.

Yet, these leading companies are still few and far between. Many mine owners and operators continue to manage sustainability with a long list of initiatives that generally lack cohesion, strategic inclusion, prioritisation, or clarity on how each one addresses the challenges. Often performance is reported through Global Reporting Initiative standards and sustainability reports, and while the tone of these reports is usually positive, sustainability continues to be a huge challenge. The lack of sustainability is at the core of some recent problems and incidents within the industry. An example is the Marikana incident in South Africa. Was it only a labour dispute or the result of deeply rooted frustrations about poor housing and lack of community development? Violent and destructive strikes such as Marikana will continue unless there is better community engagement and development. A fragmented approach to sustainability will not yield optimal results. Making a lasting impact on the mining industry requires an all-encompassing sustainability strategy'.[17]

Furthermore, globalisation calls for an urgent need of Sustainable Development in mining sector. Many problems are no longer local, regional or national concerns. There is growing evidence that ecosystems are seriously out of balance, posing new and poorly understood risks. The problem of poverty is inseparably linked with global ecological problems which cannot be solved unless people have a stake in the outcome and the resources to manage the problems. Governance systems have not yet emerged for the reality of a globalised economy. The new world information regime ushers a growing awareness and expectations across the globe.

IS MINING SUSTAINABLE?

Mining, it is generally perceived, is inherently unsustainable due to its negative impact on ecology, environment and society. It destroys the biophysical environment, its contributions to human well-being are uneven and not significant, often overwhelmed by the social and economic damage it inevitably inflicts.

According to R. Rajaram and K. Parameswaran,[18] the belief of unsustainability is based on three commonly misunderstood observations that no single mine can operate forever, mining resources are finite and non-renewable, some historic mining practices have had negative impacts on the environment that are no longer acceptable. Responses to these misconceived prepositions are that no single mine can operate for ever is not relevant because Sustainability encompasses many values than the continuing availability of the resource being developed. Indeed, it is the very fact that mineral development will end someday that makes the integration of those others sustainability considerations in to the mining process highly appropriate. In that sense, manufacture of the many advanced machineries are not durable and everlasting. But that technology is not unsustainable. Many have longer lifecycle. Many metals, once produced, are capable of being recycled, smelted time and again to their original elemental form and refined to the most demanding specifications. Metals like copper, gold and silver can be recycled or fabricated. Metals, through recycling, are infinitely renewable and are, therefore, useful to future generations, this is essence of sustainability. Mining operations must be conducted in a manner that is protective of the human health and the environment by minimising impacts.

Mining contributes to sustainable development by creating Human, Social and Financial Capital. As regards Human Capital, mining provides facilities for education and training. Trained and educated people can play a constructive role in both production of wealth and development of social institutions and building of societies. So far as social capital is concerned, mining contributes to community wellbeing and stable social order. Mining creates Financial Capital by generating employment through creation of jobs and self-employment opportunities.

It is possible for mining to contribute to sustainable development through enhancing the benefits while mitigating the negative impacts both when mining is taking place and subsequently as well as improving stakeholder participation in the management of the resources problem. The mining process is extremely demanding on resources such as water, land, and energy. The ecological and social degradation that results from the mining process prevents the industry from complying with the "Four System Conditions for Sustainability" as per The Natural Step. The current practices of minimising impacts and enforcing regulations do not address the systemic sustainability challenges within the process[19]. Mining companies should; Create conditions in which individuals and groups can develop their potential, enrich lives, advance technology, Provide an adequate standard of living for those who do not have enough, in area of influence and Stay within the capacity of ecosystems to absorb change[20]. "Anything we utilise has either been grown or mined. The most useful, sturdy and long lasting of things, such as concrete, steel and copper wire, are made from material that is mined. Metals are not naturally occurring in forms that are readily useful. The cost and effort to extract and refine elements such as iron and copper are so high that metal has always been recycled which is possible because the properties of metals are not lessened by nature over time, unlike wood and other "farmed" substances. The bible speaks of turning swords into ploughshares and that is what even industrial societies do today. For example, roughly 50 percent of all copper "consumed" in the United States is recycled. As for steel, it is estimated that in the US, 90 percent of steel used in manufacturing is recycled after an average life for each product of 22 years. There are no government mandates that force this on people, it is simply just good sense on account of the fact that it is cheaper to melt a metal item and remake something new with it than to search for a new copper mine, develop it, dig out the copper, refine it and then make the new item. In this very real sense mining is infinitely sustainable; the needs of future generations will be met over and over again by the metal that past generations have found, extracted and refined. Before the metal was found, extracted and mined, it was useless to humankind and there was nothing to even evaluate as sustainable. The fact that the metal has been removed from the ground and no longer exists in that place is not a sign of non-sustainability; rather it has created a permanent and unending supply of copper that will be recycled and reused ad infinitum. The obvious question here is to whom should we look to determine what present needs are and how can we know what future generation's needs may be so that we can ensure we do not compromise them. What seems to be implicit here is that these same needs and abilities must be determined by those in government who, in order to ensure that society is sustainable, need to force us to act according to their assessment of what future generations' needs may be. Somehow metal mining seems to be one of those activities generally regarded as unsustainable.[21] "There are various ways to sustainably use the minerals. 'Recycle and reuse' are the most Sustainable ways to use non-renewable mineral resources. In this direction, an important initiative viz. "pollution prevention pays" programme was started by Minnesota mining and manufacturing company. Under the programme, they have adopted four pronged strategy such as; Redesigning equipment and processing Using few hazardous raw materials, Identifying toxic chemical outputs and Making more non-polluting products.

There are many views and counterviews, points and counterpoints regarding 'sustainability of mining'. The term 'mineral', according to Saleem H. Ali, "is often used to define various naturally occurring materials extracted from the geological crust of the Earth. Yet, for questions of sustainability, we need to differentiate metallic minerals from non-metallic ones, because 'renewability' for operational purposes will be defined by the level of entropy (or disorder) that mineral use will generate. The question of renewability from a chemical perspective is simply one of expending enough energy to bring back the material from a higher level of entropy to allow for reuse or recycling. Energy needed to counter the entropy created by the mineral's use is the main metric to evaluate whether that material's use is sustainable or not. From an operational perspective, metallic minerals are used in lower levels of entropy. That is why we are usually able to recycle them, whereas with minerals like coal, the use itself converts the material to such a high level of entropy (in the form of carbon dioxide) that it is essentially non-renewable. If we can design products that can retrieve minerals in usable form with relatively low energy expenditure and restorable environmental impact (particularly if the energy utilised for recycling is from renewable sources), then mineral usage is indeed sustainable. From an economic perspective, the extraction process of a finite resource from the Earth's crust can still lead to sustainable development so long as the capital generated is invested in building a diversified economy. This 'weak sustainability' aspect also applies to fossil fuel-extraction economies, thus deeming them non-renewable but sustainable. Indeed, the natural resource base of some areas may necessitate mineral profits as the catalyst for a longer term development path. Rather than a simplistic rejection of minerals as 'non-renewable,' environmentalists must be willing to grapple with the chemical, ecological and economic nuances of material extraction".[22] Jamie Kneen is of the opinion that "Mining is inherently unsustainable: It is destructive to the biophysical environment, and its contributions to human well-being are uneven and often overwhelmed by the social and economic damage it inevitably inflicts. Mining must be drastically scaled back, not expanded. Where it is undertaken, it must be carried out carefully and conscientiously. There is a theory that although itself unsustainable, mining growth can contribute to sustainability by providing raw materials for cleaner, more efficient and more prosperous human societies and generating wealth and employment that can serve as a 'bridge to sustainability.' This theory, however, depends on fulfilling a set of conditions that are unrealistic given the global context: a profit-driven economic system, flawed accounting that treats collective and ecological goods as externalities, tax avoidance and financial secrecy, and weak governance relative to the power of extractive corporations. To the extent mining can be seen to have contributed to development, it is at a considerable ecological and human cost. The more comprehensive and thoughtful the plans for 'mining for development,' the clearer it is that such objectives are unattainable. Regarding the path toward sustainability, we must reduce demand, reuse manufactured items and recycle materials. We must integrate the true costs of extracting and processing raw materials into decision making. We must place strict limits on where and how mining occurs. We must mine deposits at a slower pace to minimise environmental and socioeconomic disruption and maximise benefit. We must be able to identify ecologically and culturally sensitive areas as 'no-go' zones. Indigenous peoples must be able to exercise free, prior and informed consent and mining must be subject to participatory and democratic decision making. We must not build mines that are likely to require 'perpetual care' — for example, for heavy metal or radioactive contamination. We must place a real value on our precious geological resources, leaving them in the ground until they are truly needed and then extracting them with great care and respect".[23] JankiJul emphatically argues that "Of course mining is not sustainable. Minerals do not grow back. I agree with Salim Ali that there must be more re-use and recycling as that should reduce extraction. But re-use and recycling do not make mining sustainable. They could help mining companies to keep going. In that sense the mining industry might be sustainable. But that is very different to an assertion that mining itself is sustainable. Mining continues because the true cost of extraction and manufacture (destruction of the environment, damage to communities, exploitation of workers etc) is passed on to people who do not have the power to resist and protect themselves. That is unjust. We have to stop it. We also have to get past the fairyland theories of economists who believe in perpetual growth with finite resources. As human beings we are incomplete if we do not have the right relationships with the life that exists around us. John Harrington Jul had a different point of view. "The Starship Enterprise had technology that meets the criteria suggested by Salim Ali for recycling. It was called a replicator. We don't, as yet, have those on earth. We do, however, have many life-supporting environmental services such as the provision of oxygen and clean water. These services are usually diminished by the extractive activities required for mining. If all the costs, social and environmental, of mining were internalised to the price of the refined metals, perhaps we would be more judicious in our use of mined materials. As it stands, we now replace our smart phones and tablets well before they've reached the end of their economic or material life. Mining in today's economy can not be sustainable". Mark B Mar emphasised that EPA defines sustainability as "Sustainability creates and maintains the conditions under which humans and nature can exist in productive harmony, that permit fulfilling the social, economic and other requirements of present and future generations". Defining sustainability in this way, the question "Can mineral extraction be sustainable?" can have answers that are black AND white, and also many shades of gray. It somewhat depends on the scale one considers. From the scale of a single mine, the answer is clearly: No. Extract minerals from a site and pump them into our economy, and those minerals are gone from that site, never to return. The longer the refined mineral serves some purpose in the economy, the better from a "gray area" sustainability view. It's utility can be improved by ensuring that it's used for durable goods or is recycled ad infinitum, thus reducing the amount of new mining that must be done to meet our needs. But from the perspective of a single mine, mineral extraction is inherently unsustainable.

Across the broader economy and global environment, the question of scale becomes one of resource abundance, resource demands, and time. Given a long enough time scale, humans would theoretically extract every extractable resource from the face of the earth. Thus, closing the loop on resource use is critical. Paul Hawken wrote about this in The Ecology of Commerce. Waste from one process equals food for the next. A sustainable economy cannot create waste as we traditionally have in the past. Is it time to stop all new mining development because mineral extraction is unsustainable? We can't do that now. Our economy would grind to a halt. Our population is growing (a huge problem that is completely unsustainable), and demanding things like planes, trains, automobiles, computers, cell phones, roads, buildings, and a seemingly infinite array of things that depend on mined resources. In order for our economy to be sustainable, we need to proceed with carefully managed mining for the time being. But we also must close the loop and treat waste from one process as food for the next; and we need to stabilise human population (preferably at a lower level than present). Someday, presumably, we'll have enough mineral resources in durable, recyclable, and Recycled goods that we won't need to mine new minerals". He further went to say that "This posting was typed on a computer which was manufactured with mined mineral resources. The pixels on your screen are 100 percent recycled pixels from post-consumer waste. It's very much possible to make mining more environmentally sustainable. All we have to do is to develop and integrate practices that reduce environmental effects resulting from mining operations. Think about underground mining technology that has a low energy consumption rate instead of working with machines that are not fuel efficient. Sustainable mining is basically mining that meets the present demand without compromising the future generation's needs. We can make mining environmentally friendly". According to David MacKay, "The extractive industry of mining is one of the most sustainable industries in existence. Unlike "renewable" agriculture and logging, metals once mined, remain in human use forever as they are recycled again and again to be made available to all future generations. For example, it is estimated that nearly all of the gold ever mined throughout human history is still in use. What isn't in use has been lost in shipwrecks and the like.

Today the entire economy of the copper and copper alloy industry is dependent on the economic recycling of any surplus products. Nearly half of the worlds copper comes from recycled scrap, 50 percent of which (25 percent of the world total) comes from the US. It is similar for steel; about 60 percent of US steel usage comes from recycled scrap. In Mexico, a growing country, a lower portion comes from recycling and the world's copper mines are needed to make up the difference. Mexico currently imports about 30 percent of its steel, most of that from the US".[24]

It is pertinent to mention two relevant reports in this connection.

The London Declaration in May, 18-23rd 2001; specifies that[25] (1) A moratorium on new large-scale mining projects in Greenfield areas of Asia, Africa and Latin America; (2) Companies must clean up the terrible damage caused by their past and current mines, without drawing on public funds, and be held morally, legally and financially responsible for their misdeeds; (3) The World Bank/IMF cease funding of industry-initiated mining codes which are imposed on the governments of Africa, Asia-Pacific and Latin America; (4) Mandatory higher standards in all mining; (5) That surface and subsurface rights of indigenous peoples and all mining-affected communities be unequivocally respected and enforced, as well as their right to veto unacceptable projects. We seek solidarity from civil society and specifically from development and environment NGOs, in response to the global outcry from communities affected by mining. We ask these organisations: (1) To ensure that mining affected communities are fully informed in advance on all aspects of mining and minerals projects and empowered to speak for themselves in response; (2) To refuse to participate in initiatives, such as MMSD, which are primarily spearheaded by the industry to serve its own purposes; (3) To advocate for politically and legally enforceable measures that will hold the mining industry accountable, above all, to mining and exploration-affected communities.

The report "Rich lands poor people- is Sustainable Mining Possible" consisting of 356 pages prepared by New Delhi based Centre for Social Science and Environment (CSE) inaugurated in Odisha during December 2007 highlighted, inter alia, restriction on export of Minerals, Environmental Protection, use of both low and high quality minerals, water conservation in mining areas, Proper rehabilitation, compensation to displaced families and mining affected communities in mining areas along with arrangement for Jobs at least one per displaced family and others. The report goes on to recommend a range of policy initiatives that could help India meet this challenge. Some of its main recommendations include recognising people's right to say 'no' or in other words, mining should not take place without the consent of the people; independent, impartial preparation of Environment Impact Assessment (EIA) reports; disallowing mining in forests; framing stronger mine closure regulations; and "doing more with less – a key to sustainable development". Regarding sustainable mining, in India, 'State of India's Environment' reports by the New Delhi based Centre for Social Science and Environment (CSE) points out that mining cannot be sustainable or truly environment-friendly: one, because all ore bodies are finite and non-renewable and two, because even the best managed mines leave "environmental footprints". According to Chandra Bhushan, CSE's associate director, "It is about ensuring that mining is conducted in an environmentally and socially acceptable manner". The report says, "Odisha, for instance, accounts for 7 percent of India's forests and 11 percent of its surface water resources – it also holds 24 percent of India's coal, 98 percent of its chromite and 51 percent of its bauxite. Mineral industries are, naturally, flocking to the state. But for all its mineral wealth, the state performs poorly in terms of human development indicators. The state has a Human Development Index (HDI) of 0.404 – worse than that of Kerala, Tamil Nadu, Andhra Pradesh and West Bengal. The CSE report points out that Odisha's per capita income has actually declined during the second half of the 1990s – precisely the period when the state went on an industrial overdrive. All the mineral-rich districts of the state feature in the list of 150 most backward districts of the country. In Keonjhar, the most mined district in the state, 62 percent of the population lives below poverty line. In Koraput, the bauxite capital of India, 79 percent live below poverty line. Statistics indicate that the income from mineral extraction rarely benefits the regions from where these minerals come – in fact, poverty is increasing in many of these districts."

It is a well known fact that mining as an extractive industry inherently depletes a stock resource and metal recycling and efficiency can only to some extent postpone exhaustion, but cannot make mining sustainable. Hence, the concept of "weak or quasi-sustainability" has been coined i.e. 'mining can be considered to contribute to sustainable development if its economic benefits outweigh social and environmental costs, and if mining revenues are invested in building sustainable industries, enterprises and productive capacities. The "weak sustainability" principle posits that different forms of capital (natural, human, physical) are substitutable, although, in fact, the substitutability among them is not great. Activities can be considered "sustainable" if the overall stock of capital is at least not diminished and preferably augmented. This definition suggests that mining can contribute to sustainable development, but only if it gives rise to long-term net benefits (environmental, social, or economic) that equal or exceed the values that existed prior to exploitation. To arrive at the "net", all social and environmental costs and all external costs must be subtracted from the benefits. Since these costs are rarely accurately calculated, it can be hard to claim a positive net value. In addition, the 'Trickledown Theory'—that some fraction of the benefits accrued by the recipients of royalties, profits, and taxes eventually trickle down to the impacted people—is aspirational.

Over the past two decades, the mining industry has attempted to improve its image in terms of mine safety, environmental restoration, and community relations. Through ICMM and PDAC, the concept of sustainable mining has been promoted throughout the industry. Twenty years ago, no one at a mine talked about sustainability; today there are vice presidents responsible for sustainable development. Yet, sustainable mining, as defined by the industry, follows the model of "weak sustainability" that allows trade-offs among economic, social, and environmental responsibilities and is less stringent than the concept of Responsible Mining'.[26]

PRINCIPLES

Ecology, economy and society or what is generally called as socio-economic and environmental sustainability are the basic components of sustainable mining. Some authors are of the view that sustainable mining can be attained through "Systems Thinking" and "The Natural Step" framework. Systematic Approach to Sustainability should be one that contributes to a sustainable society. The mining industries need to constantly monitor its actions and process towards adopting Sustainability Principles. They have to constantly supervise and continuously initiate actions in eliminating its contribution to *systematic increases* in; concentrations of substances from the earth's crust and that produced by society, physical degradation of nature through overharvesting, introduction and other forms of modification, conditions that undermine people's capacity to meet their needs. Canons of 'system awareness' that is applying a systematic approach to the mining process requires that the elements of each system of sustainability should be identified and the interconnectedness between the socio-economic and environmental elements in the system should be established. Besides identifying missing links and gaps between current best practices and a desired vision of sustainability, this method will help in creating new opportunities for improvement.

The mining companies should invariably apply and measure the change in Socio-Economic and Environment systems in the entire mining process and assess the feedbacks in mining Life Cycle, Environmental Quality and Human Health. This process leads to a feedback process in resolving and addressing the outstanding issues of sustainability by identifying the indicators like minerals contained within Tailings, recycling facilities, landfill facilities, in-use of products under mining Life Cycle, the quality of air, water, and land under environment and the physical, mental and cultural features under the human health. In the process, many impediments and major systemic barriers that contribute towards social, economical and environmental adverse impacts within the mining process like problems in Post-Closure Reclamation, Acid Mine Drainage (AMD), Production and Accumulation of Waste, inadequate safety norms, etc. can be avoided or eliminated and a viable natural environment, community nurturing and well being as well as resource efficiency can be ensured. This will go a long way in achieving the goal of sustainable mining system. However, there is a limited guidance for mine operators to put sustainability framework and theory in to action on the ground.[27]

"Application of sustainability principles to mining is inherently challenging, as mining is the act of removing and consuming a limited resource. A common sustainable mining framework is focused on reducing environmental impacts of mining. Strategies for assessing the sustainability of mining operations include measuring, monitoring, and working to improve various environmental performance metrics, and these are used to determine whether a mining operation is sustainable. The key metrics for environmental sustainability in mining relate to efficiencies in resource consumption, minimizing land disturbance, pollution reduction, as well as closure and reclamation of exhausted mine lands. Another sustainable mining framework transitions from the emphasis on the environmental footprint of mining operations to responsible management of non-fuel mineral resources throughout their entire lifecycle, including use phase and end of life, with attendant implications for reducing the quantity of mined material and preserving reserves for future generations. The assessment of sustainability in mining should encompass a systems view of mined materials in society, emphasizing existing environmental sustainability metrics from mining operations as well as including "circularity" or "lifecycle" metrics to assess the sustainability of production and extraction for the long term".[28]

According to Chinamyi Shalya, "two main pre-conditions for achieving sustainability are good governance and self-regulating mining enterprises which are economically viable, financially profitable and technically efficient. Sustainability principles have application for all stages of mine lifecycle – exploration, mine planning, construction, mineral extraction, mine closure and post-closure reclamation and rehabilitation. These principles include elements such as intra and inter-generational equity, the precautionary principle, scientific mining, management of environmental and socio-economic impacts, creation of substitute capital in the form of social and physical infrastructure and stakeholder engagement.[29] Besides, safety, resource efficiency, environmental management practices and community engagement are important aspects of sustainability. "For both ethical and business reasons, a mining operation should aim to prioritise safety. Characteristics of safe mines include a commitment to risk management; appropriate attitudes and behaviours; reporting systems need to be in place; a focus on education and training; and a focus on processes and equipment (Laurence, 2005). Adopting environmental management practices on leading mine sites makes excellent business sense. Unless steps are taken in the planning and operational stages to protect environmental values, long-term liabilities such as acid mine drainage, may result. According to Agnimirh Basu of CSE, important environmental requirement for a mining project is a comprehensive environment assessment. A mine needs a 'social licence to operate'. Unless the community is engaged and supportive of a mining operation, opposition and confrontation may ensue. Hence, implementation of effective and transparent engagement, communication and independently verified reporting arrangements with all stakeholders are very important". Key Sustainable Mining Principles are to: 'implement and maintain ethical business practices and sound systems of corporate governance, integrate sustainable development considerations within the corporate decision-making process, uphold fundamental human rights and respect cultures, customs and values in dealings with, implement risk management strategies based on valid data and sound science, seek continual improvement of health and safety performance, seek continual improvement of environmental performance, contribute to conservation of biodiversity and integrated approaches to land use planning, facilitate and encourage responsible product design, use, re-use, recycling and disposal of waste products, contribute to the social, economic and institutional development of the communities'.

Sustainable Principles also include; leadership, collaboration, corporate Social Responsibility, communication etc. The Miner of the Twenty first Century must know how to monitor emissions and protect the environment, how to restore a mine site back to other uses and how to get along with the neighbours which are balance of three elements i.e. techno-economic issues, environmental issues and socio-political issues. The miners, before mining; should plan for the neighbourhood, during operation; should protect the environment and after operation; should plan for closure viz. mitigate/restore the site, sustain the communities by providing alternate employment, sustain the company, implement rigorous Environmental Impact Assessment process in six areas such as; land degradation, deforestation, subsidence, air pollution, water pollution, soil pollution.[30]

Some authors like Mayank and others have discussed about 'Sustainable Mining Framework' (SMF) and Sustainable Development Framework (SDF) for the Mining Sector. The SDF provides guidance for the mining companies to improve performance on environmental and social aspects, however, over time it can also become the common benchmark against which all mining operations may be evaluated in terms of their comparative performance on sustainable development terms,[31] as analysed Later.

Schneider Electric has discussed about the evolution of a sustainability programme. He outlined various steps for preparation of a Sustainability Programme viz. '(i) Develop Sustainability Strategy, Determine corporate profile, Set goals and objectives, Governance structure, (ii) Implement Reporting Programme' Develop management plan, Establish measurement and reporting structure, Operations and Products / Supply Chain, Set baseline with third-party verification, (iii) Evaluate Footprint and Prioritize Actions, Identify portfolio opportunities, Commit to targets', Create performance metrics (KPIs), (iv) Implement Optimisation Activities, Reduce (raw material, energy, water, waste), Renew (renewable energy, recycle, treat and reuse), Restore (offset impact, manage credits) Communicate Results and Performance, Employee and stakeholder engagement / functional teamwork, Voluntary / mandatory reporting, Investor / Customer scorecards, Supplier engagement and performance — Sustainability Strategy, Common Sustainability Challenges, Connecting vision with execution', Cost/benefit metrics for sustainable initiative decision making, and Determination of materiality integrated Sustainable Development Strategy'.[32] Regarding sustainability in the mining Industry, he is of the opinion that the small 5% segments who want to avoid sustainable development issues really avoid Sustainability Issues Well Established with Good Management Common Sense, Contextual Approach. The 80 percent who develop practices based on common sense in response to particular context, which in some cases results in good work but is generally not benchmarked to international standards and can be highly variable in the quality of both delivery and outcomes. The 15 percent who are well managed, reporting material issues, showing improvements, and are recognised as leaders.[33]

However, process of sustainable development is very rigorous and difficult to achieve but once achieved, the rewards are immense. The production, consumption and living patterns will be consistent with ecological balance, there will be; dramatic reduction of the numbers of people living in poverty, greater opportunity to develop individual and group potential and more open, cooperative, transparent and consensual systems of governance'.[34]

Key Challenges

There are a number of challenges mining industries and mine owners face today. 'Sustainability is a big word. Fourteen letters, six syllables, and a broad library of books and pamphlets, sustainability has become a catch-all for business activities related to an array of factors from air, water, and land to community relations. The concept of sustainability creates pressure to stop activities such as coal-fired power generation and to alter other behaviours such as fishing, farming, and most major industrial processes. Governments, communities, non-governmental organisations (NGOs), shareholders, and individuals are all calling on industry to promote a more sustainable world. However, the calls for sustainability vary widely according to location, situation, regulations, and population—and they tend to overlap. This makes it difficult for industry to respond. Does sustainability start with the community or the environment? With water or air? With frogs, bats, people, the supply chain, or assets? Mines have a lasting impact on a community's air, water, and land. Despite the commodity crisis, more mining companies are making sustainability a core element in their business strategies'.[35] Although there are a number of tools and guidelines that contribute to sustainable development, there is no consistent, integrated approach to supporting the national mining industry.[36] Sustainability a high sounding word with great expectations. Now, 'increased awareness and expectations flow from the new world information regime. Governance systems have not yet emerged for the reality of a globalised economy. The problem of poverty is inseparably linked with global ecological problems which cannot be solved unless people have a stake in the outcome and the resources to manage the problems. There are growing evidence that ecosystems are seriously out of balance, posing new, serious and poorly understood risks. Urgency for sustainable development stem from the very concept of globalisation which means that many problems are no longer simply local or national concerns. They have wider ramifications and severe repercussions.[37] Innovations and R&D activities are key to and basics of sustainability. But there is a glaring lack of these activities by the mining companies. Many people now say mining is at a critical juncture. They are talking about "the mine of the future" and 'the introduction of crucial innovations such as automation and mechanisation (robots) and the extensive use of data analytics to optimise mining performance. Sustainability can be a catalyst to these changes as it addresses all facets of the mining value chain, from input sourcing and development of local supply and renewable energy to delivering a sustainable product to the market. Besides, mining portfolio shifts add to the detriment of sustainability. Amid the commodity crisis, a trend has emerged: large companies are selling non-core assets to smaller firms that, in some cases, have less mining experience. These divestments can be related to the portfolio (focusing on fewer but more profitable assets), regulation (governments requiring a higher level of local ownership), and sometimes even the mine lifecycle (selling the mine before its end of life). This increases the industry's sustainability risk. Smaller mining firms are potentially less focused on and may lack the skills and resources needed for socially and environmentally sound mining. Any subpar practices will tarnish the industry's already tainted image, which could have severe medium- and long-term implications. In response, governments might opt to tighten the rules. The big mining houses would be wise to work with government and smaller mining firms to address these issues and support sustainable mining.[38]

There are instances of 'frequent occurrences of collapsed pits, less regard for the disposal of waste into river bodies, poor service conditions to staff of small-scale mining companies and poor corporate social responsibility to mining communities' which are avoidable challenges in sustainability of mining. A group of three researchers have assessed the challenges to Sustainable Mining through a case study of Li Tong Mining Company in Ghana during 2015. Also a study was conducted by The Ghana Chamber of Mines during 2013. These studies revealed that management and staff of the industry have little knowledge in sustainable mining practices whilst others perceive a low senior management support. The high cost of sustainable operations and many other challenges are key de-motivational challenges to the adoption of sustainable practices. The researchers found a positive effect of the challenges on their inability to practice sustainability and recommend a periodic rigorous training for all staff and management of all small scale mining companies. Some of the key and outstanding challenges illustrated by them are enumerated as under.

Lack of expertise and technical Know-how

Sustainability is quite a new concept and its implementation will need experts so that its full benefits will be realised. Africa and Ghana in particular lack sufficient experts in this field. The cost-benefit trade-off of sustainability must be well calculated in order to provide a cost effective sustainable products and services to the consumer with little or no negative impact on the economy, environment and the society. According to the Sustainable Procurement Taskforce (2006), without sustainability training, the motivation of procurers reduces and delivery of sustainable outcomes suffers.

Internal Stakeholder Resistance

Stakeholders are those who can affect or are affected by the achievement of the company's objectives (Freeman, 1984). An organisation's stakeholders may be either internal or external. Internal stakeholders include the owner, manager and employees of the organisation whilst suppliers, customers, the society, the government, shareholders and creditors are best grouped under external stakeholders of an organisation. According to Amy Robinson (2008), a successful sustainable procurement project requires greater stakeholder collaboration. If stakeholders are unable to perceive the benefits of sustainability, it becomes difficult for them to buy- into sustainable programmes and they turn to create a massive resistance. The resultant effect is that implementers of sustainable projects become de-motivated and eventually stop implementation.

Organisational Internal challenges

Organisational internal challenges are key to the promotion of sustainable development practices. Frequent closures and strikes by labour unions and apathetic attitudes by the management hamper sustainable practices.

Political and Legal Challenges

To ensure that suppliers or producers of products and services practice sustainability, it is of much importance that the government takes the lead in enacting laws to regulate the activities of producers of products and services so that the ordinary consumer will be safeguarded. The flood that occurred and claimed about 152 lives in Ghana which was attributed to the environmental unsustainability of some suppliers of fuel who built service stations in waterways (www.dailyguideghana.com, accessed June, 2015). The mining communities have seen worse environmental and social effects from mining companies who are not committed to environmental sustainability. These companies destroy large acres of farmlands, trees, pollute and divert river paths, create and leave large pits which endanger lives and fail to reclaim lands without being punished because of weak laws and weak institutions to enforce laws. This results in many deaths and serious poverty in most mining communities. The public procurement act (2003) which serves as a guideline for procurement in Ghana throws very little light on sustainable procurement. The result of poor government leadership leads to poor awareness of sustainability to producers or suppliers of goods and services in the nation.

Economic Challenges

1. Cost of implementing sustainability

There are many barriers to sustainable project implementation.

According to the findings of a survey conducted in about 800 businesses in New Zealand, management time and cost stand out as the major challenges (Lawrence et al., 2006). The cost of training, doing market research to source for sustainable suppliers of inputs, cost of purchasing machinery which use less of energy or are less dependent on non-renewable energy, cost of disposing waste sustainably, etc. are all major concerns for management of businesses.

2. Green wash and Absence of Eco-labels

Green wash is the practice of making false, misleading and/or deceptive claims about the environmental practices of a company or the environmental attributes of its goods and/ or services (Sustainable Procurement Guide, 2013). After perceiving the benefits of sustainability, management of organisations who recognise the barriers associated with sustainable production turn to use such bad device to create a false perception of their products and services to their purchasers and the public at large. Businesses who engage in this act are able to sell their products at a very low price to obtain a competitive advantage over businesses practicing sustainability. This results in the demotivation of businesses involved in actual sustainable production or operations. Green wash also introduces confusion in the minds of purchasers as to which products are sustainable or not. Purchasers who are unable to realise the benefits of buying sustainable products become disappointed as they are unable to obtain value for their investment.

Eco-labels are designs provided by organisations who provide certification to products and services. They are meant to protect the purchaser and the consumer (www.ecolabelindex.com/, accessed February, 2015), however, many developing countries do not have enough of such organisations who provide certification and eco-labels to suppliers of products and services. This makes the control of less quality products flooding the market very difficult for the few certification agencies to the disadvantage of the purchaser and the consumer.

3. Complexity of comparing cost and value of products

The quality of a product or service is perceptual to many researchers and purchasers irrespective of the level of details of specifications provided. Sörqvis (2001) agrees with the level of complexity in measuring the quality and cost tradeoffs of services and products. This problem poses a lot of challenges to consumers who seek to achieve value for money which may not be realised as a result of fake or poor quality products flooding the market. Businesses, purchasers and consumers are therefore unable to know which products will exactly meet their satisfaction with respect to the level of investment. Campanella (1999) suggests that the best way to make quality more tangible is through quality costing.

4. Social Challenges

Habit and the difficulty in changing human behaviour

Irresponsible disposal of waste by consumers and some businesses is a habit difficult to deal with. Many organisations and consumers are reluctant to separate their waste before disposal. Some poisonous waste find their way into river bodies which pose a lot of threat to health and safety. Many consumers in Africa fail to check expiry dates of products before use whilst others feel they will be sidelined by their society should they bring such companies and individual consumers who do not practice sustainability to the lime light for them to be punished.

5. Lack of Awareness of the Impact of Sustainable Procurement

Many suppliers and consumers have little or no knowledge of the impact of sustainable procurement. This accounts for the lack of support for sustainable projects and the less patronage of sustainable products by consumers. According to findings from a research conducted by Marita, R. (2011) awareness and demand for sustainable business practices are perceived as low. This is a key contributor to the reluctance of consumers and even management of some organisations to support sustainable projects and products.

6. Poverty

Many consumers are unable to patronise sustainable products as a result of their high cost. Poverty causes suppliers and workers to accept bad terms and conditions that do not show fair trade. Suppliers are willing to accept wages that are not fair compensation for their products or services and others who have poor hygienic working conditions still continue to give services for fear of being laid off. In some developing countries such as Ghana, many mining companies and cocoa farmers use child labour which is a big challenge to the sustainability of the society. Such children are deprived of quality education, health and safety which affect their productivity in the future.[39]

There are, however, immense challenges that threaten the very sustainability of our global society, let alone individual businesses or developments. A few of the most important of these challenges—population growth, clean energy supply, fresh water availability, and global climate change. Thus, common challenges facing the mining industry are; Access to Energy, Health and Safety, Access to Capital, Volatility of commodity prices, Environmental footprint etc. As resources in some areas become scarce or depleted, companies are forced to push new frontiers of exploration.[40]

Sustainable Development Framework (SDF) and Mining

International Council of Mining and Metals (ICMM) and International Union for the Conservation of Nature and Natural Resources (IUCN) have developed some principles and guidelines to take care of sustainability as well as development requirements in mining operation. However, ICMM, in agreement with all its members/companies have developed ten Key Principles of the SDF as a part of their performance obligations as ICMM members for continual improvement in their performance and contribution to sustainable development.[41] These ten principles are reproduced as under. It is intended that by implementing these principles, investments in the mining and metals sector should be financially profitable, technically appropriate, environmentally sound and socially responsible.

1. Implement and maintain ethical business practices and sound systems of corporate governance:

   • Develop and implement company statements of ethical business principles and practices that management is committed to enforcing. • Implement policies and practices that seek to prevent bribery and corruption. • Comply with or exceed the requirements of host-country laws and regulations. • Work with governments, industry and other stakeholders to achieve appropriate and effective public policy, laws, regulations and procedures that facilitate the mining, minerals and metals sector's contribution to sustainable development within national sustainable development strategies.

2. Integrate sustainable development considerations within the corporate decision-making process:

   • Integrate sustainable development principles into company policies and practices. • Plan, design, operate and close operations in a manner that enhances sustainable development. • Implement good practice and innovate to improve social, environmental and economic performance while enhancing shareholder value. • Encourage customers, business partners and suppliers of goods and services to adopt principles and practices that are comparable to our own. • Provide sustainable development training to ensure adequate competency at all levels among our own employees and those of contractors. • Support public policies and practices that foster open and competitive markets.

3. Uphold fundamental human rights and respect cultures, customs and values in dealings with employees and others who are affected by mining activities:

   • Ensure fair remuneration and work conditions for all employees and do not use forced, compulsory or child labour. • Provide for the constructive engagement of employees on matters of mutual concern. • Implement policies and practices designed to eliminate harassment and unfair discrimination in all aspects of our activities. • Ensure that all relevant staff, including security personnel, are provided with appropriate cultural and human rights training and guidance. (ICMM SD Framework – Final Principles Date: 29 May 2003 International Council on Mining and Metals Page:3 Document Ref: C 020/290503) • Minimise involuntary resettlement, and compensate fairly for adverse effects on the community where they cannot be avoided. • Respect the culture and heritage of local communities, including indigenous peoples.

4. Implement risk management strategies based on valid data and sound science:

   • Consult with interested and affected parties in the identification, assessment and management of all significant social, health, safety, environmental and economic impacts associated with our activities. • Ensure regular review and updating of risk management systems. • Inform potentially affected parties of significant risks from mining, minerals and metals operations and of the measures that will be taken to manage the potential risks effectively. • Develop, maintain and test effective emergency response procedures in collaboration with potentially affected parties.

5. Seek continual improvement of health and safety performance:

   • Implement a management system focused on continual improvement of all aspects of operations that could have a significant impact on the health and safety of our own employees, those of contractors and the communities where we operate. • Take all practical and reasonable measures to eliminate workplace fatalities, injuries and diseases among our own employees and those of contractors. • Provide all employees with health and safety training, and require employees of contractors to have undergone such training. • Implement regular health surveillance and risk-based monitoring of employees. • Rehabilitate and reintegrate employees into operations following illness or injury, where feasible.

6. Seek continual improvement of our environmental performance:

   • Assess the positive and negative, the direct and indirect, and the cumulative environmental impacts of new projects – from exploration through closure. • Implement an environmental management system focused on continual improvement to review, prevent, mitigate or ameliorate adverse environmental impacts. • Rehabilitate land disturbed or occupied by operations in accordance with appropriate post-mining land uses. • Provide for safe storage and disposal of residual wastes and process residues. • Design and plan all operations so that adequate resources are available to meet the closure requirements of all operations.

7. Contribute to conservation of biodiversity and integrated approaches to land use planning:

   • Respect legally designated protected areas. • Disseminate scientific data on and promote practices and experiences in biodiversity assessment and management. • Support the development and implementation of scientifically sound, inclusive and transparent procedures for integrated approaches to land use planning, biodiversity, conservation and mining.

8. Facilitate and encourage responsible product design, use, re-use, recycling and disposal of our products:

   • Advance understanding of the properties of metals and minerals and their lifecycle effects on human health and the environment. • Conduct or support research and innovation that promotes the use of products and technologies that are safe and efficient in their use of energy, natural resources and other materials. • Develop and promote the concept of integrated materials management throughout the metals and minerals value chain. • Provide regulators and other stakeholders with scientifically sound data and analysis regarding our products and operations as a basis for regulatory decisions. • Support the development of scientifically sound policies, regulations, product standards and material choice decisions that encourage the safe use of mineral and metal products. ICMM SD Framework – Final Principles Date: 29 May 2003 International Council on Mining and Metals Page: 5 Document Ref: C 020/290503.

9. Contribute to the social, economic and institutional development of the communities in which we operate:

   • Engage at the earliest practical stage with likely affected parties to discuss and respond to issues and conflicts concerning the management of social impacts. • Ensure that appropriate systems are in place for ongoing interaction with affected parties, making sure that minorities and other marginalised groups have equitable and culturally appropriate means of engagement. • Contribute to community development from project development through closure in collaboration with host communities and their representatives. • Encourage partnerships with governments and non-governmental organisations to ensure that programmes (such as community health, education, local business development) are well designed and effectively delivered. • Enhance social and economic development by seeking opportunities to address poverty.

10. Implement effective and transparent engagement, communication and independently verified reporting arrangements with our stakeholders:

    • Report on economic, social and environmental performance and contribution to sustainable development. • Provide information that is timely, accurate and relevant. • Engage with and respond to stakeholders through open consultation processes.

The Government of India, however, realised that the issues on adverse social and environment impacts always loom large not only in India but across the globe with different proportions. That aside, economic impacts, mass displacement and other associated problems remain unresolved for years together. For developing a holistic sustainability approach in Indian mining sector, erstwhile planning commission constituted a High Level Committee under the Chairmanship of Shri Anwarul Hoda, Member, planning commission in the year 2005. The committee was to review the National Mineral Policy recommended, performance of mining vis-à-vis sustainable development taking into account the global trends in sustainable developments and introduce best practices in implementation of environment management. The High Level Committee made an intensive study on the impact of mining operations with the need to develop principles in mining, best practices and reporting standards which can be measured objectively. Taking into account the global trends and international practices, the committee recommended to develop a Sustainable Development Framework (SDF) suitable for India. The SDF comprised of guidelines on mining principles, reporting initiatives and good practices. Ministry of Mines, Govt. of India commissioned ERM India Pvt. Ltd. to develop a SDF for the Non Coal, Non Fuel Mining Sector in India.

The principal objective of SDF was to fully achieve sustainable development in mining sector by addressing the outstanding issues. It contains regulatory measures, guidelines at different stages, different yardsticks, measurable outcomes/achievement, reporting standards, quality assurance as well as recommendations of best practices. The key elements pertaining to sustainable development incorporated in the SDF approach are not already in the regulatory regime. The existing policies vis-à-vis profile and performance of mining companies and corporations of India display a wide diversity. They must be compliant to the stipulations and strive to improve over a period of time. According to the report, "the framework approach is a flexible one in that allows for the achievement of sustainable development objectives without being too prescriptive and formulaic. It takes into account the biggest issues facing the sector in the context of existing laws and regulations and defines a set of principles that collectively progress the sector towards sustainable development".

Ministry of Mines, Govt. of India adopted eight principles of SDF, reproduced as follows, for the Non Coal, Non Fuel Mining Sector in India, which form the core of the Sustainable Development Framework for India.

1. Incorporating Environmental and Social Sensitivities in decisions on leases:

   This principle integrates sustainable development concepts at the earliest phase of the mining lifecycle. The underlying philosophy of the principle is to categorise mineral bearing areas based on an environmental and social analysis taking a risk based approach. At the bidding stage the categorisation of lease areas into High and Low risk will allow the investors to take business decision with the knowledge that the cost and uncertainties of getting approvals as well as operations in high risk areas will be significantly higher than the low risk areas. It will also allow regulators to put additional commitments at an early stage for environmental and social performance. This principle allows for the government to balance environmental and social interests of the nation, with mining priorities in the longer term;

2. Strategic Assessment in Key Mining regions:

   Understanding that mining activities occurs in clusters which have impacts at a regional level, undertake a strategic assessment of regional and cumulative impacts and develop a Regional Mineral Development Plan based on assessment of the regional "capacity" at periodic intervals. Creating an institutional structure to own and implement such plans in key mining regions and taking critical decisions on mining, new leases, allocation of resources, and even possible moratorium on mining to ensure more sustainable planning and development in such regions;

3. Managing impacts at the Mine level impact through sound management systems:

   The key elements of this principle are impact assessment of key environmental, social, health and safety issues, development of management framework and systems at the mine level and continual improvement of the same on the basis of international standards on a self driven basis. A key elements is disclosing performance on environmental and social parameters to external stakeholder at every stage of the project lifecycle.

4. Addressing Land, Resettlement and Other Social Impacts:

   This principle demands a comprehensive assessment of social impacts and displacement of mining projects at the household, community and mining region level, and management commitment to address those impacts through mitigation measures and management plans.

5. Community engagement, benefit sharing and contribution to socio-economic development:

   This principle seeks commitment to regular engagement with the local community as well as sharing of project benefits with the affected families. It is rooted in the principle of sharing profits with the affected communities already provisioned for the in draft MMDR Act awaiting approval. It dovetails the social impact management of project operations with the CSR initiatives being undertaken and looks at an integrated approach to mitigate impacts and improve local livelihoods and living conditions in the neighbourhood areas/communities.

6. Mine Closure and Post Closure:

   Mining operations must prepare, manage and progressively work on a process for eventual mine closure. This process must cover all relevant aspects and impacts of closure in an integrated and multi-disciplinary way. This must be an auditable document and include a fully scoped and accurate estimate of planned cost of closure to the company. The cost estimates must be adequately provisioned to cover national, regional and local legal and regulatory requirements for closure; and must also include the cost of servicing all agreements/commitments made with stakeholders towards post-closure use;

7. Ethical functioning and responsible business practices:

   This principle underlines the need for ethical business practices and a strong sense of corporate responsibility among mining companies. It recommends companies to go beyond legal compliance; and

8. Assurance and Reporting:

   This principle seeks mining sector stakeholders to assess their performance against this SDF and demonstrate continual improvement on this performance over the life of the project. It requires this performance to be reported in a structured manner in a Sustainable Development Report to be disclosed in the public domain as well as to regulatory agencies to consider during approval processes. At the very least, the SDF provides guidance for the mining companies to improve performance on environmental and social aspects, however, over time it can also become the common benchmark against which all mining operations may be evaluated in terms of their comparative performance on sustainable development terms. The SDF will need to be used by mining companies to demonstrate commitment to sustainable development, and may be submitted to regulators at the time of seeking clearance or renewal or extension. It should also be used by regulators to evaluate the mining company's commitment to achieving environmental and social goals. Investors and financers may use this to assess risk and could additionally use it to demand better performance of the associated mining operations. Once this SDF is accepted, its use can be determined through more focused consultations and seeking consensus.

The challenging task, more than formulation, is the implementation of SDF principles through proper initiatives. Some of the actions required are as follows:

Promulgation of Acts/ Rules: Important components of SDF principles may be transformed in to rules making it mandatory for the mining companies to put them in their action plans. Strict regulations will have a standing impact on its effective implementation.

Coordination and Collaboration: Proper coordination between and collaborative efforts among the Departments concerned are required for regularly monitoring and reviewing the performances of SDF by the mining companies / corporations. Constant and continuous monitoring will make them alert as well as responsible.

Concurrent Evaluation: Evaluation of performances will identify the lapses to fix the accountability so much so that the defaulters can be taken to task. Besides, stumbling blocks and hindrances in the process can be taken care of. Mid course corrections, regulatory criteria and statutory clearances can also be revised accordingly.

Engaging Local community: Local people are the major stakeholders in operation of the mining activities. They will constantly watch the implementation of social and environmental safety measures initiated by the company. Civil societies/NGOs operating in the area can be entrusted to drive forward the SDF agenda by ensuring the company's accountability and mandatory disclosure of performances.

The long run expected outcomes, as spelt out in the report are:

> • "Reduced environmental and social conflicts in areas awarded for mining; • Greater clarity for all concerned stakeholders, on risk levels of mining lease areas; • Potentially reduced delays in obtaining clearances (environmental, forest) for mines; • Improved protection of high risk areas in terms of environment and social considerations; • A Regional Mineral Development Plan for selected mining areas and addressing key regional and cumulative impacts of mining through coordinated and collective action; • Opportunity for clustering of small operators to become more competitive, and compliant; • A robust E&S Management framework in mining companies enabled with continual improvement systems guiding sustainable development of mine and which is commensurate to risk category of mineral bearing area as enunciated under Principle; • A disclosure process that provides stakeholders with relevant and timely information, and allows issues to be raised in engagement forums; • Opening up of illegal mining activities to intensive stakeholder scrutiny as geo-spatial information regarding mining activity will be a published periodically; • Intensive use of geo-spatial and geo-scientific information at mine level for assessment, planning, management and monitoring of the mining sector; • Stronger monitoring and assurance systems and processes; and • SDF reporting on governance and ethical practices".

SUSTAINABLE MINING INITIATIVES (SMI)

Sustainable initiatives are required to implement the SDF in letter and spirit. ICMM and World Gold Council (WGC) together developed important activities/initiatives which mining companies need to implement for ensuring sustainability in mining operations. They have 'taken some of the best practices across the mining spectrum and compiled a list of the most important initiatives mining companies should begin implementing immediately because aside from profit margins and safety records, the key theme in the mining industry is sustainable initiatives. Not only are they vital to surrounding communities in which companies operate but they are vital for the prosperity of the world'. Key initiatives are; Implementing Community Development Programmes, Generation and Use of Renewable Energy Resources, Maintaining transparency, Upholding human rights, Innovative closure and reclamation operations etc.

Mining companies provide jobs and create investment opportunities for the affected and surrounding communities. But one of the best objectives they can accomplish for surrounding areas is building community development programmes which can have a wider impact on the societies they operate in. According to the WGC, "these programs should be designed to meet local needs and priorities, and include investments in sectors including education, entrepreneurship and healthcare". The mining companies/corporations should launch a slew of initiatives on community inspired programmes like building road connections and communications, housing for weaker sections and low income communities, healthcare centres. Mining companies, instead of heavily pressurising the available energy sources, should endeavor to generate and use the renewable energy resources. Many smart companies nowadays are taking systematic initiatives like building efficient solar parks, bio-fuel, generation of natural gas by using mine waste and tailings etc.

A responsible miner should maintain transparency in all its practices. Sustainability will be a distant dream without operational transparency. Transparency builds the much needed trust between the mining company and all its stakeholders. According to the report, "tips for addressing transparency in the metals and mining sector are: (i) Report on corporate sustainability performance in a standardised way is an essential piece of the transparency puzzle. (ii) Be responsible for directly disclosing information about operations to customers and surrounding communities.(iii) Provide responsible sourcing schemes to help define what it means to manage the supply of a given material or product. Take into account the relevant environmental and social benefits and impacts". Transparency is, therefore, imperative for forming a balanced social contract leading to sustainable development. The basic foundation for every mining company/ corporation operating in all countries of the world is respect for fundamental human rights. ICMM and World Gold Council adopted this as a key initiative for sustainable development and it's one of the biggest principles. According to the World Gold Council, "Respect for local societies and human rights are of special importance for responsible gold-mining companies. Many companies undertake their own social responsibility programmes, as well as sign up to international standards such as the Voluntary Principles on Security and Human Rights". Many mining companies are implementing an array of new initiatives and innovative practices on mine closure, reclamation and wind up operations which are vital elements of any mining considering the impact on host communities. The mining company should leave a positive impact and sustainable legacy for all its host communities after operations have ended. This requires a strategic planning, designing and effective implementation as well as transitioning mine operations to post-mining and then planning for the socio-economic and environmental challenges in future. Well designed mine closure programmes, Development of eco-centres, eco-parks and eco-tourism in mine sites are some of the initiatives taken by mining companies.

Federation of Indian Mineral Industries (FIMI) have launched SMI during 2009 "to mainstream sustainability in the mining sector and to counter the negative perception of mining. The overarching aim of SMI is to promote the tangible and measurable concepts of sustainable mining, assist the mining sector to inculcate sustainable practices in mining operations and tackle the industry-wide challenges of environmental management and socio-economic development. SMI assisted the Ministry of Mines to implement SDF at individual mine level and has been an integral part in developing the Star Rating of Mine scheme and over the years has evolved itself as an important encouragement for the inclusive and responsible growth of the mining sector. SMI has been effectively working with the mining industry to integrate sustainable practices and bring out the optimistic aspects of mining. It has grown in stature for leading the sustainable movement in the Indian mining sector and serves as a driving force for change and continual improvement on issues relating to mining and sustainable development in India. SMI engages with all stakeholders including government, industry, communities, civil societies, academia etc. to positively influence Indian policy and regulations to promote sustainable mining".

Global Mining Initiative (GMI) and Mining, Minerals and Sustainable Development (MMSD)

SMI were launched in order to counter the adverse impact of mining activities which profoundly transform landscapes escalating environmental crisis and socioeconomic patterns of affected communities. Mining companies and corporations have been increasingly promoting sustainability initiatives to make public aware the minerals' many benefits. There were, sometimes back, debates and deliberations on 'Sustainable mining'. 'Sustainable' has been narrowed to mean basically 'profitable' over a period of up to about 20 years. This phrase 'Sustainable Mining' began to be used from 1999, when the world's 10 biggest mining companies met to launch a project they called 'Mining, Minerals and Sustainable Development' (MMSD, part of their Global Mining Initiative), to see how the mining industry could 'contribute to the global transition to sustainable development'. But the long-term vision how to plan to live sustainably in the long term – is completely absent from the corporate perspective. Global Mining Initiative (GMI), promoted by CEOs of large mining companies was an important effort in this direction. 'At the core of the GMI was the Mining, Minerals and Sustainable Development (MMSD) project, which gathered over 150 individuals and organisations to understand the role that the sector could play in sustainable development'. The project resulted in the publication of important reports, despite different hurdles. Number of studies followed on the heels of the MMSD project, as authors and analysts attempted to understand the implications of sustainable development to mining organisations and operations.

"The Mining, Minerals and sustainable development (MMSD) initiative was full of promise for a new era. It was envisaged that the participation of multiple stakeholders in decisions about the sustainable use of the earth's resources would help design a sustainable future for an industry that was seen as the most challenging to global ecological sustainability".[42] Environmental sustainability can be better achieved through meaningful engagement of the NGOs, Civil Societies, Corporate bodies and local administration. Global Mining Initiative (GMI) envisages effective management of stakeholders with a focus on indigenous and local communities. Participatory approach, conflict resolution mechanism and Global Reporting Initiative (GRI) by the multinational mining corporations play a vital role in MMSD.

In the national context, mining sector in India, holds the promise of economic development. It is the backbone of India's prosperity and growth contributing significantly to the economy both directly and indirectly. "India had a vision to incorporate sustainability in this sector at least almost a decade ago, even before the 2015 United Nations Sustainable Development Goals (SDGs) became the recognised standard for sustainability. But the implementation of that vision has failed to gain momentum and now, as India looks at increasing mineral production, sustainability in the extractive industry is back in focus. Following the SDF document in 2011, sustainable development became the buzzword in the lexicon of the Indian government. Beyond the SDF for the Indian mining sector in 2011, the recent policy measures proposed or adopted by the government of India focus on this too. For instance, the National Mineral Policy (NMP), 2019 emphasised that environmental, economic and social considerations must be taken into account as early as possible in the decision-making process so that mining is financially viable, socially responsible, environmentally, technically and scientifically sound, uses mineral resources optimally and ensures sustainable post-closure land uses".[43] The NMP stipulates that 'The government shall set a benchmark against which all mining operations may be evaluated in terms of their comparative performance on sustainable development framework and enforce commitment on part of the mining companies to adopt sustainable development practices for achieving environmental and social goals'. The National Resource Efficiency Policy, 2019 of India also emphasised sustainable development of minerals. It envisaged that 'while the attainment of all the SDGs requires to a large extent the sustainable management and use of earth's natural resource base, no fewer than 12 of the goals refer directly to resources and the environment as fundamental to their achievement'. The principle of resource efficiency forms the base of SDG -12 i.e. Sustainable Consumption and Production. The policy enunciates that 'resource efficiency has enormous potential for cost savings from reduced material use, reduction in social conflicts due to mining, increased job opportunities, reduction in climate change and environmental degradation. It noted that it could lead to a reduction in conflicts and displacement due to mining, lead to job creation in recycling sectors, preserve resources for future generations, reduce ecological degradation and pollution, greenhouse gas emissions, avoid industrial and solid waste generation and provide opportunities for the restoration of water bodies. Even India's Intended Nationally Determined Contribution (INDC) submitted to the UN in 2015 focused on clean coal, renewable power and resource efficiency to usher in sustainability. At the heart of all the documents that address sustainable development in the mining sector is the idea that the mining activities will be in conformity with environmental and forest laws, including waste management rules. However, ironically, the problems currently caused by mining – displacement, health issues, land conflicts, social inequality and poverty etc. that the government plans to address by incorporating sustainability in the mining sector – are in fact a result of years of unchecked mining activities'.[44] Hence, focused interventions to check illegal and unregulated mining can ensure un interrupted sustainable mining practices.

A case study was undertaken to take a closer look at the problems of mining sustainability by covering iron ore mines of eastern India. The future generation being deprived of mineral resources will be able to enjoy the same stream of benefits from better quality of natural resources; this is the underlying principle of sustainable mining practices. This policy goal can be only achieved by reinvestment of some parts of the profit earned from the sale of mineral in natural, human, and social capital at the regional level. Case study result reveals that implementation of similar policy have yielded encouraging results in one iron ore mine.[45] Responsible mining industries in India are jumping, taking the bull by the horns and implementing sustainable mining practices. They have adopted Corporate Social Responsibilities (CSR) as a part of their operational policy principle. National Aluminium Company (NALCO), for example, is providing a better quality of life for communities residing in the vicinity of its plants. The Company has comprehensively addressed the problems of rehabilitation of displaced families with adequate compensation, housing and employment to the extent feasible. Creation of infrastructure in the surrounding villages for communication, education, healthcare and drinking water gets priority in the periphery development plans of the Company. Community participation in innovative farming, pisciculture, social forestry and sanitation programmes apart, encouragement to sports, art, culture and literature are all a part of its CSR commitment as a responsible corporate citizen. As a policy, NALCO has been allocating one percent of its net profit of the year for peripheral development activities of the succeeding year. Focus areas of key project interventions with hand holding support for a long time sustainable CSR Plan are: • Alternative Livelihood Means • Drinking Water and Sanitation • Construction Work • Medicine, Health Family Welfare • Education • Electricity/Energy • Vocational Training • Plantation • Adoption of Villages. Mineral Exploration Cooperation Limited (MECL) under the CSR, stand for raising and improving the Socio-economic status of women and weaker section. For the purpose, it provides training to acquire new skills for the career development etc. MECL also propose to provide sewing machines for the use of Mahila Mandal of the village, stitching training classes for women belonging to SC, ST and indigenous communities of the society of the villages nearer to MECL projects for improving the socio-economic condition of women belonging to weaker section in the rural areas.

Conclusion

Minerals, it is argued, are essential for sustaining the needs of around eight billion people on the globe and, thus, serve as the building blocks of modern society. More and more population on the earth strive to improve their style and standard of living, mining, therefore, has come to play a more vital role. 'Given the importance of mining contribution to nation's development both in terms of GDP and employment generation, sustainability has become a key enabler for the mining sector and hinges upon efficient use of resources with adequate environmental management and social upliftment measures in mining jurisdictions'. But, as yet, a consensus as to how to make mining compatible with sustainability has not been built up. Many development champions consider the nature of the mining industry as unsustainable due to the finite nature of minerals. This very premise puzzles the minds of the advocates of sustainable development. As a global NGO campaign has argued: 'Mining is inherently unsustainable - it requires the depletion of non-renewable natural and cultural resources'. According to Young and Septoff, in many cases, mines can be operated more responsibly, with reduced negative impacts. But a truly sustainable global society will take fewer minerals from the earth each year. 'Generalisations like the one above are rhetorically strong, but factually contentious. There are a number of issues and uncertainties involved in the extraction of minerals that call for contextual and careful considerations about the long-term viability of mining'. Sustainable wealth creation can ensure a sustainable mining with its main objectives of promoting socio-ecological sustainability i.e. development of human societies and indigenous communities and the environment. No mine, according to A. Frederick, will continue to produce indefinitely; but indigenous societies and undisturbed ecosystems can prevail for long periods of time and can continue to improve over time. Promoting sustainability in mining operations is a key to ensuring a sustainable society and a green environment'.

Many mining companies in Europe and South Africa have started Sustainability initiatives for increased responsibility in the mining sector. Sustainability initiatives aim to make the extraction of raw materials more responsible through the introduction of international standards. The biggest question is how sustainable mining initiatives be made successful and how can the challenges in mining sector be more successfully addressed? The results of a research study[46] 'Mapping Sustainability Standards Systems for Mining and Mineral Supply Chains,' are worth noting in this connection. The study shows that "the large number of initiatives, their diverse thematic and regional focuses, and the application of differing high standards in certain areas result in duplication, gaps, and a lack of efficiency. The study identifies these gaps, for example, in issues such as financial security for the closure costs of mines and the use of renewable energies or water extraction by the mining industry. On the other hand, the study points out that harmonisation and mutual recognition of standards of different sustainability initiatives would reduce auditing costs and avoid varying requirements for the mining industry - e.g., on the part of raw material customers. International reference standards such as those of the ILO (International Labour Organisation) and the Organisation for Economic Co-operation and Development (OECD) are particularly helpful here. In addition, civil society and local communities are not sufficiently involved in decision-making processes in affected countries. Ultimately, strengthening oversight of the raw materials sector in the countries themselves can and should also contribute to implementing measures and reducing the effort required for implementation. After all, a country's governance capacities are key to the effective implementation of environmental and social standards. However, it requires further efforts by all actors along raw material supply chains to make the sector more responsible and to achieve a sustainable contribution to improving governance in countries that produce raw materials". Mining and mineral extraction has to go on despite the fact that mining operations from setting up to expansion to closure, entails immediate, short, medium and long term management issues as well as other challenges with significant impact on the society, economy and environment. Hence, sustainability approach must be adopted for obviating the adverse impacts. The importance and advantages of sustainability principles need no reiteration. Mining companies, therefore, have no choice but to practice it earnestly and in letter and spirit in order to survive in the competition. But, its implementation is not that smooth. It is utmost important for mining companies to forge a meaningful balance and effective trade-off between local/regional vis-à-vis national interests in ecological, economical, social and political issues.

End Notes & References

1. John Locke, The Phrase was adopted in 1690.
2. Dr. Gro Harlem Brundtland, *"Our Common Future",* 1987.
3. www.cannybuyer.com/guidebook.
4. Meshach Awuah- Gyawu, Samuel Brako, Emelia Darko Adzimah, Ghana, *"Assessing the Challenges to Sustainable Mining: A Supply Chain Perspective, A Case of Li Tong Mining Company in Ghana".*
5. Forum for the Future, *Buying a Better World: Sustainable Public Procurement,* 2007.
6. Mayank, *"Sustainable Development Framework for the Mining Sector".*

7. www.sa-green-info.co.za › portal › article ›
8. www.bgs.ac.uk › mineralsUK › mineralsYou ›
9. sumas.ch › 5-examples-of-sustainable-development, Jun 15, 2020.
10. miningfocus.org › metal-mining-and-sustainability
11. Burger, ASMR Newsletter, Fall 2012.
12. Elmira TajvidiAsr, RezaKakaie, Mohammad Ataei, Mohammad Reza Tavakoli Mohammadi, *'A review of studies on sustainable development in mining lifecycle'*, 2019.
13. www.mining-technology.com › future-of-mining ›, Mar 29, 2016.
14. www.downtoearth.org.in › blog › can-mining-be-sustainable, Mar 29, 2016.
15. Elmira TajvidiAsr, RezaKakaie Mohammad Ataei, Mohammad Reza Tavakoli Mohammadi, *op.cit.*
16. www.technology-group-study-of-peristaltic-hose-pump-technology-in-precious-metal-mining
17. Prabhash Gokarn, Head Raw Material Projects Tata Steel, Mining Engineers' Association of India (MEAI), Bhubaneswar Chapter, Workshop At Sukinda Chromite Mines, Tata Steel, 28th Feb 14.
18. R. Rajaram and K. Parameswaram, *"Sustainable Mining Practices – A global perspective"*, Great Britain, 2005.
19. Vijay Kumar, Lecturer in Mining Engg. at Govt. Polytechnic, Bicholim-Goa, Feb 14, 2017.
20. Prabhash Gokarn, *op.cit.*
21. (a) Meshach Awuah- Gyawu, School of Business Management, Department of Procurement and Supply Chain Management,Christ Apostolic University College, Kumasi-Ghana (b) Samuel Brako, Faculty of Business and Management Studies, Department of Procurement and Supply Chain Management,Takoradi Polytechnic, Takoradi- Ghana (c) Emelia Darko Adzimah, Faculty of Business and Management Studies, Department of Procurement and Supply Chain Management, Kumasi Polytechnic, Kumasi–Ghana- *"Assessing the Challenges to Sustainable Mining: A Supply Chain Perspective, A Case of Li Tong Mining Company in Ghana"*.
22. Saleem H. Ali, Director, Centre for Social Responsibility in Mining, University of Queensland, Australia.
23. Jamie Kneen, Communications and Outreach Coordinator, MiningWatch Canada.
24. David MacKay http://www.withouthotair.com
25. partizans@gn.apc.org
26. Robert Goodland, *'Responsible Mining: The Key to Profitable Resource Development'*, 2012.
27. David Laurence— *"Establishing a sustainable mining operation: an overview."*
28. Miranda, R. Gorman, David A. Dzombak; *"A review of sustainable mining and resource management: Transitioning from the lifecycle of the mine to the lifecycle of the mineral"*.

29. Chinamyi Shalya, *Sustainable mining – Challenges of Polish mines.*
30. Prabhash Gokarn : Tata Steel, *Op.Cit.*
31. Aggarwal, Mayank: *"Despite an early vision for sustainability in mining, is India's progress faltering?"*, 24 July 2020.
32. Schneider Electric – *'Sustainability in Mining: Technologies and Processes for Data Aggregation, Management, and Reporting'*– SME 2013.
33. *Ibid.*
34. Prabhash Gokarn : Tata Steel.
35. Soziale Verantwortung, *"Mining takes on the sustainability challenge."*
36. Katarzyna Pactwa, Justyna Woźniak, Andrzej Strempski; "*Sustainable mining – Challenge of Polish mines."*
37. Prabhash Gokarn : Tata Steel.
38. Soziale Verantwortung, *"Mining takes on the sustainability challenge."*
39. (1) Meshach Awuah Gyawu, (2) Samuel Brako and (3) Emelia Darko Adzimah; *"Assessing the Challenges to Sustainable Mining: A Supply Chain Perspective,* (A Case of Li Tong Mining Company in Ghana)", European Journal of Business and Management, ISSN 2222-1905 (Paper) ISSN 2222-2839 (Online), Vol. 7, No. 25, 2015.
40. www.mdpi.com, *'Some Challenges to Sustainability'* – MDPI
41. ICMM SD Framework – Final Principles Date: 29 May 2003 International Council on Mining and Metals Page:2 Document Ref: C 020/290503
42. Ruth Phillips, *"NGOs in a sustainable relationship for sustainable mining?The Australian NGO perspective on what happened after the MMSD initiative"*, Third sector review 18 (1), 171-193, 2012.
43. Aggarwal, Mayank; *'Despite an early vision for sustainability in mining, is India's progress faltering?'*, 24 July, 2020.
44. *Ibid.*
45. Sinha, S. and R. Banerjee, *"Mining, sustainability and policy initiatives*", Journal of Mines, Metals and Fuels 54(12):459-462, December 2006.
46. Adelphi, together with the Federal Institute for Geosciences and Natural Resources (BGR), the Federal Environment Agency (UBA) and the Center for Social Responsibility in Mining (CSRM) at the Australian University of Queensland, has compiled the results of three research projects and prepared recommendations for various actors along with raw material supply chains and for political decision-makers in a short paper as part of the BGR series *"Commodity Top News"*.

Chapter – 4

Responsible and Ethical Mining: Principles and Practices

This chapter introduces the concepts of Responsible and Ethical Mining, their background and an overview. There is a thumbnail picture on the Principles and Practices with an analysis on the challenges of Responsible and Ethical Mining. There are flashes of global scenario and international experiences. The chapter ends with a concluding note.

Responsible Mining

The term 'Responsible Mining' is relatively a new concept and generally refers to mining operations in a more responsible and sound manner so that negative impacts can be avoided or reduced to a maximum extent possible. Responsible mining first appeared in an article entitled "Re-inhabitory Mining" and next in another article titled "Ecological Mining". The term "Responsible Mining" is also claimed as having been formulated by Ranil Senanayake of the International Analog Forestry Network and Brian Hill of the Institute for Cultural Ecology. Responsible mining is enshrined in the Philippine law since 1995 though exactly not in same connotation. According to the Philippine Mining Act of 1995, "It shall be the responsibility of the State to promote their rational exploration, development, utilisation and conservation through the combined efforts of government and the private sector in order to enhance national growth in a way that effectively safeguards the environment and protect the rights of affected communities". Responsible Mining was, thus, a composition of three elements – environmental protection, economic growth, and social equity. With one of these elements missing, consider your company next on the hit list.[1]

The Initiative for Responsible Mining Assurance (IRMA) defines responsible mining as an environment where the mining industry respects human rights and the aspirations of affected communities; provides safe, healthy and respectful workplaces; avoids or minimizes harm to the environment; and leaves positive legacies. Marta Miranda, David Chambers, and Catherine Coumans suggest the following four components as central to defining responsible mining: • Acknowledging the need to preserve ecologically and culturally significant areas. • Ensuring environmentally responsible mine development • Ensuring that mine development results in benefits to workers and affected communities • Ensuring that appropriate corporate governance structures are in place. In Mongolia, a series of multi-stakeholder meetings convened by The Asia Foundation in 2006 brought together representatives from civil society, government, industry and academia to agree on the following definition of responsible mining: Responsible Mining is a comprehensive and transparent minerals activity respecting the rights of all stakeholders, especially of local people, environmentally friendly and free of human health impacts, embracing the best international practices and upholding rule of law whilst generating a sustainable stream of benefits. According to the report of Mongolian River Resources in 2008, 'Responsible Mining is a complete set of activities in the minerals sector, respecting the rights of all stakeholders, including local communities; environmental friendly; having no negative effects on human health; open; respecting the rule of law; and which contributes Sustainably to the benefit of the nation. It further states that the principles of responsible mining are: (i) Ensuring multi-stakeholder engagement; (ii) Transparency and openness; (iii) Upholding law and its enforcement; (iv) Responsibility for the safety of people and the environment; (v) Investment in future development; (vi) Ensuring fruitful productivity and efficiency; (vii) Humane and ethical; and (viii) Based on advanced and modern technology'.

Various discussions, deliberations and dialogues on the concept of 'responsible mining' were held during 2018 and 2019. According to Carmel V. Abao, the document, 'Discussion Paper on Mining: Consolidated Notes from the 1st and 2nd Round Table Discussions (RTDs) on Responsible Mining' were held on November 27, 2018 and December 11, 2018 to find out a common understanding and mutually acceptable articulation of what 'responsible mining' is. The first RTD focused on large-scale mining operations while the second one discussed small-scale mining. The results of these two dialogues were then presented during a third RTD held on March 17, 2019. The first two dialogues resulted in an initial definition of responsible mining as “one which contributes significantly to economic growth while at the same time ensuring that the impacts to the environment are remediated toward sustained productivity of the land after mining and the host and neighboring communities are developed into self-reliant communities beyond the life of the mine”. This definition was said to be “founded on the threefold elements of economic viability, technical feasibility and social acceptability”. Moreover, the first two dialogues resulted in a claim that “for mining to be responsible, it has to meet the fundamental tenets of economic growth, environmental protection and social development or loosely translated as ‘profit, planet and people,’ all with the end goal of sustainable development. The three ‘Ps’ are to be treated as equals, but with ‘planet’ as the primus inter pares (first among equals), signifying that primordial consideration is to be given to mining’s responsibility to protect the environment”. A fourth “P” — policy — has also been deemed necessary because mining “should be legally compliant”. In the third RTD, the need was emphasised “not just for a law but a robust policy framework that integrates rights and privileges, taxes, penalties, tree-cutting permits, resilience cost and benefit, wealth accounting and valuation, island ecosystem and biodiversity”. The issues of labour and mining workers were also raised. The third RTD also resulted in several discussions about the important role that ‘government’ plays in mediating the various interests in the mining industry. The government as the main duty-bearer has to ensure that policies are agreed upon democratically and implemented effectively.[2] Responsible Mining, thus, refers to advocacy to reform mining activity, as well as to a marketing strategy used by mining companies to promote their operations as environmentally or socially sound.[3]

The contribution of mining to socio-economic progress and development cannot be understated. Minerals are foundation for the creation of wealth and progress but at the same time, mineral extraction and related downstream processing continue to be associated with significant negative social and environmental impacts. Mining operations are highly detrimental for the environment and associated with severe social impacts such as those upon the health and welfare of workers.[4] ‘To date, such impacts are addressed by means of governance and innovation approaches, that contributed each to substantially reducing the negative sustainability impact of mining and moved the sector to higher responsibility. Note that the term responsible mining is often preferred over sustainable mining. This is based on the formation of mineral resources over geological time scales such that human extraction by far exceeds natural formation. As a result, there is an exhaustion of the resources in the long or even short run limiting the resources' availability for future generations. Responsible consequently refers to mining practices that are performed in a sustainable manner while ignoring that a mineral deposit is exploited'.[5] Moreover, the scope of responsibility is understood in a broad sense, as defined by van den Brink et al. as "the management of social, environmental and/ or economic sustainability".[6] Although the definition of responsibility given transcends the boundaries of social sustainability in a narrow sense, it nevertheless focuses on a comprehensive contribution to the welfare along the SC. However, a SC always starts with raw material production such as in agriculture, forestry, or mining. These basic economic actions are the prerequisite of today's often global SCs.[7] Moreover, mining has been found to be one of the most impactful economic operations with more than 30 Millions of formal and informal miners.[8] This social relevance is complemented by long-lasting and far-reaching environmental degradation of mine sites and surroundings as well as resource consumption, at least if the mine is poorly managed and operated.[9] Responsible mining is thus vital for "improving the welfare of individuals and communities through employment, consumption or participation" as called for by the OECD.[10] However, the dominance of other goals, such as economic performance, short-term profits, insufficient knowledge but also price pressure and increasing demand at the global mineral market, has long impeded the diffusion of more responsible mining practices.[11]

Responsible mining has multiple definitions and dimensions, ideas and approaches. Responsible mining demonstrably respects and protects the interests of all stakeholders, human health and the environment, and contributes discernibly and fairly to broad economic development of the producing country and to benefit local communities, while embracing best international practices and upholding the rule of law.[12]

Certain principles are pronounced in the Cape Town Statement on Geo-ethics. 'Responsible mining concerns the principles and ethics of sustainable development applied to the exploration for and exploitation and use of economic mineral resources, including the entire value chain, from studies, exploration, and extraction to processing, refining, waste management, mine closure and rehabilitation. In particular: (i) It is about concrete commitment to managing the economic, social and environmental challenges related to mineral resources development, to build a system capable of ensuring/promoting responsible extraction of minerals and developing a proper alignment of the corresponding benefits at local, regional, national and global scales. (ii) It is about how to build trusting and transparent relationships with society in general and with the actors more directly involved that allow a fruitful involvement of local communities and government authorities in the creation of sustainable benefit for all parties. (iii) It is a way to minimise and mitigate environmental impacts related to water, biodiversity and land. (iv) It meets and tackles climate change issues through implementation of innovative technologies across the value chain, but also by producing the supplies of minerals and metals needed in low-carbon energy systems. These principles expressed through the Cape Town Statement on Geoethics are essential for responsible mining. It is clear that there is the expectation that any geoscientist, working in this area will make her or his responsible contribution'.[13] Hence, Responsible Mining calls for proper checks and balances, strict rules and regulations, empathy and ethics to minimize and control the hazards and negative impacts ultimately leading to maximisation of benefits for all stake holders. The term responsible also means capacity for taking moral decisions and accountability, liable to legal review and penalties in case of fault besides being honest, reliable, sound thinking, trustworthy decisions and good judgment based on accurate information. Responsible mining does not only require actions and commitments from mining companies, but is likewise dependent on the active and constructive engagement and involvement of all actors (including governments). They all have a responsibility to be well informed, transparently updated and fully aware of all aspects of any mining activity that touches them in any way. Social and environmental responsibility is integral and rests on all actors involved.[14]

Ethical Mining

Mining companies adopting and maintaining business ethics in their operations right from sourcing to closing may be termed as ethical mining. Kerry Clarke-Potter have discussed about ethical mining in supply chain management of various products and ethical sourcing.[15] According to him, "Your mobile phone, your car, your clothes. All of these things came from a supply chain. Your mobile phone might be sitting in your hand, right now, but at some point, Cobalt, Lithium and Tungsten were all mined for its creation. Do you know if they were responsibly mined? Was there child labour involved? Forced labour? Did it fund conflict? Supply chains are everywhere, and so are concerns surrounding supply chains, particularly mining ones. Are they ethical? The ethical problems of supply chains have attracted various criticism and responses. For good reasons: in mining supply chains, it is difficult to track mineral sourcing down to every supplier. This is a problem. Minerals like Cobalt, Lithium and 3TG are integral components in our mobile phones and cars. They are central to modern life. Yet frequently, they are sourced unsustainably and unethically, with the use of child labour and modern slavery. There are three main things that should be considered when sourcing minerals; i.e. Sustainability, Ethics and Economic impact. These all form 'responsible mining practices'. Not only are the three vital to the protection of our environment, but they are also fundamental in our community and economy. Nobody wants murky supply chains – no, not even companies. They, and we, want insight into supply chains so we know that minerals can be sourced in a responsible way".

Some renowned companies are taking proactive initiatives towards ethical mining. In 2017, Amnesty International praised Samsung and Apple for their transparency. These companies have maintained their supply chain and ensured that their materials are sourced ethically. Kerry Clarke-Potter cited the example of mining in Democratic Republic of Congo where no efforts are taken to promote ethical mining. Forced and child labour are prominent, workers are subjected to long days in poor conditions, many facing incredible weakness and illness and there is mercury pollution. Exposure to mercury in the processing of ores leaves workers vulnerable to poisoning. Besides, some Small-scale and artisanal mining groups make economic exploitations. They often force the existing communities to work. Companies provide a reliable source of income and these people become reliant on their economic power. These communities will plunge into poverty, if they won't work.

Mining companies bearing responsibilities on environmental, economic, social and governance issues are doing ethical mining operations but companies/corporations only concentrating on business and profit without discharging any socio-economic and ecological commitments are simply unethical. Hence, ethical mining and responsible mining are very much correlated and may, sometimes be used as coterminous. However, some key principles of ethical mining are: Refraining from illegal mining and illegal practices, Shouldering socio-economic and environmental responsibilities and commitments, Respecting human rights, Absence of bribery and corruption in operation, Maintaining Transparency and good governance.

Ethical mining is to a great extent concerned with human rights-based approaches (HRBAs) to social and environmental issues. "On the latter aspect in particular, it is striking that, even though human rights instruments still take up environmental issues only scarcely, human rights-based approaches to global challenges such as climate change have gained momentum".[16] Mining companies should, therefore, pay a due respect to principles of human rights. 'Development' gradually came to be seen as a holistic notion, encompassing economic, social, cultural, political and environmental dimensions'.[17] Ethical mining is an integral part of inclusive mining and development. Inclusive development follows UNDP's human development approach and integrates the standards and principles of human rights: participation, non-discrimination and accountability.[18] This was emphasised in the 1986 UN Declaration on the Right to Development which described the right to development as "an inalienable human right by virtue of which every human person and all peoples are entitled to participate in, contribute to, and enjoy economic, social, cultural and political development".[19] Environmental (protection) concerns were added to this list by the 1992 Rio Declaration on Environment and Development.[20] Human rights are both substantive objectives in themselves, but also prescribe a certain process for achieving those objectives. This process should among others be inclusive/non-discriminatory, participatory and provide for accountability.[21] These components, therefore, form integral parts of all HRBAs to mining development. According to Anton and Shelton, human rights and environmental protection can be seen as "based on fundamental different and ultimately irreconcilable value systems".[22]

There is a very conflicting issue in ethical mining. Mining is necessary, but it has some negative impacts. From the practical points, mining operations are unethical – in terms of environmental degradation, ecological destruction, pollution, safety of employees, land acquisition, displacement etc. It is, therefore, not a question, just to accept that mining is unethical but necessary. Hence, there should be a balancing act or trade-off between the two. "Mining is going to have to exist: to have a green economy, we will need certain metals such as copper, nickel, cobalt, aluminium, lithium, etc. We believe it is the responsibility of the mining companies to operate in the most ethical manner possible, and to be held accountable. --- There is just no global consensus on the issue of ethics. You might make an argument that mining thermal coal is bad, but there is just no consensus that says that diamond mining is bad, for instance. --- Various considerations key to company's ESG credentials. Socially speaking, it is the company's social licence to operate, its employee health and safety performance/programmes and its human rights policies that we are most concerned with. Finally, from a governance perspective, the considerations are similar to other sectors, ranging from bribery or corruption, policies and business ethics to management incentivisation, etc. From an ESG perspective, everything has to come back to 'net', as in the 'net benefit' of any action should be positive.- - - - It comes down to information. The more accurate information we have on the whole process, the better the decisions we will make as to whether something is worthwhile. This process has two layers. Firstly, there has to be some form of regulatory oversight. Regulators have a duty to remain one step ahead to ensure that companies' activities do not cost society as a whole more than they contribute. Within the second layer, the key decision-makers are corporates, capital and consumers. Companies are fundamentally capitalist and should be free to operate within regulations, but as the awareness of, say, environmental concerns become more widespread, the company-consumer feedback loop means that profits will be driven to companies who operate ethically. --- There are also considerations around the ESG track record of the potential acquirers and whether the assets will continue to be operated to a high standard and with transparency. It's worth pointing out that listed coal companies tend to trade on cheaper multiples than other industrial mining companies. It's interesting to see this development being reflected in prices, and it could be cited as an example of the market being efficient with respect to ESG considerations. In terms of the decision to not invest in coal assets, there are other factors to consider – many communities rely on coal mining; and in the event of a mine closure, ensuring sustainable transition, decommissioning and rehabilitation of the mine site. There are multiple dimensions to the issue".[23]

Genesis

Mining is the building block and economic foundation for developed and developing countries. With the exponential growth of Population, increase in per capita income and rapid urbanisation, demand and necessities for more buildings, cars and consumer products grows. This, in turn, generates an enhanced need for mined products and more so in modern societies this has become essential components for the physical and social infrastructure.

The very process of mining operation is complex and intensive and it causes massive socio-economic and environmental changes. Mining adversely affects the social structure and economic condition of the local communities, induces displacement, causes conflict and degrades environment and ecology. It has many other challenges to tackle. But if managed responsibly and implemented ethically, mining could have multifarious advantages in the form of employment generation, job creation, building new and improving existing infrastructure and guaranteeing increased income and improved quality of life.

Looking at the history of transformation of the mining industry in the past decade i.e. in the 1990s, the mining industry across the globe was suffering from an existential crisis and the legacies of a number of major mining projects had triggered a significant, and increasingly global, environmental movement against mining. David Humphrey in 'The Remaking of the Mining Industry' in 2015 observed; "Historically, mining contributed to environmental degradation, displacement of populations, worsening economic and social inequality, and armed conflicts. This negative correlation came to be known as the 'Resource Curse'. Unwanted pollution and undesirable social impacts resulting from mining, contributed to the industry being labelled as a 'problem industry'". Since then, the concepts of Responsible and Ethical Mining gained momentum and mining companies/corporations gradually started taking initiatives on Responsible and Ethical Mining practices.

Why Responsible and Ethical Mining?

Mining's contribution to the economy and development in general and society and community in particular need no reiteration. Then the obvious question would be why Responsible and Ethical Mining? What is its Need and Necessity? Why do a government or mining company need to implement Responsible Mining? Mining is essentially required for boosting a country's economy. But at the same time, mining operation leaves a trail of adverse consequences on earth and environment, people and planet. Mining is, among many, an issue that continues to divide the people. On the majority side of that divide are people who sees mining as damaging and has irreversible consequences. This acknowledgement of mining's threats to the environment, alongside with the closure of mining ventures, the call for mining companies to start practicing responsible mining has become louder.[24]

The challenges of mining are enormous calling for cautious approach and responsible handling. The adverse impacts should be minimised if not completely banished, local communities should not feel deprived, infrastructure should be upgraded, and at same time profit should also be attractive. Therefore, Responsible Mining is widely emphasised to deal with these outstanding issues.

The basic objective of development should be the distribution of the benefits, wealth and resources equitably and reduction of the adverse impacts on the local people. Mining definitely benefits and positively affects the host countries but with a dent on the environment and cost to the local communities if there is lack of proper management of operations, effective dealing of relationships and efficient handling of local resources, communities, ecology and the environment or in other words this necessitates responsible practices by the mining companies with strong ethical compass. Geography and geology determines the occurrence of mineral deposits. Mining cannot choose locations that are logistically, socially, environmentally or politically optimal, appropriate or 'friendly'. This means that companies may have to deal with circumstances that could pose ethical challenges including: the relationship with local communities, position in the landscape/environment, relationship with local and national governments, weak governance and associated increased risk of corruption and bribery. It is necessary to deal with these challenges in a responsible way'.[25]

Responsible mining plays a vital role in the society. Besides generating jobs, it can help in averting the country's economic crisis. Foreign Direct Investment, in addition, continues to capitalize in the countries' mining industry and thereby provide economic benefit for the country as a whole. Therefore, miners, geoscientists and engineers will need to responsibly navigate the different aspects, upscale their capacity and upgrade skills on how to maintain and deal with local communities and related social issues. The big question before the countries, nowadays, is not whether they should mine the minerals or not, but how to mine responsibly. Responsible mining is essentially propoor and proprogress. Development of roads, electricity, other infrastructure and advancements due to mining will benefit the public at large. Besides, after exhaustion of mining resources, there are chances that mining operations may wind up. In that scenario, responsible mining is crucial in guaranteeing all affected stakeholders – most especially the people – that with a reconditioned ecosystem, they can still go back to what they consider as their homes but this time, with new know-hows on more advanced living. Mining responsibly is not entirely about economic gains, but this practice is also geared towards taking care of what made it possible in the first place – the environment and the public.[26]

This is in line with what Miranda et al[27], stated as 'the principles of responsible mining which are sustainable development, equity, participatory decision-making, accountability and transparency, precaution, efficiency and polluter responsibility'. According to Robert Goodland, there are two outstanding reasons why should mining become responsible? First, mining is damaging communities and ecosystems worldwide. Our energy and industrial activities have released enough carbon into the atmosphere to damage the climate. Industrial waste has so polluted the oceans that they are becoming less alkaline (they acidified by 30 percent over the last few decades).[28] Lower ocean pH is already reducing the size and weight of oyster and crab shells and corals as calcium carbonate becomes less available.[29] Besides, big companies often excavate huge open pits, which may be abandoned after mining. Also abandoned are the commonly toxic and extensive tailings lagoons retained by impermanent dams, which often fail in severe weather. Open mining pits and tailings lagoons can remain risky for decades or longer. Mine closure procedures are improving in the best-run corporations and in countries with prudent regulations and enforcement. However, in weakly governed countries, mine closure bonds are often still incommensurate with prudent costs and communities can be left in the lurch when companies leave or sell their holdings. Vast literature supports the claim that mining operations are causing increasing social and environmental disruption.[30] Second, Mining corporations that adhere to responsible mining principles can reduce local conflicts, prevent environmental impacts and even improve their profits. Responsible mining can actually offer companies bigger and quicker profits, with no conflicts with surrounding communities and with fewer impacts than irresponsible mining. Mining can provide significant local employment, economic opportunity and government revenue. According to the first annual corporate responsibility report of Centerra company, 2010 the benefits of responsible mining include 'reduced labour shortages though investing in local education and skills training, more consistent production as a result of a healthier workforce, less likelihood of conflict by building better relationships with local indigenous people and artisanal miners, better access to lower cost services and supplies through regional business development and faster access to financing because of lower perceived risk by equity markets'. Socially responsible mining corporations will have few, if any, conflicts with Indigenous Peoples and communities surrounding the mine site. Absence of conflicts and fostering of the consensual approach will avoid lengthy delays and will accelerate the permitting process. Achievement of social responsibility will benefit best-practice corporations and severely hamper slow adapters. Corporations that earn the people's and government's trust will ultimately be rewarded with a higher stock price. Ethical investors will shun corporations in conflict with communities and government, thus depressing stock prices. Mining companies choosing not to become responsible face scrutiny of their human rights (e.g., slavery) and labour (e.g., child labour) records, as well as their environmental prudence. Shocking headlines exposing irresponsibility can damage the mining companies share price.[31]

Responsible mining provides many basic materials for present generation and future generations. But because minerals are public assets, decisions about their exploitation must be transparent, participatory and subject to informed scrutiny by civil society. Problems arise from the gross asymmetry of power between rich and knowledgeable mining corporations and impacted people who are unprotected by weak governance. Best-practice responsible mining seeks to redress this imbalance.[32]

Responsible mining's main postulate is that mining should not damage life-support systems. Proper dialogue with and coordination among all stakeholders is a basic prerequisite. According to Christian S. Monsod, mining is an issue of social justice, because its impacts on agriculture, rivers and coasts are mostly borne by the poor.[33] Moreover, the present burden is also shifted to posterity. Future generation will face extensive and expensive problems without adequate resources to cope with. It is, thus, an issue of 'intergenerational justice'. Hence, mining operations should ensure maximum benefits to all stake holders with the minimum socio-environmental impacts. From the vantage point of environmental protection, responsible mining may seem like an oxymoron given the obvious toll of mining on natural resources, especially arable land, water and forests. The nature of this industry thus goes against the principle of preserving and cultivating land and nature for present and future generations.[34] And responsibility in operation can only ensure the sustainability in mining. Since mining, as a driver of economic development can never ever be done away with, it must be sustainable and in order to be Sustainable, mining has to be Responsible. Hence, ensuring sustainability in mining is the most important justification for adopting responsible practices in mining operations. In other words, responsible Mining paves the way and leads towards Sustainable Mining. Therefore, the catch phrase in 'sustainable mining' always is and should be 'responsible mining'. It has been realised that although some degree of disturbance is inevitable even in the best managed mines, according to Miranda, nearly all negative social and environmental impacts are avoidable if companies would operate according to the best possible standards. The main challenge here is how to ensure that mining communities benefit from the presence of mining companies and to reduce the negative effects of mining in such communities. Agencies charged with ensuring responsible mining need to make sure the right thing is done to ensure sustainability while deriving the advantages of mining. These are the responsibilities and ethical practices of both the government, mining companies and all other stakeholders and this is what has been referred to, in recent years, as 'Responsible and Ethical Mining'. Mining has to be responsible and ethical for maintaining its sustainability. There are strong linkages between responsible and ethical mining and sustainable development of mining sector.

Principles and Practices

There are a number of guiding principles, best practices and standard norms on responsible mining discussed by various authors on the subject and prescribed by international agencies and authenticated regulatory organisations. It is not just the absence of responsible and ethical mining which may cause irreversible damages to the planet's ecology and environment. It is not also synonymous or coterminous with 'rational mining'. The very terms "ethical" and "responsible" means taking or capable of taking a moral, honest, reliable, and trustworthy decisions based on good judgment and sound thinking and hence, there is an element of morality, responsibility, accountability, answerability and liability in decision as well as action. Many mineral rich countries do not have any legal definition for fixing yardsticks on how responsible mining can be operationalised, practiced, measured and complied with, some other countries developed stringent audit criteria for responsible mining based on socio-economic and environmental parameters. Mining has linkages with other sectors of the economy and a bearing on different factors of production. Hence, related acts and regulations have also been promulgated for effective implementation of strict laws on responsible mining. However, an universally agreed responsible mining model has not yet been developed. Some miners are doing cost-benefit analysis of mining operations using Total Economic Value (TEV) as model. Ecosystems services are being fully accounted in weighing the cost as against the benefit of mining in a particular area. The benefits are calculated in terms of employment, social infrastructural development programmes, taxes, forward and backward linkages. The costs include air and water pollution, land degradation, agricultural resources, coastal and marine resources, forest resources, health impact, tourism, soil erosion etc. But in the cost-benefit analysis, it is difficult to calculate or account for the benefits accrued, such as the free education, free healthcare, roads and bridges built, school buildings, healthcare facilities, and livelihood projects as provided by mining companies. Hence, in responsible mining, it is at times not possible to measure whether the costs outweigh the benefits or vice versa.

Many mining companies sincerely desire to be responsible miners but actually not adopting the responsible practices. Juan Pablo Duque[35] once remarked, 'Wanting to be socially responsible is different than having to be socially responsible.' Some companies have better approach on Corporate Social Responsibility (CSR) in relevant areas like environment and ecology, community and infrastructural development, health and safety, good governance and management etc.

In the Annual International Convention of the mineral exploration industries of Canada known as Prospectors and Developers Association of Canada's (PDAC), four relevant ideas pertaining to responsible mining emerged. They are; (1) Mining companies should do things that don't always make business sense. For example, better access to clean water was the community's highest priority. It sure would have been the better thing to do in terms of building long term support in the community. Taking action that doesn't appear to be in the best interests of the business creates a very high level of authenticity and good will, even among the most challenging stakeholders such as advocacy groups. (2) Mining should be made much more transparent. Most mining operations take place in very remote places. Before the Internet age, this meant that mining companies could operate with relative impunity. Not anymore. We're in the "age of transparency" and corporations have to be accountable to a degree that was never anticipated. "Stories about mining operations can now be spread worldwide at a moment's notice". Mining companies that are operating responsibly (or are on a track to improving their operations) should share what they're doing through blogs that are open to employers, advocacy groups, indigenous people, investors, and other stakeholders; give outsiders a window into mining operations through webcams and videos of operations; and ensure that mineral development agreements and CSR performance are accessible to all stakeholders. A strong link between technology, environment and people need to be established. (3) Application of engineering discipline to solve social problems. According to the American Society of Civil Engineers (ASCE), "engineers have a leading role in planning, designing, building and ensuring a sustainable future. Engineers provide the bridge between science and society". Engineers in mining companies are trained to solve complex problems, many of which are related to social issues such as better access to clean water, a resource that is vital for mining companies and for communities. Engineering is a profession with a social purpose and mining companies have yet to leverage the social value of their own engineers. (4) Acknowledging the fact that mining has a social purpose. Mining companies have a distinct social purpose and, in many cases, are directly responsible for the well-being of indigenous people, often in the absence of local government policies and programmes. Participants at PDAC agreed that existing voluntary guidelines related to corporate social responsibility such as GRI and ISO 26000 are important but don't capture the qualitative outcomes of corporate social investments in education, health, environment, and local culture. Responsible mining companies should develop meaningful social objectives, understand the ways in which social outcomes intersect with and support their business goals, implement a social continuous improvement process, and communicate outcomes in a clear and transparent way. These responsibilities are good for business and are the right things to do.[36]

Robert Goodland has discussed the eight principles on Responsible Mining.[37] According to him, following best practices means fully espousing all relevant policies and procedures as set out in OECD's Guidelines for Multinational Enterprises,[38] plus the eight principles as listed below plus the guidance regarding 'No Go Zones'. Together, they provide voluntary principles and standards for responsible business conduct in areas such as employment and industrial relations, human rights, environment, information disclosure, combating corruption, consumer interests, science and technology, competition, and taxation. These principles can also apply to small and artisanal mining, but those types of mining deserve special treatment through training, education, poverty reduction, health and safety, and alternate job creation which are not included under the eight principles. These eight and three other principles (i.e. Principles of No Go Zones, Compensatory Offsets, Principles of Precaution) are discussed below.

Social and Environmental Assessment

Environmental and Social Impact Assessments (ESIAs) is a mandatory provision for mining operations in many countries. This process begins as soon as project's engineering and pre-feasibility planning starts with standard procedures.[39] ESIAs are generally carried out by concerned companies through independent Technical Support Institutions (TSIs) or Third Party Consultants (TPCs) under government's guidance and supervision. The basic objective of an ESIA is to determine the economic feasibility of the project by using full-cost pricing after internalising all socio-environmental costs after which a mitigation plan is designed to do away with or minimize adverse social and environmental impacts and taking care of the displaced and affected communities. However, utmost precaution should be taken while selecting the TSIs/TPCs. Their technical expertise, experience and competence must be verified and they should be unbiased as well as reliable. The most effective mechanism is for a small team or panel of social and environmental experts (PoE)[40] to help the proponent select the most appropriate ESIA team.

Transparency and Honesty

Mining companies should be very honest and maintain transparency in their activities and aptitude. It is their responsibility to adopt zero tolerance for corruption and foster an open information sharing system, particularly with communities. All stake holders including 'potentially impacted people, both women and men, must fully and openly participate in or be meaningfully involved throughout the approximately two-year ESIA preparation period, from stakeholder identification, through the PoE review, ESIA report, Impact/Compensation Contract, to restoration, rehabilitation, and monitoring. Awareness and receptiveness of citizens have increased tremendously nowadays with the advent of Internet, social networking and mobile phones and in this age of transparency, they should actively participate in all activities. The responsible mining companies should be very much accountable and answerable for all their omissions and commissions.

Acceptance by the Stakeholders

Grounding of mining companies and its operations should not go ahead and become critical without full consent of all stakeholders including managing body and employees of mining company, indigenous people, local communities and the local or provincial government organisations that grant permits, levy and receive taxes, royalties as well as the stock and share holders of the mining company. If people and communities don't want or resist, responsible mining companies never force mines on them. All mining companies should establish legitimate, independent, and representative citizens' advisory councils (CACs)[41] as called for by IUCN[42]. 'Even after free prior informed consent (FPIC) has been achieved, active citizen engagement in the EIA process is necessary. Companies should follow some degree of corporate social responsibility (CSR). The best corporations aim to ensure that all potentially impacted stakeholders actually welcome a project because the risks are slight; compensation is great; and job training, employment, and local procurement are attractive. FPIC, as mandated by the UN's Declaration on Indigenous Peoples[43], is the best practice' and mining companies should effectively negotiate with the indigenous peoples.

Food Production

Land and mining are scarce natural resources on earth. Mining operations which tend to deplete or degrade available resources in areas of scarce land or water should not be allowed. People and the planet are utmost priorities and prioritisation of minerals over them is hardly desirable. Mining operations normally cause decrease of water level and production of food in adjoining areas. Water resources have multiple domestic uses in order of priority viz. water supply to households, irrigation, power generation, fisheries, fodder and livestock raising, industrial use and the last one is mining. Mining must not dampen agriculture and production of food crops through various types of irrigation.

Compliance with International Standards

Corporate and Social Responsibility (CSR) is the guiding and abiding principle which a responsible mining company should comply with as well as follow invariably upholding all international socio-environmental agreements, standards and best practices. Most of the companies have adequately staffed dedicated units to look after the implementation of CSR. CSR seeks to ensure that the interests of all stakeholders about social and environmental impacts are met in corporate policies and projects. In a narrow sense, CSR means operating a business in a manner that accounts for the social and environmental impacts it creates. CSR is a form of voluntary corporate self-regulation such that the company actively complies with the spirit of the law, ethical standards, and international norms.[44]

Robert Goodland has cited a number of international environmental and social agreements relevant to mining, such as: African-Eurasian Waterbird Agreement (AEWA, 1995), Basel Convention on the Control of Transboundary Movements of Hazardous Wastes and their Disposal (1989), Cartagena Protocol on Bio safety to the Convention on Biological Diversity (2000), Convention Concerning the Protection of the World Cultural and Natural Heritage (1972), Convention on Biological Diversity (CBD, 1992), Convention on Climate Change (UNFCC, 1992) and Kyoto Protocol (1997), Convention on International Trade in Endangered Species of Wild Fauna and Flora (CITES, 1973), Convention on the Conservation of Migratory Species of Wild Animals (CMS or Bonn Convention, 1979), Convention on Wetlands (Ramsar, 1971), Convention to Combat Desertification (UNCCD, 2007), International Treaty on Plant Genetic Resources for Food and Agriculture (Plant Treaty, 2004), Montreal Convention on Substances that Deplete the Ozone Layer (1987), Nagoya Protocol on Access to Genetic Resources and the Fair and Equitable Sharing of Benefits Arising from their Utilisation to the Convention on Biological Diversity (2010), Rotterdam Convention on the Prior Informed Consent Procedure for certain hazardous Chemicals and Pesticides in International Trade (1998), Stockholm Convention on Persistent Organic Pollutants (2001), UN's Protect, respect and remedy framework (see Ruggie, 2011), UNESCO Biosphere Reserves (1968), Vienna Convention for the Protection of the Ozone Layer (1985), Vienna Convention on the Law of Treaties (1969) etc. which principled mining corporations are aware of and comply with the standards. According to him, 'one of the main goals of CSR is the Social Licence to Operate (SLO) which is formalised in free prior informed consent (FPIC). The concept of a mine to have a 'social licence to operate' is emerging within the hard-rock mining community and acceptance of the definition and application is broadening. SLO is a comprehensive and thoroughly documented process in which local stakeholders identify their values and beliefs as they participate meaningfully throughout the environmental and social impact assessment process of the proposed project, from scoping through mine closure and restoration. The SLO usually includes three goals for successful development: first, maximize advantages of the mining project to local national economies; second, minimize damages to communities, environment, and socio-cultural assets; and third, strengthen the capacity of impacted communities and their advocates (civil society) and governments to participate meaningfully in the mining process.'[45] Social licence[46] must be earned and then maintained. It is dynamic and non-permanent because beliefs, opinions, and perceptions are subject to change as new information is acquired. A social licence is usually granted on a site-specific basis: hence, a company may have a social licence for one operation but not for another. The bigger the impacts of a project, the more difficult it becomes to earn the social licence. A mining company wanting to relocate an entire village faces a much bigger challenge.[47]

Prequalification or Certification of Potential Mining Permit Seekers

The responsible mines should engage in an Independent Certification Regime to be funded by the payments from the mining companies. Prequalification or certification of potential bidders on governmental work is a mandatory provision in most of the countries. Through this mechanism, only the selected better mining companies those have CSR standards, E&S policies with in-house cells /units and better quality track records are being encouraged and weak companies are automatically eliminated in the process. Reputed independent, unbiased and neutral organisation(s) can take up third-party review, monitoring, conformity assessment, certification etc. and could be responsible for ESIA contracting, guidelines, international standards, and quality review. This will be funded by corporate contributions to international bodies and not by the company itself.

Insurance and Performance Bonds

In order to fulfill and comply with contractual obligations in time, insurance and performance bonds mechanisms should become standard in responsible mining. The insurance and bonds should be adequate enough to cover accidents, any noncompliance and the future exigencies. If sometimes after closure of a mine, leakages start in a toxic waste pond, the cleaning-up, repair and restoration liability would be financed from the reclamation bonds. This also holds good in case of other unforeseen contingencies. Similarly, the provision of transferability of insurance and bonds in case of taken over or sale of a company should be clarified in advance. Hence, a realistic calculation of all short-term and long-term future costs of a mine should be done and accordingly provisions made in insurance and bonds.

Royalties, Taxes and Fees

Costs-Benefits analysis of a proposed mine should be made appropriately taking in to account all the pertinent parameters i.e. the social and ecological impacts as well as payment to governments in the form of royalties, taxes, fees etc. towards the cost and the net profit and others, if any, in the benefit side. Net profits accrue to the company, royalties, taxes etc. to the government and social and ecological impacts to the beneficiaries. All the internal and external costs and benefits should be properly identified and allocated even though it is difficult to make a proper estimate of environmental and social costs of a mining project. This task can be assigned to an expert organisation or third party independent institution. Responsible mining companies should be very transparent and honest in making payments to government, preventing corruption,[48] mitigating adverse impacts caused to the communities and scrupulously follow the 'Extractive Industries Transparency Initiative'(EITI).

Reasonable compensation to the affected people, provision of social infrastructure in adjoining areas, making good the environmental hazards, preventing ecological damages and impacts on land (agricultural, homestead, grazing etc.), water, (fresh, marine, catchments etc.) and air, conservation of eco-systems and mine closure funding for repair and restoration activities, community training and skill development for alternative avocations or jobs and providing sustainable livelihood income generating schemes are the best practices and standards of a responsible mining. Along with socio-ecological risk-bearing capacity, a responsible mining company should be well equipped to combat post-closure catastrophic damages and other eventualities.

It is a well known fact that certain costs, adverse impacts and hazards cannot be compensated and, therefore, mining companies should ensure that the compensation and benefits to the affected people should always outweigh the loss sustained can be achieved by ensuring that the benefits accruing to the potentially impacted people clearly exceed the costs and impacts. Best-practice mining companies should agree to contribute at least one percent of profits (or revenues, or even better, of the value of the minerals extracted) to environmental conservation efforts in their area of operation, as proposed in IUCN's World Conservation Congress Resolution 4.085 of 2008: establishing the one percent Earth Profits Fund.[49]

No-Go Zones

Responsible miners should avoid operation in certain extremely valuable areas, unless and otherwise in extraordinary cases, the potentially affected communities accord full consent for the project. They are socially or environmentally sensitive areas[50] and off limits to mining, mainly; indigenous peoples reserves, conflict zones, fragile watersheds, special biodiversity habitats, cultural properties etc.. These areas should have special considerations in mining regulations[51] and be excluded from Mining operations as 'No Go Zones'. The utility and importance of these areas should not be jeopardised by any extractive industries.

Indigenous Peoples reserves are areas in which Indigenous Peoples, tribes, ethnic minorities, forest dwellers and other vulnerable groups have their settlement and livelihood for generations but without any legal authority on the land. Once displaced or dispossessed, they never regain their earlier status and ease of living, however successful may be the resettlement or rehabilitation programme. Conflict zones are conflict-prone areas or areas of overt social and armed conflict. Such conflict zones in most cases exacerbates conflict and national governments should either prohibit or postpone a mining project in these areas or issue a transparent guideline on whether industrial mines should be grounded in conflict-prone areas and if at all it is allowed to operate, under what circumstances and terms and conditions.

Areas providing critical sources of water, Fragile water-regulating systems, glacier, watersheds that conserve water for irrigation and agriculture, ecosystems those help for conserving riparian services, all water catchments above or feeding into irrigation systems and areas protecting a dependent project downstream should be prohibited from mining. Besides, mining should not be permitted or banned or avoided in mountainous zones, areas vicinity to any source of water, active seismicity or geological faults with the risk of leakages of toxic lagoons and rupture of mine wastes, steep slopes, areas prone to landslides or mudslides, areas subject to very high rainfall and areas with a wide swath either side of possible hurricane or cyclone paths. These areas are 'No Go Zones' for extractive industry, need protection and should be off limits. Any mining on small islands, like Indonesian law, should be illegal, because a big mine on a small island is likely to damage water supplies and marine livelihoods by dumping wastes.[52] It must also be strictly prohibited to mine anything in most important and precious areas like the areas of high biodiversity and endemism, rare or endangered species, rare habitats, and intactness e.g., coral reefs, mangroves, tropical rain forest, remaining old growth forests, biological hotspots, wetlands, and wilderness,[53] and all conservation units such as; national parks, state or provincial parks, UN Biosphere Reserves, UN World Heritage Sites, areas scheduled for inclusion in the national system of conservation units, protected forests, most mangroves, old-growth tropical forests, UN Ramsar Convention wetland sites as well as their buffer zones. Responsible and best practices are to avoid mining in these areas. Green forest areas which are vital for greenhouse gas (GHG) emissions and sequestration function, conservation of biodiversity, attenuating floods, providing water in the dry season and community livelihoods should not be denuded rampantly and mercilessly for the purpose of mining. As far as practicable, deforestation should be avoided, or reversed and must be more than replenished by tree plantations or regeneration and reforestation. Averting the worst consequences of human-induced climate change is a "great moral issue" on a par with slavery.[54] Besides making an agreement with the mining company on a heavy price for GHG emissions, Government should also impose a huge tax on all carbon emissions to generate sufficient fund for substantial measures to combat climate risks. Responsible mining should not destroy green forests and become carbon-neutral as much as possible. Similarly, highly valuable and important cultural properties like archaeological sites, old battlefields, ancient religious sites, areas of sacred groves, petro glyphs, geo glyphs or rich fossil sites are No Go Zones for responsible mining. The category of areas discussed above are non-replenishable, irreplaceable, and highly vulnerable in nature due to their ecosystem services and socio-environmental benefits.

Compensatory Offsets

The term 'offset' is a consideration or amount that diminishes or balances the effect of an opposite one and 'compensate' is used to mean measures to counteract adverse impacts. However, 'offset' is often used interchangeably with 'compensate'.

Mining in no-go zones are, in exceptional cases, unavoidable and has to be negotiated after exhausting the best possible options or alternatives. In view of long term development perspectives or greater public interest or overriding compulsions, a compromise must have to be forged. There may be such other situations in prohibited areas and in these cases, the adverse impacts or costs of mining should be fully compensated or the offsets are such that there should not be any net loss. The higher compensatory offset than 'no net loss' is the best practice of responsible mining.

The best available options in any development project are to first, prevent, second to minimise, and third to restore or mitigate any harmful impacts. In unavoidable circumstances, mining can go ahead in off-limit areas, but the company should initiate adequate palliative steps to avoid adverse impacts to the maximum possible extent and hence, offsets and other compensation are the last option, if the other three cannot be accomplished. Compensatory offsets are usually conservation measures intended to compensate for unavoidable environmental and social harmful impacts caused by implementation of development project, after appropriate preventive, minimisation, and mitigatory measures have been taken so that a clear net gain can be achieved. Compensatory offsets, besides bolstering the company's social licence to operate, also enhances the company's goodwill and trustworthiness between all the stakeholders. Carbon offset for a carbon neutral mining company could be tree plantations or rehabilitation or regeneration of degraded forest to neutralise the amounts of greenhouse gas emissions (GHG), the project expects to emit during its lifetime. The offsets should expand carbon sequestration capacity. Likewise, biodiversity offset should conserve an area greater or at least, similar to the tract lost to the mine and in ecosystem service. Any compensatory offset for biodiversity loss should leave the environment far better off than pre-project period. Ecosystem services are the benefits or functions that people, including businesses, derive from ecosystems. They include four types of function: (1) provisioning services for food, freshwater, timber, fibers, medicinal plants; (2) regulating services such as surface water purification, carbon storage and sequestration, climate regulation, protection from natural hazards; (3) cultural services such as natural areas that are sacred sites and areas of importance for recreation and aesthetic enjoyment; and (4) supporting services such as soil formation, nutrient cycling, and primary production.[55] The area is better off if there is "informed agreement of stakeholders that the proposed offset is more extensive in area, greater in environmental value (less disturbed, less damaged, more biodiversity, greater environmental service value), higher in ecosystem function, and under a more secure level of protection, such as by financing in perpetuity".[56] The best practices in responsible mining is not just to replace the lost area with a similar or an equal amount of land elsewhere which may not be sometimes adequate. 'In practice, research has shown that 1:1 offsets have often failed to deliver similar ecosystem services if they are small isolated tracts, not contiguous to other habitats and thus susceptible to the "edge effect," which reduces reproduction in small conserved plots. Rather than "no net loss," the accepted practice is increasingly to offer a "net benefit". Thus, the area or value of the tract lost is multiplied by a "small multiple" of more than one, commonly three, and less than ten. The small multiplier must be big enough to compensate fully and unambiguously for all areas lost to the project'.[57] Ultimately debates over compensatory ecosystem values and the "equivalency" of ecosystem gains and losses are usually reduced to establishing a "compensation ratio," a number that establishes the number of compensatory hectares required per hectare of ecosystem impacts.[58] The U.S. 1972 Clean Water Act (and the U.S. 1990 Clean Air Act) use a rule of thumb of a 3:1 ratio for wetland banking, although this would now be considered minimal.[59] Hence, compensation ratio should be fair enough to generate a net gain for the loss sustained as also in case of "Compensation for Social Impacts" which the mining company pays to adversely impacted people in terms of money and other benefits. The compensation amount should have been publicly agreed upon or transparently negotiated in the Impact-Compensation Contract (ICC) or Impact-Benefit Agreement (IBA)[60] that are intended to compensate the adversely affected local communities and indigenous peoples for the loss of their livelihoods, traditional domains and ancestral properties due to mining. IBAs or ICCs may include training and employment in the project, local procurement and services to the project, revenue sharing, environmental provisions, health and safety, reclamation procedures, scholarships, apprenticeships, cross-cultural training, relationships between communities and mine employees, control of employees behaviour, penalties and incentives, and dispute resolution.[61]

Principles of Precaution

Principles of Precaution originated with the famous quotation 'Quod dubitas ne feceris' which means 'When in doubt, don't do it'.[62] There are various other precautionary principles such as; an ounce of prevention is greater than a pound of cure, better safe than sorry, err on the side of caution, look before you leap, first do no harm than think otherwise etc.. ESIA mandates the analysis of alternatives—better, safer, cheaper ways to do things—and the development of "cleaner" products and technologies. One powerful alternative is simply slowing down in order to learn more about potential harm—or doing nothing—the "no project" alternative.[63] Generally, the responsible companies avoid the risk of damages to areas of historical and cultural significance or high-conservation-value by not mining there. Precautionary measures should be initiated even in uncertainty, if there are risks of social or ecological damages because neither the people or communities nor the government can wait till the certainty of adverse impacts or till the apprehensions are fully established. Precautionary principles are normally the risk-aversion measures.

Some other authors have discussed the principles of responsible mining from various viewpoints. According to Sara Bice, the concept of responsible mining is based on five key principles or pillars: holistic assessment; ethical relationships; community-based agreements; appropriate boundaries and good governance. Together, these pillars circumscribe global best practice and innovative ideas to catalyse new and improved approaches to a sustainable mining industry. Responsible mining must be comprehensive in its approach and in the concerns it encompasses from exploration to closure, from environment to community, from corporate governance to financial transparency. For mining to be responsible, companies must engage meaningfully and ethically with communities. Responsible mining, therefore, brings together strong impact assessment methods and mitigation planning; principled relationships with local communities which prioritise equal community involvement in agreement-making; and sets boundaries for corporate activities and roles in states and communities within a framework of good governance. These five pillars hold the potential to transform the mining industry and its impacts to undergird a viable, ethical and accountable mining industry. The author argues that these practices are critical to the future viability and social acceptability of the global mining industry and draws on a range of case studies, including from Australia, Canada, Central Asia, Papua New Guinea and west Africa. Mining can have negative environmental and social impacts, but can also be responsible. However corporations have little impetus to act responsibly without being held to account by an informed and active public, and by strong institutions and governments which not only create but also enforce legislation.

Sara Bice discussed the five pillars of responsible mining in seven chapters. Chapter 7 reviews the five pillars to establish theoretical propositions for how they can be better incorporated into corporate practice. The journey towards responsible and ethical mining has been decades in the making, and will continue for decades to come. It requires a belief in possibilities, a willingness to question the status quo, ingenuity and tenacity.[64] Anshari & Isnalita are of the view that ‘number of scientific publications, according to Web of Science*(*a Clarivate Analytics Product), that correspond to the search terms “low-grade ore”, “sustainable mining”, “zero waste” and “mineral waste metal recovery” on environmental to social aspects, its general effect on the extraction processes was more strict legislation with regard to residue containment and treatment and the need for the development of less polluting technologies. Furthermore, an optimisation of the mining and metal extraction industrial processes according to sustainability principles is required'.[65] But Jeroen Spooren et.al. are of the opinion that 'given that transactions in investment primarily involve the private sector, self-regulation is the primary means through which irresponsible mining practices could be prevented. However, government intervention also plays an important role in incentivising responsible investment and reducing harmful practices'.[66] Noma Nyembo, however, observed that 'built on strategies developed in Latin America and the Caribbean in 1996, governments have been challenged with remaining transparent, providing stable institutional arrangements, investment certainty, and fiscal responsibility whilst mining companies have been charged with increasing consultation with affected people and mitigating negative social, economic and environmental impacts. Drawing on guidance from these initiatives and relevant literature a code to modern day mining practices can be organised around 10 main themes including: Legal frameworks, economic and fiscal policies, institutional reforms, CSR and participation of affected peoples, access to mining activities, ongoing obligations, regulatory aspects, ancillary licenses and permits, investment contracts, and environmental and social matters'.[67] According to Goodland, 'Mining corporations that follow the eight principles, abide by all international environmental agreements, strive to achieve best practices, and avoid No Go Zones would become industry leaders. Many mining corporations already have adopted responsible mining rhetorically. The best mining corporations will put it into practice'.

Key principles and best practices of responsible mining are outlined by IRMA, ARM, RMF and some important index of responsible mining has also been developed in the form of RMI. Besides, ICMM has developed some Principles which are used as a benchmark for responsible mining by member companies. All these are highlighted in the discussion that follows.

Initiative for Responsible Mining Assurance (IRMA)

Early 2000 witnessed a spate of international standards and frameworks for and mining industry echoing a broader trend of global civil regulation. 'Documents including the ICMM '10 Principles for Sustainable Development', the OECD's 'Guidelines for Multinational Enterprises', the Initiative for Responsible Mining Assurance (IRMA), the Extractive Industries Transparency Initiative (EITI), and the Global Reporting Initiative's (GRI's) Mining and Metals Sector Supplement punctuated a cultural turn in companies' approaches to corporate social responsibility'.[68] During 2008, a resolution supporting Responsible Mining was passed by the International Union for Nature Conservation (IUCN) and its Extractive Industry Responsibility Initiative (EIRI) was established.[69] Prior to that, the Initiative for Responsible Mining Assurance (IRMA)[70] was founded in 2006, by a "coalition of non-government organisations, businesses purchasing minerals and metals for resale in other products, affected communities, mining companies, and labor unions"[71] with decisions as to how best to follow the mitigation hierarchy of "avoid, minimise, restore, and offset". Bringing these stakeholder groups together is truly a social innovation and as a result IRMA is regarded as the MSI with "the most comprehensive and legitimate catalogue of requirements".[72]

Minerals covered under IRMA, 'range from coal, aluminium and dimension stones as examples for mass raw materials to gold and diamonds produced in small amounts only. Some address several mineral raw materials if these share common characteristics, for example, the conflict minerals tantalum, tin, tungsten and gold (3TG) or cyanide use in gold and silver production. Its scope i.e. the goals and focus on groups of minerals, company scales, geographical regions, supply chain parts or—as in the case of the "International Cyanide Management Code for the Manufacture, Transport, and Use of Cyanide in the Production of Gold" (Cyanide Code)—a mining practice. The MISs' goals vary and extend from strengthening the social licence to operate through commitment and reporting (e.g., ICMM), to professionalisation of and improvements in ASM (e.g., Fairtrade, Fairmined), providing information for more sustainable purchasing decision making (Bettercoal or Fairstone), ensuring compliance with the Dodd–Frank Act for downstream actors of the supply chain (e.g., the Conflict Free Smelter Initiative, CFSI) to making a positive sustainability impact in mining and processing.[73] On 28th June, 2018, the IRMA released the Standard for Responsible Mining, a precedent-setting global certification programme for industrial-scale mine sites as a Multi-sector effort to certify social and environmental performance at mine sites globally. The Standard for Responsible Mining's best practice requirements for mining include elements such as health and safety for workers, human rights, community engagement, pollution control, mining in conflict-affected areas, rights of indigenous peoples, transparency in revenue payments from companies to governments, and land reclamation once mining is done. The Standard is not simply a pass/fail system, instead focused on transparency, where different levels of performance are recognised and continuous improvement is encouraged, but where certification is still available for those industrial-scale mine sites meeting all major best practice requirements[74]. According to Aimee Boulanger, Coordinator of the IRMA, "The Standard for Responsible Mining responds to the power of markets to create greater incentive and leverage for protection of the environment and the communities who live closest to mines. It will offer shared value to corporations who seek to make a profit, while also offering an ethical value chain to their customers". IRMA standard report released on June, 2018 Contains Preamble, Introduction to the IRMA Standard, Principles and Objectives, Scope of the IRMA Standard, Chapter Structure, Language, Basis for Certification, Continuing Improvement, Flagged Items, Associated Documents and Materials, Collaboration with Related Standards and Certification Systems, Business Integrity Requirements, Chapter 1.1 to 1.5 highlighting Legal Compliance, Community and Stakeholder Engagement, Human Rights Due Diligence, Complaints and Grievance Mechanism and Access to Remedy, Revenue and Payments Transparency and Planning for Positive Legacies Requirements. Besides, other chapters are Chapter 2.1- Environmental and Social Impact Assessment and Management, Chapter 2.2- Free, Prior and Informed Consent (FPIC), Chapter 2.3- Obtaining Community Support and Delivering Benefits, Chapter 2.4- Resettlement, Chapter 2.5- Emergency Preparedness and Response, Chapter 2.6- Planning and Financing Reclamation and Closure, Social Responsibility Requirements, Chapter 3.1- Fair Labour and Terms of Work, Chapter 3.2- Occupational Health and Safety, Chapter 3.3- Community Health and Safety, Chapter 3.4- Mining and Conflict-Affected or High-Risk Areas, Chapter 3.5- Security Arrangements, Chapter 3.6- Artisanal and Small-Scale Mining, Chapter 3.7- Cultural Heritage, Environmental Responsibility Requirements, Chapter 4.1- Waste and Materials Management, Chapter 4.2- Water Management, Chapter 4.3- Air Quality, Chapter 4.4- Noise and Vibration, Chapter 4.5- Greenhouse Gas Emissions, Chapter 4.6- Biodiversity, Ecosystem Services and Protected Areas, Chapter 4.7- Cyanide Management Chapter 4.8- Mercury Management, and lastly, Glossary of Terms.

In the Preamble part, it is clearly outlined that 'IRMA believes that many of the negative social and environmental impacts can be avoided if mines operate according to leading practices. The Standard for Responsible Mining specifies a set of objectives and leading performance requirements for environmentally and socially responsible practice. The IRMA Steering Committee set the mission to establish a multi-stakeholder and independently verified responsible mining assurance system that improves social and environmental performance and creates value for leading mine sites. IRMA envisions a world where the mining industry is respectful of the human rights and aspirations of affected communities; provides safe, healthful and respectful workplaces; avoids or minimizes harm to the environment; and leaves positive legacies'.

The IRMA Standard is designed to support the achievement of overarching principles. It is intended; To support the application of the laws and regulations of the country in which mining takes place, or exceed host country laws in a manner consistent with best practice, To support mining company decision-making and enable communities and stakeholders to participate in mining-related decisions that affect their health, well-being, safety, livelihoods, futures and the environment, To respect human rights, and identify, prevent, mitigate and remedy infringements of human rights, To provide accessible and effective means for affected communities and individuals to raise and resolve mine-related complaints and grievances at the mine operational level, while not limiting their ability to seek remedy through other mechanisms, To increase transparency of mining related payments and provide communities and the general public with the information they need to understand and assess the fairness of financial arrangements related to mining operations, To proactively anticipate and assess environmental and social impacts; manage them in accordance with the mitigation hierarchy; and monitor and adapt environmental and social management systems in a manner that protects affected communities, workers and the environment throughout the entire mine lifecycle, To demonstrate respect for the rights, dignity, aspirations, culture, and livelihoods of indigenous peoples, participate in ongoing dialogue and engagement and collaborate to minimize impacts and create benefits for indigenous peoples, thereby creating conditions that allow for indigenous peoples' free, prior and informed consent and decision-making regarding mining development, To obtain and maintain credible broad support from affected communities, To avoid involuntary resettlement, and when that is not possible, equitably compensate affected persons and improve the livelihoods and living standards of displaced persons, To plan for and be prepared to respond effectively to industrial emergency situations, To protect long-term environmental and social values, To maintain or enhance the social and economic well-being of mine workers and respect internationally recognised workers' rights, To identify and avoid or mitigate occupational health and safety hazards, maintain working environments, To protect and improve the health and safety of individuals, families, and communities affected by mining projects, To prevent contribution to conflict, To manage security in a manner that protects mining operations, To foster positive relationships between large-scale mines and artisanal and small-scale mining (ASM) entities, To discharge Environmental Responsibilities such as; Waste and Materials Management, Water Management, Air

Quality, Noise and Vibration, Greenhouse Gas Emissions, Biodiversity, Ecosystem Services and Protected Areas, responsible cyanide and Mercury Management etc. The IRMA Standard is designed to be applicable to all types of industrial- or large-scale mining, including surface, sub-surface and solution mining, and all mined materials e.g., minerals, metals with the exception of energy fuels. Thus, "this is the first time that a set of practices for responsible mining have been designed by representatives from multiple sectors. The Standard is aimed at multiple industrial-scale commodities, ranging from building materials and automotive parts, to jewellery and electronics. It covers all the key issues deemed relevant for responsible mining, including human rights, transparency, community, labour, safety and environmental matters".[75]

Alliance for Responsible Mining (ARM)

An ARM was formed with the objective of improving the quality of life of small-scale artisanal miners by reducing the environmental and human impacts of toxic substances. It is intended that ARM would encourage the mining community to participate in designing local policies and supports the adoption of better social, business and environmental practices that comply with the Fair mined Certification Standard. ARM is working with artisanal and small-scale mining organisations (ASMOs) in some of the selected area. In order to foster community involvement in decision-making, in particular enabling the development of a common, responsible vision of the mining sector, ARM has been piloting local mining governance models since 2017.

Responsible Mining Foundation (RMF)

RMF aims to ensure that 'human rights defenders play a vitally important role in standing up against human rights abuses that can be associated with the presence of mining operations – abuses that range from forced evictions to arbitrary arrest and detention and from toxic waste spills to violations of Indigenous Peoples' rights'. It encourages continuous improvement in responsible extractives acts to guard against threats towards and killings of human rights specifically by mining and extractives industries. RMF intends to ensure that mining corporations/ companies fully respect or demonstrate a commitment to respect and protect the human rights. According to Joan Carling; "If we are to change this reality on the ground through the implementation of the UN Guiding Principles on Business and Human Rights, the bad legacy of business operations in indigenous territories need to stop".[76]

Responsible Mining Index (RMI)

Responsible Mining Index (RMI) was launched on 11 April, 2018 by the Responsible Mining Foundation. This Index is designed to assess the policies and practices of 30 large-scale mining companies, revealing how well they are performing on economic development, environmental, social and governance issues. Prof. Gilles Carbonnier observed "Mining is strategic for a whole range of resource-rich developing countries. It's a sector that can greatly contribute to the emergence of a green economy and to sustainable development, but it can also contribute to environmental degradation, social conflicts and human rights violations. The Responsible Mining Index provides an incentive for continuous improvement, highlighting where progress is needed and what best practice looks like".[77] A wide spectrum of economic, environmental, social and governance (EESG) issues related to responsible mining are covered under RMI which assesses companies from these perspectives and examines the extent to which companies are addressing these issues throughout the project lifecycle. The RMI assessment has 73 indicators applied at the company-wide level across the six thematic areas which includes Economic Development, Business Conduct, Lifecycle Management, Community Wellbeing, Working Conditions, Environmental Responsibility. Based on these six thematic areas which covers the scope and content of the RMI, it is intended to monitor how companies: (i) contribute to national-level socio-economic development in producing countries; (ii) demonstrate their commitment to ethical behaviour and good corporate governance; (iii) manage their impacts throughout all phases of their operations and plan for post-closure viability from the earliest stages; (iv) engage with affected communities and manage their socio-economic impacts at a local level; (v) provide safe and healthy workplaces, respect the rights of their workers and prevent unethical labour practices; and, (vi) assess and manage their environmental impacts in a systematic, collaborative and transparent manner. The 'Index revealed that while many companies have introduced responsible mining policies on issues such as corruption and CO_2 emissions, much progress remains in areas including monitoring how mining affects children and protecting female workers from harassment and sexual exploitation'.[78]

The International Council on Mining and Metals (ICMM)

Some specific mining principles have been developed by ICMM which should be used as a benchmark for responsible mining by member companies. The objectives of these principles are to maximise the industry's benefits to host communities and minimise the negative impacts to effectively manage social issues. The ICMM's social and economic development programme aims to strengthen social and economic development impacts including community engagement, human rights, as well as broader benefits that mining can deliver.[79] The ICMM released a revised, tougher set of Mining Principles amid increasing global pressure on mining companies to mine responsibly and rising stakeholder expectations around environment, social and governance (ESG) best practice. There are 10 Mining Principles, supported by 38 performance expectations which set out clear requirements on issues including biodiversity, gender, human rights due diligence, labour rights, local content, mine closure, pollution, resettlement and waste, to name a few. They apply to more than 650 of the ICMM members' assets in over 50 countries. However, five key Mining Principles that guide responsible mining are: Ethical Business, Human Rights, Risk Management, Social Performance and Stakeholder Engagement. These five principles the member countries to; (i) Apply ethical business practices and sound systems of corporate governance and transparency, (ii) Respect human rights and the interests, cultures, customs and values of employees and affected communities, (iii) Implement effective risk-management strategies and systems, (iv) Pursue continual improvement in social performance and contribute to the social, economic and institutional development of host countries and communities, and (v) Proactively engage key stakeholders in an open and transparent manner, effectively report and independently verify progress and performance. These Mining Principles build on the values of members while also define good practice ESG governance requirements by member companies.[80] ICMM team supported the company members to respond to the COVID-19 pandemic. The ICMM and its members have supported local response efforts, working with community members and organisations in often remote locations to identify and help meet health and other specific needs during the unprecedented global crisis.

Challenges

Mining and for that matter extractive industries can behave and operate responsibly, they can very well offset the adverse impacts on society and environment. But mining have myriads of challenges. Sheila Siegel aptly pointed out, "there is a maddening futility about speaking of 'mining,' as if it were singular or coherent. Rather, there are many mining industries, and each has its own culture, directives, structure, purpose, and pathologies. "Mining is the material basis for life, making it difficult to exaggerate its significance. George Orwell called it part of the "metabolism" of civilisation. Major divisions of history are named in accordance with their dominant mineral products: the Paleolithic and Neolithic Periods; the Copper, Bronze, and Iron Ages. More than ever, humanity relies on minerals to sustain its existence. The growth of population, speed of transportation, proliferation of electronic gadgets and games, and delivery of electricity all depend on the expansion of mining. And yet we are ready to discuss almost any other ethics before the ethics of mining. We have more faith in our capacity to restrain or end violence and war than to address the ethics of mining".[81] Environmental Challenges are one of the prime challenges of minerals and mining. The minerals and mining were included as part of global environmental and development ethics in the United Nations Scientific Conference on the Conservation and Utilisation of Resources during 1949. Minerals was one of the categories in that conference which was divided into four categories; the others were agriculture, forestry and fisheries.

But mining was left out when the postwar international environmental and development institutions were created. The minerals and mining does not figure in Agenda 21, the nonbinding, voluntary UN action plan for sustainable development that has guided environmental negotiations since the Rio conference of 1992. This had also happened earlier in the global plan from 1987, "Our Common Future," a policy manual intended to unify the international environmental and development agendas. In 1972, the Stockholm Conference on the Human Environment was held which led to the creation of the United Nations Environment Programme. The book "Only One Earth" was published on this occasion but the book devotes just few pages to mining/mineral extraction. The net outcome of 'mining's omission from environmental ethics is that as other disciplines and sectors gradually integrated concerns about sustainability into their knowledge communities, mining engineering, mineral economics and processing, geochemistry, and other sub-disciplines associated with mining have remained static. While there is much anxiety about the failure to enact the ethics of climate change or environmental health, mining does not even have an ethical roadmap. There remains no baseline for articulating and pursuing the principles of sustainable resource extraction. There was a sea change in mining and minerals scenarios and experiences of both developed and developing countries in pre-World War I and post-World War II. Most of the participating countries exhausted the essential resources and had turned to a net importer. To add to this, the world became more resource-dependent in the postwar period. Previously, only a handful of minerals were used and it was once relatively easy to count the critical minerals like iron, copper, zinc, lead, tin, coal or mercury but with the adoption of modern lifestyle, growth in demand and mushrooming of electronic industries, at least ninety metals and mineral commodities are required. "It is significant that in the cases of agriculture, forestry, fisheries, and certain other fields of resources development some progress has been made in the direction of conservation. All these are renewable resources. Yet in the case of minerals, which are not renewable, there has been practically no effort, except in time of war, to interfere with the free play of a market that is interested primarily in profits. This anomaly cannot continue indefinitely. ---- I don't know why mining vanished from environmental and development ethics. Perhaps the idea was that resource extraction would be handled in a different policy sphere, or maybe there was an assumption that managing climate, forests, biodiversity, and other ecological stresses implied an inherent reckoning with the limits of extraction".[82] However, Only in the last decade a more vocal public discourse about global resource policy has emerged and careful, internationally coordinated mineral extraction principles initiated.

Irresponsible mining makes the environment more fragile and prone to disasters. It causes several destructions, threatens the region's source of water and watersheds. In some countries, mining disasters had led to the death of rivers. A big fracture in the drainage tunnel of large pits containing leftover mine tailings led to a discharge of toxic mine waste into the river systems and caused flash floods in areas along the rivers. As a result, the drinking water was contaminated; fish and other food in the river vanished leading to biological death of a river. Some environmentalists, though not scientifically proved, blamed mining for the disastrous earthquake. Jonathan L. Mayuga has cited the case of Philippines to describe the environment related challenges. The massive destruction of ecosystems in mining tenements—the massive cutting of trees, blasting of mountains, the digging and hauling to extract mineral ores—are to blame for the siltation of rivers, as well as degradation of coastal environments that affect agricultural and fishery production. People suffer because of these. Surface mining activities may cause destabilisation of ground surface and this can exacerbate geo-hazards like landslides. Open-pit mining causes massive destruction of forest ecosystems—as forests are shaved, mountains flattened, and the digging and hauling continue, until huge bowl-like holes are formed. Besides, Mining pose threats to biodiversity. When miners cut trees and haul soil, they kill forests, which are home to threatened species. While rehabilitation is possible, the extensive damage of open-pit mining leads to irreversible biodiversity loss.[83] Fitrian Ardiansyah observed that 'it is, therefore, a huge challenge to find an appropriate solution that could balance mining development and environmental performance. A 2003 study conducted by the World Resources Institute reveals that approximately one-third of the world's active mines and exploration sites are located in intact ecosystems of high conservation values'.[84]

Many research studies documented negative environmental impacts of mining operations. It directly causes deforestation and habitat destruction in several countries. In particular, the most immediate impact on forests from mining activities is the removal of forest vegetation. In 2000, a study conducted by William Sunderlin and Sven Wunder, shows that there is a statistically significant relationship between the amount of mineral exports and rate of deforestation in Indonesia during the period of 1976 to 1980. Another study on Papua and Kalimantan, outlined that gold and copper mining activities in these two major islands had severe environmental damages on surrounding forests since 1990s. Most of the research findings reveal that mining results in adverse impacts on forest, other ecosystems and the society including contamination of rivers used for the purposes of drinking water and food supplies.

Two other related challenges of an ethical and sustainable mining are; transparency and corporate social responsibility (CSR) or Entreprise Social Commitment (ESC). "While transparency initiatives concentrate on exposing revenue transactions between the private and public sectors in extractive industry projects, corporate responsibility efforts focus on the improvement of relations between companies and communities. The transparency movement has sparked advocacy and legislative activity in the United States, United Kingdom, and Canada—the host markets for much of the world's trading of mining shares. Meanwhile, companies are dedicating more staff and resources to ensure the benefits of mine development reach communities in the form of improved services, infrastructure, and education. These twin concepts are intended to transform resource extraction from a winner-takes-all model to one in which all parties benefit" [85]. But the problem arises because there is no trade-offs or reconciliation of ESC and transparency with environment and ecological limits in mineral extraction. Gradually, when underground resources become scarce, the mining companies/corporations run for profit with less concern to ESC and transparency. This has led to the depletion of land and resources, especially the smallholder farming and mining economies upon which people depend in mineralised places resulting in endemic poverty, conflict, and ecological collapse in these erstwhile resource rich regions.

Another significant challenge in responsible and ethical mining is 'Conflicts' in areas where mineral industries operate. Protests, strikes and moments by the people including indigenous local communities disrupts the operation process severely or sometimes closure of the mines. Besides, severe conflicts often arise between the host communities and outsiders/ mining workers who use to settle in mining colonies. Local people object the use of CPRs by others. Sometimes, when underground resources get exhausted or become scarce, conflicts occur due to the pressure created by crossing a threshold of scarcity. There are cases of increasing social conflict over access to mineral resources. Conflicts between mine owners and workers are most often noticed. There are instances of the massacre of striking miners in different parts of the world. It is impossible to identify any part of the world where mineral extraction is expanding without conflicts. Many people have been killed in mining clashes.

Adoption of proper Technology and right Strategy by mining companies/ corporations pose serious challenges in resource extraction. Lack of Technology and Strategic Management may push the company to premature closure. Advanced technology, though can't overcome, can delay the inevitable process of depletion of underground resources.

Illegal mining and two other related practices of bribery and corruption pose significant challenges to responsible mining. Illegal mining has been operated in different parts of the world. Corruption and bribery has become culturally acceptable practices particularly in many developing countries. Corporations/companies indulging in rampant, unabated and illegal mining and corruption can never implement responsible mining practices. Apart from the issues discussed above, which are common to all major mining activities, the outstanding challenges and preponderant problems that appeared in the forefront are rampant and unbridled mining as well as illegal mining and corruption at different stages. Since mineral resources are finite and there is no scope of their renewal, rampant and unscientific mining are being carried out. Expression of interest basis or tender process or competitive bidding process in mining leases as per the provisions of law are not being followed resulting in illegal mining and corruption.

Weak Governance is another serious challenge in responsible mining. Mining can spark the economic development of poor regions, but at the same time it is associated with severe sustainability issues, particularly in areas with governance deficits. It is an established fact that self-regulation, suo moto CSR practices and voluntary compliance are not enough. Stake holders and societies cannot rely only on the good intentions of mining corporations. For real change to happen, mining companies/corporations need effective guidance and active support from government. 'The widespread and strengthening opposition to mining, the increasing number of jurisdictions adopting metal mining moratoria, and the soaring need for improved protection of the earth, are evidence that voluntary, nonbinding, and self-reported standards are not working adequately There must be checks and balances, mandatory regulations, government sanctions for violations and prohibitory measures. There are many cases where state governance is weak or absent and can be seen rampant irresponsible mining. This is in line with existing literature on responsible mining that often focuses on failed states and missing governance'.[86] An analytical research framework covering governance, change in practice, and diffusion reveals that the change in practice is enabled via a change in governance'.[87] This signifies the role of strong governance in implementing responsible mining practices.

Deloitte developed "Tracking the Trends 2012" which contains the top 10 trends mining companies may face in the coming year. According to him, "As mining company activity catches the international spotlight, industry stakeholders find themselves subject to higher levels of activism than ever before". Some mining companies are striving their best to implement CSR and ESC. "Although mining companies have adopted the principles of corporate social responsibility and are taking positive steps to be more responsible at a local level, many significant challenges remain. Mining companies have not developed the right recipe for sustainable social programmes. For example, funding the construction of local schools is seen as a valuable social investment, but providing ongoing support for school operations or repairs is not. (In fact, delegates acknowledged that there isn't a clear idea of what social and environmental sustainability actually looks like.) Improving stakeholder engagement won't solve the problem of government policies that are inconsistent with the UN's 'Guiding Principles on Business and Human Rights'. And on the one hand, mining companies now have to be responsible. On the other hand, CSR isn't having enough impact and operating risks are seen to be increasing".[88] What boils down from the above discussion is that responsible mining has no single or dual but multiple challenges which need strategic tackling. 'The environment is not the only thing that needs protection. One also has to protect the ancestral domain of the indigenous peoples where mining sites are often located, jobs of mine workers, and profit for mining companies. One also has to consider the overall potential of mining as a driver of economic development. Above-mentioned values and interests, however, often run in opposition to one another and the debates around mining center on the idea of trade-offs: protection of ancestral domains or economic development? Environmental protection or job creation? profit or environmental protection? jobs in mining or no jobs at all? While there are no easy answers to the question of trade-offs, processes for consensus-building have recently been undertaken by several stakeholders, to support genuine multi-stakeholder dialogue on mining policy. For consensus-building to be sustained, there must be "responsible-mining champions" in government, in the executive and legislative branches, and at the national and local levels'.[89]

Global Experiences

Let's have a bird's eye view on the global Scenario and International Experiences on responsible and ethical mining. The global mining industry generated record profits of US$133 billion in 2011. The top 40 global miners paid out dividends in excess of US$32 billion, more than the purchasing power parity GDP of many small-to-medium sized countries.[90] In the year 1998, Global Mining Industry has developed a collective conscience and drawn its attention to Ethics and Responsibility in mining. During this year, 'the heads of ten major global miners converged at the Global Mining Initiative (GMI), aiming to craft an industry-wide response to an enlarging litany of complaints and concerns. This included a series of unfortunate events which concentrated worldwide attention on the industry, especially the leaching of toxic chemicals into the Papua New Guinean Fly River downstream of BHP Billiton's Ok Tedi mine.[91] According to Dashwood, the GMI meeting spawned the ICMM and the Mining, Minerals and Sustainable Development (MMSD) project, which certain experts peg as the genesis of corporate social responsibility (CSR) in mining. Bondy observed, mining's effects and the duties of companies to prevent, mitigate and respond to those impacts, responsibility and more specifically socially and environmentally responsible business practice is a widely acknowledged component of the mining industry and corporations. They gradually linked CSR with corporate social performance, corporate financial performance and within risk management frameworks.[92] Rio Tinto, a leading group of influential miners has, in the past, appealed to 'enlightened self-interest'.[93] Mining companies, nowadays, invest billions in community investment and CSR-related programmes. Major multinationals like BHP Billiton and DeBeers annually earmark at least one percent of pre-tax profits in community investment, a figure which has become a global benchmark.[94] According to ICMM, American mines accounted for between 30 to 40 percent of total global mining between 1900 and 1960 but after 2012, mining in America accounts for less than 10 percent of the global mining industry, extracting mainly coal. In the United States today, it takes several hundred metric tonnes of ore to produce one metric tonne of copper, resulting in mountains of waste rock up to 400 hectares. Due to tapping and exhaustion of the high quality ores mostly low grade ores which hold lesser amounts of minerals have been left which ultimately results in breaking more rocks, using more chemicals.[95] Thus, more ore processed and more waste produced to realise the same amount of materials leaving environmental responsibility in jeopardy.

Sara Bice in "Responsible Mining", has cited the case of the world's largest mine pit at Bingham Canyon, Utah which is 4.4 kilometres wide and 1.2 kilometres deep. Big pits mean big mining business but commitment to business ethics and responsibilities has to be displayed. Mining is the building block of the economy of Australia. It represents the number one export industry and is the cornerstone of country's economic development. According to the Australian Bureau of Statistics 2010 and Minerals Council of Australia 2012, mining contributes an estimated seven percent of GDP and comprising almost 2,500 firms paying out US$14.97 billion (AU$18 billion) in wages to 180,000 individuals. Mining workers there earn highest average weekly salaries more than other workers. Besides their 'feel good factor', they work in good atmosphere and healthy condition.

The Argyle Diamond Mine, discovered by the Rio Tinto, a leading global mining group in the late 1970s, respected the principles of mutual recognition and obligation and ensured indigenous Australians and Aboriginal owners' active participation in decision-making, protected the heritage sites, offered employment, training programmes, business development opportunities, and imparted cultural awareness training. It is rare to find such intensive and responsive engagements by mining companies and definitely this sort of response evoke the possibilities and potential of a responsible and ethical mining industry. According to the International Council on Mining and Metals 2012, developing countries generate more than one-fifth of total global mineral production in comparison to Europe and the United States which generates only 3.5 percent and 4.2 percent respectively. Minig, today, is thus vital and serves as the life line of developing world. Heavy mineral deposits recently discovered in countries like Mongolia, Kazakhstan and the Kyrgyz Republic that hold enough potential to boost Foreign Direct Investment thereby encouraging stronger connections to the international community than ever before. These staggering mining potential and subsequent operation as well as activities put the geology and climate of these developing countries at stake.

Angola is famous for 'Blood diamonds and the Democratic Republic of Congo's has huge untapped mineral wealth. Angola has been riddled with civil war and according to Transparency International 2013, Congo ranks at the very bottom of global corruption indices. Thus, both the countries fared poorly in responsible mining. For Mongolian herders, the cost of mineral wealth is measured in water and history. In an area which receives an average annual rainfall of 80 mm per year, herders worry the country's largest proposed mine will soak up or contaminate this sparse resource. They worry that the presence of an open pit mine and all its heavy equipment accoutrements and safety exclusion zones will defeat their nomadic way of life and endanger their indigenous culture and livelihood.[96] In Papua New Guinea, the Tolukuma Gold Mine has dumped its tailings—the waste rock leftover into the local river since 1994 with the approval of the Papua New Guinean Government. This Cost-effective and Convenient method displays an irresponsible mining and catastrophic from Environment and social point of view. Mining projects in Mongolia are adopting responsible practices. The Uguuj Gold Mine Project shows an enhanced positive engagement with all stakeholders and an effective grievance handling mechanism. Community grievance handling can "range from commonly occurring, relatively minor issues to more entrenched or serious ones that can become a source of significant concern or resentment. A fundamental goal of the grievance mechanism is to solve problems early at the lowest level. A 'grievance officer' is responsible for the overall management of the system. Finally, a 'grievance and complaints committee' is responsible for authorising resolutions beyond the authority level of the resolving officer and, if necessary, escalate concerns and grievances to senior management or independent external bodies".[97] Similarly, Terra Energy which owns and operates the Baruun Noyon Uul (BNU) coal mine has 'developed an internal checklist to ensure compliance with the Voluntary Principles on Security and Human Rights across its mining operations. Besides, the company aims to respond quickly and effectively to community concerns, manages grievance mechanisms, and instil a high level of cultural understanding in its employees and develop positive community relations'.

In Philippines, mining and the administration of mining activities date back during the Spanish era, from small-scale artisanal gold mining or gold panning to large-scale, open-pit mining method. According to the country's law, minerals belong to the State and their disposition, administration, exploitation and development shall be done through licence, concession or lease. It was reorted that some companies are operating irresponsible mining which threatens the country's source of water. Besides, Marcopper mining disaster, largest mining disasters in the country's history happened in the province of Marinduque in the year 1996 that has led to the death of rivers.[98] There was also a report that a fracture in the drainage tunnel of a large pit containing leftover mine tailings led to a discharge of toxic mine waste into the Makulapnit-Boac river system and caused flash floods in areas along the river besides contaminating the drinking water. Another mining disaster was the leak from Tailings Pond 3 at the Padcal mine operated by Philex Mining Corporation in Tuba and Itogon, Benguet, on August 1, 2012. Over 20 million tons of mine tailings sediment were dumped into the Agno River and Balog Creek entirely contaminated the river water. Environment and human-rights groups have been blaming mining for militarisation, killings and human-rights abuses in mining areas. The Philippines is highly vulnerable to climate change, and irresponsible mining is making the already fragile environment more fragile and prone to disasters. Mining can exacerbate geo hazards like landslide and cause threat to biodiversity.[99] Government of Philippines, in view of these hazards cancelled the licence of many mining companies. However, in Philippines, Rio Tuba Nickel Mining Corporation (RTNMC), a subsidiary of Nickel Asia Corporation, was the winner of the First *Asean* Best Practices in Sustainable Mineral Development Award. RTNMC has also been recognised by DENR for its responsible mining practices. It not only dedicated to conscientious mining but also towards helping the environment and the community. The corporation allotted new houses, provide employment and free medical care and education to the community. They also have a Human Resource Development Programme, scholarship programme, infrastructure, household building and road constructions programme. Besides, they are very much involved in Air and Water Quality Management, Waste Management, and Research and Development.[100]

Ghana, a country famous for gold mines, is implementing responsible practices. Formation of Relationship Committee and allotment of Development Support Fund (DSF) are important milestones in that direction. A Cooperation Agreement[101] between Umnugovi Aimag and four project affected soums was instituted in April 2015. Under the agreement, the Relationship Committee and the Goviin Oyu Fund are the key implementation mechanisms for supporting affected people and engage the stake holders. The Development Support Fund (DSF) was provided to the DSF Board for the long-term development of OT and Umnugovi Aimag. Funds were allotted for implementation of programmes on the basis of suitability of proposals and their relative merit.

Of late, it was realised that deep sea mining and Extra terrestrial mining should be encouraged to put at rest the concerns of EESG issues. According to Roche, the deep sea will provide the answer to environmental and social concerns about mining by taking the process away from the communities and landscapes it effects. A Canadian mining company; Nautilus Minerals Inc. was granted the world's first deep sea mining lease in 2011 by the Papua New Guinean government. The Nautilus Company took the approval of the International Seabed Authority and started to use underwater digging and vacuum technologies to suck mineral rich 'polymetallic nodules' to the surface. According to Sonter, 'some answer that space will provide the solution, with metal-laden asteroids floated home to earth, perhaps in a parachuting extravaganza á lá Felix Baumgartner'. But, Cohen observed that 'Extraterrestrial mining smacks of science fiction. Yet NASA scientists debate its costs, not its possibility. However, such possibilities and prospects are never ending.

Conclusion

It is an undeniable fact that responsible mining means higher operational costs. Some companies may pass it on to the consumers and others may absorb it. Then a clear choice and consideration lies with the customers whether to buy a fair-trade, responsibly and ethically mined metal or to go for a cheaper one. But the decision can be much more difficult if the price jumps by 10 to 15 percent and therein lies what is called the 'end-user dilemma'. Because higher price makes no difference in quality of the product, why a buyer should seek out such product? Boyd answers, "A clean conscience. The knowledge that you're helping to preserve, rather than destroy our planet for generations to come". Kathryn Money adds; "You can feel good knowing that you are getting exactly the same quality product, but with a much lower environmental and human cost".[102] Many intellectuals on the subject put a question mark on whether 'responsible and ethical mining is possible'? But many mining companies and corporations across the globe have demonstrated and proved that responsible and ethical mining is not just a myth. It is absolutely possible and is kicking. Mining companies who put responsible mining into practice show that they continue to be the most financially viable and also practically the most socially responsible. Responsible mining is the Key to Profitable Resource Development. It is pro-poor and pro-development. Along with its massive contribution to the countries' economy, it also contributes to the progress and development of the people. If realised perfectly, it provides job opportunities to the marginalised, especially to people in remote and far-flung areas, ensures a better lifestyle for the people, preserves and protects the environment by rebuilding a better forestry and thus, effectively addresses the socio-economic and environmental concerns.

End Notes & References

1. Aliyah Jeande Muta, *"It shouldn't just be mining"*, May, 2018
2. Carmel, V. Abao, *"Responsible Mining: Planet, People, Profit, Policy,"* 2019.
3. Philip C., Sauer and Michael Hiete, *"Multi-Stakeholder Initiatives as Social Innovation for Governance and Practice: A Review of Responsible Mining Initiatives"*, December 2019.

4. Kickler, K.; Franken, G., *'Sustainability Schemes for Mineral Resources. A Comparative Overview',* revised version; DERA: Hannover, Germany, 2017.
5. Philipp, C., Sauer and Michael Hiete; *op.cit.*
6. Van den Brink, S.; Kleijn, R.; Tukker, A.; Huisman, J., *Approaches to responsible sourcing in mineral supply chains.* Resour. Conserv. Recycl. 2019, p. 392.
7. Mena, C.; Humphries, A.; Choi, T.Y., *Toward a theory of multi-tier supply chain management.* J. Supply Chain Manag. 2013, pp. 49, 58–77.
8. Zeballos, E.J.; Garry, S., *Jobs Recovery. Sectoral Coverage: An Overview of Employment Trends and Working Conditions by Economic Activity:* First Semester 2010; ILO: Geneva, Switzerland, 2010.
9. (i) Mancini, L.; Sala, S., *Social impact assessment in the mining sector: Review and comparison of indicators frameworks.* Resour. Policy 2018, pp. 57, 98–111. (ii) Murguía, D.I.; Bringezu, S.; Schaldach, R., *Global direct pressures on biodiversity by large-scale metal mining: Spatial distribution and implications for conservation.* J. Environ. Manag. 2016, pp. 180, 409–420.
10. OECD., *'LEED Forum on Social Innovations'*. Available online: https://www.oecd.org/fr/cfe/leed/forum-socialinnovations.htm, September 2019.
11. Fleury, A.M.; Davies, B., *Sustainable supply chains—Minerals and sustainable development, going beyond the mine.* Resour. Policy 2012, pp. 37, 175–178.
12. Arvanitidis, N., Boon, J., Nurmi, P. and Di Capua, G., White Paper on Responsibile Mining. IAPG - International Association for Promoting Geoethics, 2017.
13. Ibid.
14. Op.Cit.
15. Kerry Clarke-Potter, *"Why is Responsible Mining So Important?"*, Dec 10, 2018.
16. Arts, Karin (arts@iss.nl) Current Opinion in Environmental Sustainability 2017, 24:58–62.
17. *Ibid.*
18. UNDP, *'Inclusive Development'*, http://www.undp.org/content/ undp/en/home/ourwork/povertyreduction/focus_areas/focus_inclusive_development. html
19. UN General Assembly: Declaration on the Right to Development. UN Doc. A/RES/41/128; 1986:. adopted on 4 December.
20. UN General Assembly: Rio Declaration on Environment and Development. UN Doc. A/CONF.151/26; 2017: Annex 1 adopted on 1 August 1992.
21. Darrow, M., *The millennium development goals: milestones or millstones? Human rights priorities for the post-2015 development agenda.* Yale Hum Rights Dev J 2014, 15:55-127.
22. Anton, DK, Shelton, DL, *Environmental Protection and Human Rights.* Cambridge University Press; 2011: p. 119.
23. The transcript is the extract reproduced from a roundtable discussion on *'can mining be ever ethical?'*, held with Jason Mitchell, co-head of responsible investment; Mark Ashton, a portfolio manager focused on the mining sector; and Michael Canfield and Louis Rieu, co-portfolio managers.

24. Aliyah Jeande Muta, *"It should n't just be mining"*, May, 2018.
25. Arvanitidis, N., Boon, J., Nurmi, P. and Di Capua, G., *"White Paper on Responsibile Mining"*, IAPG - International Association for Promoting Geoethics, 2017.
26. Aliyah Jeande Muta, *Op.Cit.*
27. Miranda, M.; Chambers, D.; Coumans, C., *"Framework for Responsible Mining: A Guide to Evolving Standards"*, 2012.
28. UNESCO, *"The Precautionary Principle"*, Paris, France, 2005.
29. Ten Kate, K.; Bishop, J.; Bayon, R., *"Biodiversity Offsets: Views, Experience, and the Business Case"*, International Union for the Conservation of Nature (IUCN): Gland, Switzerland, 2004.
30. King, D.M.; Price, E.W., *"Developing Defensible Wetland Mitigation Ratios: A Companion to the Five-Step Wetland Mitigation Ratio Calculator"*, University of Maryland, National Oceanic and Atmospheric Administration: Solomons, MD, USA, 2004.
31. Robert Goodland, *'Responsible Mining: The Key to Profitable Resource Development'*, 2012.
32. *Ibid.*
33. Business and Biodiversity Offsets Programme (BBOP), *"Standard on Biodiversity Offsets"*, BBOP: Washington, DC, USA, 2012.
34. Carmel, V. Abao, *"Responsible Mining: Planet, People, Profit, Policy,"* 2019
35. Juan Pablo Duque is the CEO of Four Points Mining, a Colombian-British company devoted to the exploitation and exploration of gold and silver in Colombia.
36. Paul Klein, *"Four ways to make mining more responsible"*, The CSR Blog.
37. Goodland, Robert, Responsible Mining: The Key to Profitable Resource Development, 2012.
38. Organisation for Economic Cooperation and Development (OECD), *"Guidelines for Multinational Enterprises"*.
39. (i) Goodland, R., "*Environmental and Social Assessment*"; World Resources Institute: Washington, DC, USA, 2008. *(ii)* Vanclay, F.; Esteves, A.M. (Eds.), *"New Directions in Social Impact Assessment: Conceptual and Methodological Advances"*; Edward Elgar: Cheltenham, UK, 2011. (iii) Esteves, A.M.; Franks, D.; Vanclay, F., "Social impact assessment: The state of the art". *Impact Assessment, Project Appraisal,* 2012, pp. 5–44.
40. Goodland, R.; Bedy, S.; Maniates, S., "*Best Practice for Panels of Experts: Effective Independent Oversight of High Risk World Bank Group Projects"*; Bank Information Center: Washington, DC, USA, 2011.
41. Steiner, R., "*Citizens' Advisory Councils for Mining in the Pacific—Mining and Mining Policy in the Pacific: History, Challenges, and Perspectives"*; Noumea: New Caledonia, France, 2011.
42. Resolution 4.088, passed at the International Union for Nature Conservation (IUCN) World Conservation Congress in Barcelona in 2008 supporting

Responsible Mining and established its Extractive Industry Responsibility Initiative.

43. United Nations, *"United Nations Declaration of the Rights of Indigenous Peoples"*; New York, USA, 2008.
44. United Nations, "*United Nations Principles for Responsible Investment*".
45. Robert Goodland, *Op.Cit.*
46. Shepard, R.B., *"Gaining a Social License to Mine"*, Applied Ecosystem Services, 2008.
47. Sociallicense.com. *"What is a Social License"?*
48. Bauer, A., Revenue Watch International, *'Philippine Mineral Wealth for Development'*, paper presented in Mindanao Conference, Ateneo de Davao University, Davao, Philippines, January 2012.
49. Resolution 4.085, passed at the International Union for Nature Conservation (IUCN) World Conservation Congress in Barcelona in 2008 establishing the 1% Earth Profits Fund.
50. Business and Biodiversity Offsets Programme (BBOP), *"Biodiversity Offset Design Handbook"*; BOP: Washington, DC, USA, 2009. Available.
51. Dudley, N.; Stolton, S., *"To Dig or Not to Dig? Criteria for Determining the Suitability or Acceptability of Mineral Exploration, Extraction and Transport from Ecological and Social Perspectives"*; WWF International: Gland, Switzerland; WWF UK: London, UK, 2002.
52. Earthworks and Mining Watch Canada, *"Troubled Waters: How Mine Waste Dumping Is Poisoning Our Ocean, Rivers, and Lakes"*. 2012.
53. Phillips, A., *"Mining and Protected Areas"*; Mining Minerals and Sustainable Development project of the International Institute for Environment and Development: London, UK, 2001.
54. Hansen, J., *"Scientific case for avoiding dangerous climate change to protect young people and nature"*. Proc. Natl. Acad. Sci. USA 2012.
55. International Finance Corporation (IFC), *"Performance Standards on Environmental and Social Sustainability"*.
56. Goodland, R., *"Sustainable Development Sourcebook for the World Bank Group's Extractive Industries Review: Examining the Social and Environmental Impacts of Oil, Gas, and Mining"*; World Bank: Washington, DC, USA, 2003.
57. Robert Goodland, *Op.Cit.*
58. (i) King, D.M.; Price, E.W., *"Developing Defensible Wetland Mitigation Ratios: A Companion to the Five-Step Wetland Mitigation Ratio Calculator"*; University of Maryland, National Oceanic and Atmospheric Administration: Solomons, MD, USA, 2004. (ii) National Marine Fisheries Service (NMFS). *"Guidance for the Review, Establishment, Use and Operation of Conservation Banks and In-Lieu Fee Mitigation Programs"*; NMFS: Washington, DC, USA, 2011.
59. National Research Council, *"Compensating for Wetland Losses under the Clean Water Act"*; National Academy Press: Washington, DC, USA, 2001.
60. (i) Sosa, I.; Keenan, K., *"Impact Benefit Agreements between Aboriginal Communities and Mining Companies: Their Use in Canada"*; Canadian

Environmental Law Association: Toronto, Canada, 2001. (ii) Martin, S., *"Free, Prior and Informed Consent: The Role of Mining Companies"*; Oxfam: Melbourne, Australia, 2007.

61. (i) O'Faircheallaigh, C., *"Public participation and environmental impact assessment: Purposes, implications, and lessons for public policy making"*, 2010, pp. 19–27. (ii) Gibson, G. O'Faircheallaigh, C., *"IBA Community Toolkit Negotiation and Implementation of Impact and Benefit Agreements"*; Walter & Duncan Gordon Foundation: Toronto, Canada, 2010. (iii) The Impact and Benefit Agreement (IBA) Research Network website.
62. Pliny (61 AD—*ca.* 112 AD), His famous saying, many believe to be the origin of 'principles of precaution'.
63. UNESCO, *'The Precautionary Principle'*, Paris, France, 2005.
64. Bice, Sara, "Key Principes for Industry Integrity", Routledge Studies of the Extractive Industries and Sustainable Development.
65. Rahman Anshari & Isnalita Isnalita, *"Near-zero-waste processing of low-grade, complex primary ores and secondary raw materials in Europe: technology development trends"*, Feb. 2020.
66. Jeroen Spooren, Koen Binnemans, Johanna Björkmalm, Päiv Kinnunen, Article on *"Resource conservation recycling"*.
67. Noma Nyembo, A Thesis on *"Interrogating the legitimacy to enter into a social licence in the mining industry in South Africa: a community perspective"*, July, 2018.
68. Sara Bice, *'Responsible Mining'*, *Op.Cit.*
69. Resolution 4.088, passed at the International Union for Nature Conservation (IUCN) World Conservation Congress in Barcelona in 2008 supporting Responsible Mining and established its Extractive Industry Responsibility Initiative.
70. (i) Initiative for Responsible Mining Assurance (IRMA). http://www.responsible mining.net (ii) Miranda, M.; Chambers, D.; Coumans, C., *Framework for Responsible Mining: A Guide to Evolving Standards.*
71. IRMA, History. https://responsiblemining.net/about/history.
72. Kickler, K.; Franken, G., *"Sustainability Schemes for Mineral Resources. A Comparative Overview,"* revised version; DERA: Hannover, Germany, 2017, p. 31.
73. Kickler and Franken, *ibid.*
74. Press release, PR newswire, *new standard for responsible mining*, toronto, june 28, 2018 /cnw.
75. Dmitry Palekhov & Ludmila Palekhova, *"Responsible Mining: Challenges, Perspectives and Approaches"*, Nov 2019.
76. Joan Carling is the Co-convenor of the Indigenous Peoples Major Group for Sustainable Development & member of the Responsible Mining Foundation's Advisory Council.
77. Prof. Carbonnier, is the main applicant for the research project on improving natural resource governance, He delivered this opening speech at the launch of Responsible Mining Index at the Graduate 0, Geneva.

78. Hélène Piaget, the CEO of Responsible Mining Foundation.
79. Dr. Nicky Black, director of the Social and Economic Development Programme
80. Chantelle Kotz, *"ICMM paves the way for responsible mining practice"*, Mining Review Africa Issue, June, 2020.
81. Shefa Siegel, *"The missing ethics of mining"*, February, 2013.
82. An extract of the speech of Keenleyside. Keenleyside was a civil servant from Canada and a proponent of resource interdependence.
83. Jonathan, L. Mayuga, *"Is responsible mining possible"*, 2017.
84. Fitrian Ardiansyah, *"Responsible mining: Is it possible?"* Coal Asia, June 11 – July 20, 2013, pp. 96-97.
85. Jonathan, L. Mayuga, *Op.Cit.*
86. (i) Rajak, D., *"Platinum City and the new South African Dream"*. Africa, 2012, pp.252–271. (ii) Young, S.B., *"Responsible sourcing of metals: Certification approaches for conflict minerals and conflict-free metals"*, 2018, pp. 1429–1447.
87. Hofmann, H., Schleper, M.C.; Blome, C. *"Conflict minerals and supply chain due diligence: An exploratory study of multi-tier supply chains"*, 2018, pp. 115–141.
88. Paul Klein, *"Four ways to Make Mining More Responsible"*.
89. Carmel, V. Abao, *"Responsible Mining: Planet, People, Profit, Policy,"* 2019.
90. Klare, M.T., *The Race for What's Left: The Global Scramble for the World's Last Resources;* Metropolitan Books: New York, NY, USA, 2012.
91. Deegan et al., *'Banks and Ballard-1997'*; 2002.
92. Corporate Social Responsibility Initiative Report No. 66. Cambridge, MA: Harvard Kennedy School, 2014.
93. Harvey and Nish, 2005.
94. BHP Billiton 2012; DeBeers 2012.
95. Gavin M Mudd, *'The Sustainability of Mining in Australia: Key Production Trends and Their Environmental Implications for the Future'*.
96. Oyu Tolgoi Watch et al. 2012.
97. Davis, Rachel and Daniel M. Franks, *"Costs of Company-Community Conflict in the Extractive Sector"*. Corporate Social Responsibility Initiative Report No. 66, Cambridge, 2014.
98. BHP Billiton Watch. The Philippines: Supply-chain responsibility: The Sibuyan killing and Culture of Corruption. 21 October 2009.
99. Jonathan, L. Mayuga, *Is responsible mining possible?* February, 2017.
100. Aliyah Jeande Muta, *Is responsible mining a myth.*
101. Byambajav Dalaibuyan, *'The Story of the Discovery of the Oyu Tolgoi'*, 2016.
102. Kathryn, Money was the vice president of strategy and merchandising for Brilliant Earth, which uses recycled gold.

Chapter - 5

Triple Bottom Line

In development parlance, "Triple bottom line" is a pathway that meets the development across environment, economic and social (EES) needs. In mining, it means that a mining should be ecologically, economically and socially sustainable by adopting responsible and ethical practices. Triple Bottom Line is taken as synonymous with the 'three Ps' i.e. the planet, the people and the profit. According to Nethra Rajendran, "this is a framework stating that businesses should consider the economic, social, and ecological effects when making decisions. The bottom line in business was once just profit. Profit is the main driver of business and without it, it is said that there is no purpose. But the Triple Bottom Line introduces two other crucial factors: people and the planet. People being the stakeholders: employees, the wider community where corporations do business and humankind. The Planet being our environment, cities and ecosystems. The economy, society, and the environment should not be viewed as separate forces that sometimes coexist, sometimes overlap, and sometimes affect each other. Inter-sectionalist between these three forces always exist and they are forces that feed off of each other".[1]

Mining companies/corporations must adopt a holistic approach and maintain a balance or a trade-off between the earth or the planet and the people, the stakeholders or the society at large with the profit or the economy in general. A state of equilibrium can be achieved when there will be long-term environmental protection, maintenance of ecological health, safeguarding of peoples' interest, uninterrupted community prosperity and regular earning of profit resulting in sustainable development of economy. Hence, a balance is maintained when EES needs are fulfilled. It is a holistic approach which all Mining stakeholders should adopt and Mining should not cross and go below that bottom-line. In an attempt to maximise profit, mining companies must not sacrifice their responsibilities for mitigating the social and ecological problems and challenges. There is, therefore, the need for responsible and ethical mining. Responsible and ethical practices can subvert the adverse impacts of mining and fulfill the EES needs.

In triple bottom line, ecology and environment stands as the overarching force. Economy and society subsist within the all encompassing planet or the environment. Economy is dependent on and mostly driven by both the society and environment. Society, on the other hand, exists within the environment. "A healthy environment provides the resources needed to drive and supply a healthy economy and a healthy community. Everything is connected through the environment. It's time we take a holistic approach and understand the long term need to invest in our greatest asset: The Earth. We are seeing examples of how an unhealthy environment can change everything every day. In many countries, we see the trend of economy growing exponentially while the people and the planet are getting left behind"[2]. Majority of renowned mining companies and big corporations now realised the importance of responsible and ethical mining and implementing the best practices. They are trying to fulfill their commitment towards society and the environment and diffuse or minimise the negative impacts through various initiatives viz. Entreprise Social Commitment (ESC), Corporate Social Responsibility (CSR) etc.

Profit may be the prime motive but should not be the sole intention of companies' business. They have to perform their responsibilities towards ecology and society in order to survive and achieve sustainability. In the words of Ray C. Anderson, "There is not a more strategic issue for a company, or any organisation, than its ultimate purpose. For those who think businesses exist to make a profit, I suggest they think again. Business makes a profit to exist. But, surely they must exist for some higher, nobler purpose than that." Responsible and ethical mining is essential for sustainable mining. Sustainable mining, in its turn, benefits the 'three Ps', meets the ecology, economy and social (EES) needs and responds to the outstanding challenges. Triple Bottom Line i.e. those that meet the needs of society and are environmentally tolerable, economically viable as well as both socially and environmentally bearable, economically and socially equitable, can be the means for change from bad to better. Triple Bottom Line, therefore, is based on three distinct but interrelated, and intergenerational equity anchored subjects. Responsible miners doing ethical business must be constantly mindful and continuously follow the balancing, trade-offs, relationships and complementarities among and between them. They should promote basic tenets of this holistic approach and uphold the canons of this paradigm not only at the global but also at national and regional level. What is most required is that all stakeholders and key players at all levels right from the United Nations (UN) to governments, private sector, and civil society organisations have to be aware, conscious, cultured and compliant on responsible and ethical mining in terms of EES principles, policies, regulations, laws and Information, Education and Communication (IEC).

The very existence of human beings, their activities and wellbeing centres round the environment, economy or society. These are the three basic spheres and constitute a set of interrelated concepts which should form the basis of human decisions and actions.[3] Proper decisions on sustainable resource management in these three areas will bring about sustainable growth for sustainable society.[4] Examples of these include decisions on land use, surface water management, agricultural practices, building design and construction, energy management, education, equal opportunities as well as law-making and enforcement.[5] This is a win - win situations for all because natural resources are preserved, the environment is protected, the economy booms and is resilient, social life is good because there is peace and respect for human rights.[6] The three outstanding issues of sustainable mining development are economic growth, environmental protection and social equality and the concept of responsible and ethical mining rests fundamentally on environmental, economic and social responsibilities.[7] The major adverse effects and negative impacts of mining could be lessen, restored or even improved through responsible and ethical mining. The sustainability and future of mining, therefore, depends on adhering to responsible and ethical mining practices i.e. protecting and preserving the environment, social commitment in enhancing the life quality of the community and development of the area and improving the economy.

Mining corporations/companies are now moving to under developed countries and developing regions mainly for two reasons. Rich underground mineral deposits are found in these areas and resources are getting exhausted in countries of their operation. Secondly, environmental protection regulations and other laws are less strict. But as Jaskoki has pointed out, implementation of large mining projects in developing countries created many situations of vast wealth for a minority, while the large majority of the population has not seen improvement of their socioeconomic status and welfare. To the contrary, the mining impacts often degraded their living conditions. However, gradually, developing countries, the potential hosts of mining projects, are enacting their legislation for promulgation of and socio-environmental protection measures. They are insisting on, inter alia, consent from the community i.e. social licence to operate mining in their region besides procedural formalities of mining permit, performance of the corporate social responsibility and strict adherence to environmental protection regulations. There were many agreements on this at international levels. One such global congregation was held in 1992 in the name of the 'Earth Summit' at Rio de Janeiro. There was a consensus on many Principles pertaining to the aforesaid three dimensions viz. economic growth, environment, ecology and natural resources protection, social and community development. But practically, no significant changes took place after these international agreements. The abundance of natural and mineral resources has been proved to be a "resource curse" as expressed by Ross. The term resource curse according to Ross, has been used to "identify a process when natural resource wealth tends to adversely affect a country's governance and has three main harmful effects: tends to increase corruption, make authoritarian regimes last longer and trigger violent conflicts. Although there are natural resources wealth countries that fail to reach their full potential, governments can make policy decisions that help avoid some of the negative consequences of resources extraction and invest profits in social and economic development of the country. International endeavours, such as the Open Government Initiative (OGP) and Extractive Industries Transparency Initiative (EITI), are voluntary initiatives that try to beat the resource curse". However, many international organisations, according to UNEP are 'attempting to develop theoretical models and concepts to afford growth decoupling from resource mining and environmental impact from mining'.

With exponential growth of population in some countries, the future generations will require either more or at least the same natural resources as today and it is the responsible and ethical mining which could play a vital role in ensuring the availability of these resources to the posterity. No quantum of development and rate of growth can afford to utilise and exhaust available resources without devising any alternative means to provide them to future generations. Besides, the first developing automation, robotisation and technological developments in mining sector will have changing equations and far reaching societal implications. It will affect the employment scenario, conservation of environment and above all socio-economic values. It is again the responsible and ethical mining which can steer and guide the process of safety mining operation in yielding positive results in social-economic impacts, life quality, conservation, preservation and protection of the environment and long-term management of natural resources.

Mining, operated responsibly and ethically, can reverse the adverse impacts on environment. Mining has environmental impacts felt beyond mines and their vicinity that could harm the people and any living organisms if not taken up responsibly. Mining Companies have their environmental responsibility that includes the protection of land, water and air in their mining area and the near vicinity[8]. Amla is of the view that 'New and current mining projects are expected to incorporate the lessons from past mining activities and solve their detrimental collateral effects on environment and public health. As these effects often relate to contamination from tailings and acid mine drainage produced during and after mining operations, the EIA of mining projects should encompass the entire lifecycle of the mine, should incorporate the assessment of negative impacts and societal benefits also, and must incorporate the liability and the cost – and thus make funding provisions timely – for post mine environmental rehabilitation. Current mining activities need to reinforce procedures for protection of environment and public health. Mining laws and legal and technical capacity in many countries still need to be enhanced in order to include new internationally recommended good mining practices'. Some responsible mining companies are, however, adopting good mining practices in line with the best international standards. But this is limited to some companies, it may take some years to make it generalised principles across all companies.

Ecosystem integrity should be maintained, the natural environment should be made productive and resilient to support human life, used as a source of economic inputs and as a sink for waste.[9] Since the earth systems have limits for maintaining balance or equilibrium level, it's natural resources must be harvested no faster than they can be regenerated while waste must be emitted no faster than they can be assimilated by the environment.[10] The effects of climate change, for instance, provide a convincing argument for the need for environmental sustainability. Climate change refers to significant and long-lasting changes in the climate system caused by natural climate variability or by human activities.[11] These changes include warming of the atmosphere and oceans, diminishing ice levels, rising sea level, increasing acidification of the oceans and increasing concentrations of greenhouse gases.[12] While dire predictions abound, the full impacts of global warming are not known.[13] The current rate of biodiversity loss exceeds the natural rate of extinction.[14] What is clearly advisable is that, for the sake of preservation of environment, all societies must adjust to the emerging realities with respect to managing ecosystems and natural limits to growth. They have to ensure how the natural environment remains productively stable and resilient to support human life and development.[15] Another important dimension of environmental impacts remediation is evaluation of impacts of mining which many mining companies do not pay due attention to. The development of Environmental Impact Assessment (EIA) procedures aiming at identifying disruptive effects of mining to the environment, the potential effects and damages caused by mining developments in the environment are helping to foresee and estimate gains, costs, losses, and consequences of such projects.[16] 'As the environment effects often relate to contamination from tailings and acid mine drainage produced during and after mining operations, the EIA of mining projects should encompass the entire lifecycle of the mine, should incorporate the assessment of negative impacts and societal benefits also, and must incorporate the liability and the cost – and thus make funding provisions timely – for post mine environmental rehabilitation'.[17] In recent years, variations of EIA have been developed such as the Strategic Environmental Impact Assessment (SEIA), Radiological Environmental Impact Assessment (REIA) to help evaluating the risk assessment of mine operations to surrounding human settlements and for projects involving radioactive materials, such as a uranium mine or a nuclear facility. The exposure to harmful contaminants like uranium or other metals exist in many regions and exposure of large human communities to radiation and radioactive waste is not acceptable according to current radiation protection standards.[18] EIA or SEIA or REIA must be taken up by the mining companies to facilitate simultaneous, timely and appropriate implementation of remediation measures for recuperation of contaminated areas, a healthy environment and control of water, air and soil quality.

Conclusion

Inclusiveness, Sustainability and responsibilities must be the three major objectives of any business concern and economic sector, mining for that matter. Triple Bottom Line; the planet, the people and the profit should, therefore, be the prime concerns. Economy, ecology and people are not always in three overlapping circles. Our planet, the earth and environment is an all-encompassing sphere with the people and their economy within the circle. Economy guides the people's reaction, outdoor activities, investment, consumer behaviour, purchase and business action, production, distribution etc. which ultimately drives the planet's future.

End Notes & References

1. *Nethra Rajendran, "Redefining the Triple Bottom Line",* September 30, *2019.*
2. Nethra Rajendran, *ibid.*

3. Wanamaker, C. (2018), The Environmental, Economic, and Social Components of Sustainability: The Three Spheres of Sustainability: Adapted from the U.S. Army Corps of Engineers.
4. Yang, L. X. (2019), From general principles of civil law to general provisions of civil law: A historical leap in contemporary Chinese civil law. Social Sciences in China, 2, 85–91.
5. (a) Montaldo, C. R. B. (2013), Sustainable Development Approaches for Rural Development and Poverty Alleviation & Community Capacity Building for Rural Development and Poverty Alleviation (b) Porter, M.E., & van der Linde, C. (1995), Toward a new conception of the environment competitiveness relationship. Journal of Economic Perspectives, p. 9 and pp. 97–118.
6. DESA-UN (2018, April 4), The Sustainable Development Goals Report 2017.
7. Taylor, S. J. (2016), A review of sustainable development principles: Centre for environmental studies. South Africa: University of Pretoria.
8. Mannylumanao, *"The environmental effects of irresponsible mining industry" (edited).*
9. Goodland, R., & Daly, H . (1996), *Environmental sustainability: Universal and non-negotiable: Ecological applications, 6(4),* 1002–1017.
10. Diesendorf, M. (2000), *Sustainability and sustainable development. In* D. Dunphy, J. Benveniste, A. Griffiths, & P.Sutton (Eds.), *Sustainability: The corporate challenge of the 21st century* (pp. 2, 19–37). Sydney: Allen & Unwin.
11. Coomer, J. (1979), *Quest for a sustainable society* . Oxford: Pergamon.
12. Du, Q., & Kang, J. T. (2016), *Tentative ideas on the reform of exercising state ownership of natural resources: Preliminary thoughts on establishing a state-owned natural resources supervision and administration commission.* Jiangxi Social Science, 6, 160.
13. Ukaga, U., Maser., C., & Reichenbach, M. (2011), Sustainable development: principles, frameworks, and case studies. International Journal of Sustainability in Higher Education, 12(2), Emerald Group Publishing Limited.
14. UNSD. (2018c), SDG indicators metadata repository.
15. Campagnolo, L., Carraro, C., Eboli, F., Farnia, L., Parrado, R., & Pierfederici, R. (2018), The ex-ante evaluation of achieving sustainable development goals. Social Indicators Research, 136, 73–116.
16. Jain, A., & Islam, M. (2015), A preliminary analysis of the impact of UN MDGs and Rio +20 on corporate social accountability practices. In D.Crowther & M. A. Islam (Eds.), Sustainability After Rio developments in corporate governance and responsibility (Vol. 8, pp. 81–102). Emerald Group Publishing Limited.
17. Food and Energy Security published by John Wiley & Sons Ltd. and the Association of Applied Biologists. Mining and Sustainable Development F. P. Carvalho.
18. *International Atomic Energy Agency, 2014.*

Chapter - 6

Mining and Economy

Mining and minerals have a paramount significance and growing role to play in economic development of a country in general. Like any other commodity in an economy, minerals have occupied a distinctive place amongst all the economic resources. They have to be produced like any other economic commodity before they can be utilised for satisfying human needs. Minerals are indispensable not only for the well-being of man, but also for his survival. They are essential in the production of goods and services of various kinds. Minerals and mineral industries have important micro links with the economy.[1] Along with the vast array of positive contributions; mining has some negative economic impacts also. Let us examine both these aspects.

Positive Contributions

The extraction and processing of minerals and metals to provide goods and services essential to human society is as old as human development itself. Minerals and metals have brought huge benefits to society – they are vital commodities that serve as a foundation to society's material quality of life. In today's world, population growth, urbanisation, social and economic development and even demands for a green (or low carbon) economy are all contributing to increase in the demand for minerals and metals.[2] According to ICMM, '(i) Minerals, metals and the vast array of products that result, play a critical and evolving role in society, a role that will continue far into the future. (ii) Mining and metals activities, if well-managed, can provide lasting opportunities for economic growth and development. (iii) The demand for minerals and metals is growing and will continue to do so. (iv) Mining will be needed to meet this demand, even with society achieving greater efficiencies through reduction of extraneous uses, reuse, and recycling'.

A country rich in mineral resources has got an enviable status. If it can effectively harness export-oriented mineral resources, it can dictate to the international market and earn huge amount of foreign exchange. Minerals are sometimes consumed by the producers by themselves in their own industries. Most of the minerals mines, however, are transported to various mineral-based industries of the state and the country. Minerals are the base of all industries. Even some agricultural-based products like paper and textile require minerals or mineral-based chemicals at various stages of production. The machines required for their production are made essentially out of iron ore and other minerals. Thus, Minerals and mineral deposits are at the foundation of industrial development in any country.[3] However, mining is a key contributor to economic growth and improved material quality of life. Mining's long lifecycle, its potential for engagement and employment of both skilled and unskilled labour as well as the it's products catering the needs of industries ultimately results in poverty alleviation across the world. With responsible management leading to the sustainable mining can contribute immensely to economic development. Mining leads to economic advantages such as higher tax incomes for states, improving Services and increasing employment in local communities and raised demand for goods and services makes the economy of local communities thrive. According to the World Bank, Mining Sector Results Profile; many countries view the mining sector as a key engine of economic development. Ample evidence exists that a country that adopts modern mining legislation and an enabling environment can attract private sector investment in mining exploration and production. This, in turn contributes to increased tax revenues, export earnings, employment opportunities, infrastructure development especially in rural areas, and transfer of technologies to the host countries. In many countries with mineral resources, there is a close link between efforts towards poverty reduction and the performance of and framework for the mining sector. Mining sector generates employment and business, thus leading to economic growth. "Minerals and metals produced through mining are used directly for infrastructure, technologies, energy management and agriculture; and used indirectly for livelihood generation, education, healthcare, water and sanitation, creation of sustainable cities, promoting innovation, positively contribute to the SDGs etc."[4] Besides generating income and creating opportunities for growth for lateral or downstream businesses, mining can contribute to economic growth and sustainable development in a variety of ways, both directly and indirectly, at micro and macro level. At micro level, mining sector plays a vital role in poverty reduction, employment promotion, income generation thereby increasing per capita income, boosting local economy etc. and at macro level, it has a significant contribution in enhancing Gross domestic product (GDP), investment, exports and foreign exchange revenue of a nation, propelling tourism, ensuring low carbon economy, promoting sustainable development besides other miscellaneous contributions.

It is pertinent to indicate here that ten case studies that were conducted by ICMM and the consultancy firm Oxford Policy Management (OPM) using the ICMM Mining Partnerships for Development Toolkit, give valuable context-specific insights into the macro-economic contributions of mining. This, in turn, provides the basis for assessing a broader range of impacts on national economies using export and production data alone. The case studies are: Brazil, Chile, the Democratic Republic of Congo, Ecuador, Ghana, Guinea, Lao PDR, Peru, Romania and Tanzania. They clearly show that contributions are multifaceted within countries. Mining provides vastly different levels of contribution in relation to the different macro-economic aggregates of most economies. Contributions also vary across and within countries. For example, the contribution of mining to government revenues and employment can differ quite significantly from one country to another depending on a range of factors. The major findings from the case studies reveal that mining boosts foreign direct investment (FDI), total national investment and exports, enhances net foreign exchange earnings and government revenues, gross domestic product (GDP) and provides various miscellaneous economic benefits.[5] Some of the major contributions of mining to the economy and economic development with illustrations of countries across the world[6] as well as national and regional experiences are discussed below.

Gross Domestic Product (GDP)

It is revealed from the case studies that mining typically provides only a modest direct contribution to a country's GDP and the various income modules that are components of GDP (typically around 2–4 percent of national totals). This relatively low number is partly explained by the fact that developing host countries often lack the industrial base to supply the sophisticated mining technology used in modern mines. As a result, many upstream value additions or GDP contributions for that matter, such as mineral beneficiation, take place outside the host country. However, in more mature industrial economies such as Romania and Brazil, the multiplier effects on GDP can be quite significant - one dollar of economic activity in the mining sector can generate three dollars or more of economic activity elsewhere. Thus, the total GDP direct and indirect contribution of mining is much larger than the "typical" 2–4 percent referred to above.

Romania

For Romanian regional development policy, the effects of regional disparities across the country remain among the most pressing issues. There is a clear "centre to periphery type" approach creating a structure delivering faster growth for those regions immediately surrounding the capital, driven by investment that prefers developed areas, labour migration and government intervention in its more developed regions in order to achieve a higher national growth rate. Regions within Romania suffering with poor economic development are continuously searching for new opportunities for growth aimed at reducing as much as possible the gap between themselves and other more developed ones. One such example is Rosia Montana, where there has been mining activity for two millennia. There is a compelling case where mining and tourism could become joint instruments towards a vastly improved regional development. The modern mine proposed by Gabriel Resources in Rosia Montana through its Romanian subsidiary, Rosia Montana Gold Corporation (RMGC), could remedy the heritage and environmental issues and also become a tourist attraction during its operation as the largest operation of its type in the EU, potentially one of the largest world-wide. RMGC plans to start developments in tourism during the lifespan of mining activities, from restoration and refurbishment of Roman galleries and other heritage monuments, to cleansing the water course in and around the mine. Once mining operations cease, long-term opportunities created by the rehabilitated landscape shall be put for tourist use. Mining have created directly and indirectly, large-scale business spin offs ensuring the viability of services and facilities needed for tourism. Contemporary mining could therefore, become a visitor attraction and act as a key attraction for educational visits for university students studying mining, economics, planning, etc.

Poverty, Employment and Income

Mining sector has a tremendous impact on poverty reduction, employment promotion and income generation. Drawing on economic data available on mining sector contributions to national economies, several conclusions can be drawn regarding the sector's contribution to the economy and poverty reduction in low and middle income countries. There is a growing economic contribution by the mining and metals industry to many low and middle income countries. The rapid expansion of mining activity in recent years has made many low and middle income countries more reliant on mining for generating foreign exchange earnings. The rising importance of mining challenges both the critics of mining and its supporters to improve their understanding of the mechanisms that allow mining to have increasingly positive developmental effects. Mining in recent years has been the single most dynamic component of many poorer countries' total productive activity. Thus, it has become a potential source of both direct and indirect incomes and a potential catalytic force for faster overall economic growth. In many countries, the mining and metals industry can and should be recognised as an important potential contributor to the critical policy objectives of both job creation and poverty reduction. Smaller, low or middle income countries score relatively highly in the Mining Contribution Index (MCI). This is because of the high ratio with respect to the three variables used to calculate the index: share of mineral exports in total exports, rate of growth in mineral exports relative to other sectors and mineral production values as a percentage of GDP.

Employment and wages

Positive findings have been obtained in relation to mining's contribution to employment and wage earnings. New direct jobs created by large mining companies are normally well-remunerated compared to prevailing national average income levels, though the numbers of such jobs are relatively small – rarely more than 1.5 percent of total national employment. Analysis from the case studies shows that mining may be successful in generating indirect employment in the supply chain, as well as induced employment as the salaries of direct employees or supplier employees are spent within the wider economy. Such employment multiplier effects can often be significant. One direct mining company employee may correspond to three or four employees elsewhere in the economy. In poorer regions where mining is the dominant activity, these indirect employment effects can be especially important. Analysis from the case studies shows that mining may be successful in generating indirect employment in the supply chain, as well as induced employment as the salaries of direct employees or supplier employees are spent within the wider economy.[7] The Geneva-based International Labour Organisation reported that in 2010, there were 1.5 million people employed in the mining sector in developed nations, and 2.2 million in developing/emerging nations. The number of people working in mining jobs dropped significantly during the 2008 recession, although many of these jobs have since returned. In addition, artisanal and small-scale mining employs an estimated 25 million people worldwide, indirectly supports more than 150 million people. Mining takes place in more than 100 countries, with more than 50 regarded by the World Bank as "mining countries" because of the importance of mining to their exports, domestic markets, or to employment.

The formal mining industry not only creates mining jobs but also generates employment indirectly by stimulating demand for goods and services. This generates income, earnings at all levels and enhances purchasing power. Mines spend millions of dollars on equipment, maintenance, food and other services and often use local contractors and suppliers. This creates jobs indirectly in processing and manufacturing mined goods, transportation, and providing equipment and services to the mining industry. All industries will create these types of spillover effects, known as multipliers, but the mining sector generally creates a higher multiplier effect than other sectors. The multiplier effect can be significant but varies depending on the amount and type of outsourcing done. Generally the amount of indirect employment generated by a mine exceeds its direct employment. For example, case studies found that for every job directly created by the Yanacocha mine in Peru, 14 jobs were indirectly created. In addition, the number of indirect jobs increases over time and can stimulate local entrepreneurship and new business development. Mining is an important economic activity in many developed and developing countries. The formal mining sector employs more than 3.7 million workers. In addition, more than 20 million people work in artisanal and small-scale mining (ASM). Despite its importance, there is a misperception that mining is not a significant contributor to global employment. People often reach this conclusion by considering only direct employment or by comparing mining to other sectors. However, such a narrow view fails to consider the many jobs created in the support industries that provide goods and service to the mining sector.

Small-scale mining provides employment for about 13 million workers and their families worldwide, in particular, in countries such as Bolivia, Brazil, Colombia, Tanzania, Burkina Faso, Ghana, Congo, India, Indonesia, and China. Commercial-scale mining provides employment and skills transfer to workers and can also be an important source of social services to remote communities. All forms of mining can be accompanied by the growth of small and microenterprise activity, providing supplies and related services to mining companies, miners, and their families, and generating substantial further incomes.

Canada

Canadian-owned mining firms are active internationally, with nearly 1,000 exploration companies in more than 100 countries and more than 4,300 mineral projects at various stages of development. Canadian mining companies employed more than 169,000[i] workers in Canada and abroad in 2011. Mining also remains an important source of employment in Canada.

In 2010, one out of every 50 Canadian workers was directly employed in mining, resulting in 308,000 jobs. In 2009, this sector accounted for 2.1 percent of Canada's total employment yet contributed 2.7 percent ($31.9 billion) of Canada's total GDP, down from 3.2 percent ($39.8 billion) in 2008. Mining also generates significant indirect employment in Canada in both rural and urban areas. Statistics of Canada does not break down employment data in mining-related supporting activities (e.g. exploration, contract drilling, and transportation). However, the Mining Association of Canada reports that 3,223 companies supply goods and services to mining companies such as accounting, environmental consulting, legal and technical advice, and finance. They also note that the Canadian mining industry also supports many jobs in shipping, rail and transportation. A 2011 analysis of the economic impacts of mining in British Columbia found that 21,112 people were employed directly in mining (2 percent of BC's labour force), with an additional 16,590 jobs indirectly created.

Foreign Direct Investment (FDI)

In most cases the contribution of mining to the case-study countries' total FDI has been very high – typically more than half of total annual FDI. This contribution is especially significant in low income countries, pointing to their ability to attract mining investment even when FDI into other sectors appears unattractive. The emerging evidence also indicates that mining FDI combined with increased foreign exchange earnings can be a catalyst for some countries to improve their credit ratings and attract long-term loan funding from abroad (e.g. Ghana and Tanzania). There are, however, some significant exceptions to this pattern. Brazil, for example, is one of the world's most successful emerging market economies in terms of attracting FDI (in excess of US$50 billion per annum in recent years). Yet FDI into the mining sector represents a very small share of the total FDI received by Brazil – partly because the economy is more diversified and partly because a significant proportion of Brazil's recent investment in mining is attributable to domestic investment by its largest mining company, Vale.

The main areas of mining's large, direct economic benefits in most low and middle income countries are shown in Table 6.1. The table demonstrates that the greatest benefit mining affords to these national economies is not government revenue, but rather the contribution to foreign direct investment (up to 90%). It may be noted that the other four segments' contribution (foreign direct investment, exports, government revenue and national income measured by GDP) accrue to national governments. Only the smallest contribution, employment, can be said to directly benefit local communities. The greatest benefit mining affords to these national economies is not government revenue, but rather the contribution to foreign direct investment.

Table 6.1: Macroeconomic contributions of mining to low and middle income economies

Sl. No.	Segments	Percentage Contribution
I	II	III
1	Mining Employment	1 – 2 % of total employment
2	Mineral Income	3 – 10 % of total national income
3	Mineral Taxation/Revenue	3 – 20 % of government revenue
4	Mineral Exports	30 – 60 % of total exports
5	Foreign Direct Investment (FDI)	60 – 90 % of total FDI

Source: ICMM (Various years) and OPM (Various years).

Note: The percentages are not additive but indicate the range of stand-alone contribution of each segment.

Total national investment

Mining makes a very significant contribution to national investment totals especially when mining activity is building up from a low base (e.g., Tanzania 1999–2006). In the case of Brazil, Vale's investments alone are the equivalent of almost 5 percent of total annual national investment.

Tanzania

Tanzania is endowed with abundant mineral resources of international value, including gold, diamonds, salt, gypsum, gemstones, iron ore, natural gas, phosphate, coal, nickel, cobalt and tanzanite. The country's major gold fields are located in Geita, Musoma, Tarime, Chunya and Canda. Although records indicate that mineral exploration and exploitation in Tanzania began in the 1880s following the establishment of the German administration, there is evidence suggesting that local people, using traditional methods of mineral prospecting, produced minerals centuries before the establishment of the colonial administration. Hilson reports that as many as 40,000 years ago, regional hunter-gatherers exploited obsidian and chalcedony rock for stone implements and weapons, and used iron ore for painting. The government of Tanzania instituted a new Mining Act in April 1998 that is conducive to foreign investment. Mineral production increased by 51 percent in 1991 and 24 percent in 1992, mainly in response to the implementation of the Trade Liberalisation Policy in 1985 and the enactment of the National Investment Act in 1990. Tanzania has a series of older mines, such as Williamson Diamond Mine at Mwadui and the Kiwira Coal Mine in Tukuyu, as well as a number of newer operations, most of which have been established in the Lake Victoria Gold Fields (LVGF).

As indicated in the Tanzania Economic Survey, industry liberalisation has been a major reason behind marked increases in national mineral production. Notable achievements include increased gemstone and gold production, which, between 1984 and 1991, increased from 400 to 29,600 and 39,500 to 3,851,000 tons, respectively. Currently, Tanzania ranks third in continental gold production behind South Africa and Ghana. Rises in mineral production has increased the contribution of the mining sector to national Gross Domestic Product (GDP), which rose from 1.1 percent in 1989 to 2.3 percent in 2000. However, overall, mining contributes a relatively small share to national GDP, suggesting that the Tanzanian government, despite its successes in attracting foreign investment, has allowed incoming mining companies to export the bulk of extracted and processed product. In fact, findings by Tauli-Corpuz, Akabzaa, Darimani, Jones and Awudi confirm that mining has provided marginal contributions to the communities surrounding operations.

Exports

From a global production perspective, the most significant players in terms of mining and metals are large developed countries and middle income countries. Countries that are currently home to the largest mineral and metal resources fall into two major categories: traditional producers such as Australia and the US and major emerging economies including Brazil, Russia, India, China, and South Africa (BRICS). The top 20 countries in terms of the absolute value of their mineral production have a very limited overlap with the top 20 countries in terms of mineral export contributions to total exports: the latter group are mostly small and developing countries. Globally, according to International Council on Mining and Metals (ICMM), 2010, the value of Global minerals production was four times higher than in 2002. At the same time there are 40 countries with significant dependence on mineral exports. During this period, growth in value has been significantly greater than growth in world gross domestic product (GDP). Thirty of these are low or middle income countries and many have low Human Development Index (HDI) Scores. Evidence from ICMM case studies in Chile, Ghana and Brazil illustrates that mining areas have enjoyed stronger poverty reduction and social development performance than non-mining areas. Mining activities has significant contribution to the economies of 212 countries in its sustainable development. According to CHR Michelson Institute (CMI) Report captioned "Socio-Economic effects of Gold mining in Mali", the mining has a strong positive impact on the national Economy in terms of its contribution to exports, GDP and Government Revenue. It has resulted in setting up of mine-induced local development projects and contributed significantly to enhance employment salary levels. The World Gold Council study commissioned by Price Water House Coopers puts the direct economic impact of Gold mining at some $210 billion plus globally more than the GDP of most countries without taking in to account its indirect-impact. Gold mining industry has the potential to transform the economies of emerging nations in particular. Gold mining supply the base materials for an enormous secondary sector in regaling, Jewellery, investment and industrial applications which employ enormous numbers of people globally.[8]

The mining industry is a major force in the world economy, occupying a primary position at the start of the resource supply chain. However, its role in contributing to the national economies of different countries varies greatly and is neither well documented nor well understood. Another perspective on mining's contribution to the world economy is depicted in Annexure 6.1. It presents a simple summary of mining production values in the 20 most significant mineral producing countries in terms of production value and share of global production value in 2010. It compares the 2010 values and shares of production in each country (columns 3 and 4) with the corresponding values and shares for the year 2000 (columns 5 and 6), and the change in production values between 2000 and 2010–2010 (column 7). It indicates the relative significance of mineral production by expressing it as a percentage of each country's GDP (columns 8 and 9) and shows a corresponding value for reliance on mineral exports (column 10). It is dominated by the major developed country producers and by large emerging market producers including the five BRICS (Brazil, Russia, India, China, and South Africa) as well as other important emerging economies (e.g. Chile, Indonesia and Mexico). This contrasts sharply with the ranking of countries by their reliance on mining exports presented in Annexure 6.2 in which lower income countries figure much more prominently.

An assessment of global trade data indicates that in 2010, there were 40 countries that could be defined as relying on non-fuel minerals for over 25 percent of their merchandise exports according to OPM, 2011. Of these, 75 percent are low and middle income countries. This number represents a considerable rise over time: in 1996 there were only 29 mineral reliant economies and, as on 2005, the numbers were 33. Dramatic changes in this indicator have occurred since 2005 when commodity prices started to rise rapidly. In the years to 2010, eight additional countries have entered the ranks of mineral reliant countries including Bolivia, Burkina Faso, Georgia, Ghana, Guyana, Lao PDR, Montenegro and Somalia. Annexure 6.2 presents a list of the 20 countries with the highest mineral export contributions as a percentage of total merchandise exports in 2010. It shows that over time, more relatively low income countries are becoming increasingly reliant on export revenues from minerals as the main source of foreign exchange earnings. Many of these countries have low Human Development Index (HDI) scores, which draws attention to the potential for productive sectors, such as mining, to make a contribution. Case-study work of ICMM has found that in some countries (for example Chile, Ghana and Brazil), mining areas have enjoyed stronger poverty reduction and social development performance than non-mining areas. Half of the non-fuel mineral reliant countries are in Africa and some of these – for example, Tanzania, Ghana and Mozambique – have become significant exporters of oil and gas as well as non-fuel minerals. This trend is of rising significance for many aspects of policy in the affected economies and warrants greater and more regular attention. Above all, it confirms the proposition that in so far as production and income generation are critical forces in poverty reduction, mining has an increasingly significant role to play.

Case studies confirm the high contribution mining makes to exports for many countries. The contribution of mining to total exports in 2010 amounted 19 to 78 percent in Different countries as indicated in Table 6.2.

Table 6.2: Contribution of mining to total exports

Sl. No.	Name of the country	Percentage of contribution of mining to total exports
I	II	III
1.	Democratic Republic of Congo	78.3
2.	Chile	66.0
3.	Guinea	65.2
4.	Peru	63.0
5.	Lao PDR	45.0
6.	Tanzania	41.0
7.	Ghana	25.0
8.	Brazil	19.0

Source: ICMM.

Ghana

The mining industry remains the backbone of many economies in the developing world. Its resurgence in Ghana since 1989 was driven by the global paradigm which emphasises private sector-led development as the engine of economic growth in developing countries. The historical importance of mining in the economic development of Ghana is evident in the country's colonial name, Gold Coast (Akabzaa, T. and Darimani, A., 2001). Ghana earned $2.5 billion from mineral exports in 2007. As many as 20,000 people are employed in large-scale mining whilst 500,000 are employed in the small-scale sector. Mining contributes about 7 percent of Ghana's total corporate tax earnings, 41 percent of total exports, 12 percent of revenue collected by the Internal Revenue Service and 5 percent of Gross Domestic Product (Ghana Chamber of Mines, 2008). Gold dominates the mining sector and the country is Africa's second largest gold producer after South Africa.

Foreign exchange revenue

Commercial-scale mining can be an important source of foreign exchange and fiscal receipts for governments. When managed well, the net foreign exchange and taxes generated by mining can be used by governments as an engine for overall economic growth and as a source of financing to support national budgets for social-sector and poverty reduction programmes. Substantial fiscal impact from mining can be found in countries such as Chile, Mexico, Peru, Botswana, Ghana, South Africa, and Papua New Guinea.

Net foreign exchange earnings and Government revenues

The mining sector generates significant foreign exchange earnings. A substantial proportion of these do not, however, enter the national economy as they are used by mining companies to import goods and services during construction and operation. Nevertheless, the case studies point to a significant net foreign exchange contribution in the operational phase, even when imports are factored in. During the construction phase the high-import content is typically almost fully accounted for by the inflows of foreign equity and loan capital. The time sequence of the various different foreign exchange effects are well illustrated by the examples of Tanzania and Romania. There is no economic study group – including from within the World Bank or the International Monetary Fund – that systematically captures the important role of the extraction and production of mining, minerals and metals in the economies of each country of the world. Until now, this data limitation has prevented a country-by-country assessment of the varied macro-economic contributions of the mineral sector to national economies. As a basis for discussion and a first step towards providing a long-overdue and continuously updated set of data on mining's overall economic contribution, ICMM, with Oxford Policy Management (OPM), has developed a preliminary index, which ranks countries by the importance of mining and metals within each national economy i.e. the Mining Contribution Index (MCI). The MCI is calculated based on aspects of mining and metals contribution to national economies where country-by-country data exists.

All countries are ranked on this MCI score and grouped under ten categories. A comprehensive table with data by country is provided in Annexure 6.3. This demonstrates the relative significance of mining's contribution in each national economy i.e. MCI ratings by country (from 100 to below 10). The MCI provides a reasonable first approximation of the relative importance of mining and metals to each national economy. However, as highlighted by the case study evidence above, there are many more direct and indirect potential contributions from the mining sector.

The findings suggest considerable country-by-country variation in terms of the contribution of mining to total government revenues, and the allocation of these revenues between the national and sub-national levels. However, the full fiscal contribution of mining projects does not occur until well after project start-up, particularly in terms of income taxes paid by companies. This is due to long lead times as well as the accelerated capital write-downs that are commonly used to incentivise investment.

Miscellaneous Economic benefits

The value of mining's contribution depends on the overall value of production and the payments to production factors including financial capital, equipment, land and labour. As compensation for these factors of production, mining companies pay suppliers of goods and services, wages and salaries to workers, and interest to loan providers. Companies must also pay a minimum risk-adjusted return to shareholders (whether foreign or domestic) in the form of dividends or capital appreciation – to reward them for investing in risky, long-term mining assets.

The surplus value that remains once all of these factors of production have been compensated is referred to as the rent from mineral resources. In broad terms, these rents are allocated between the host government and company investors in a manner depending on the fiscal regime in place. Greater integration of mining, through various forms of linkages, holds promise for increasing mining's contribution to national economies at each stage of the value chain. The extent and nature of these contributions, however, will vary depending on country-specific circumstances. For example, in many poorer countries, the absence of quality suppliers prevents mining companies from procuring domestically. The composition of production value is an important guide to the economic benefits that mining could potentially make to the national economy.

Development of Local economy

Large mining operations can be found to invest substantially in local economic development, through providing training, public services such as education and health, and public goods, such as clean water, transport, energy, and infrastructure.

Europe

Mining has a significant role in promoting Europe's regional development. According to Oxford Policy Management Institute, International Council of Mining and Metals, "Regional Development Planning is an approach to making long-term plans for a country's regional development, which links feasible private sector initiatives in support of coherent and integrated, productive and social infrastructure to enable economic diversification of the region. ******** Mining can create significant economic benefits – the direct benefits that come in the form of income and employment as well as the indirect benefits stimulating local economic activity". Europe is rich in raw materials and exploiting them can only have a positive impact on competitiveness. With about 30 million EU jobs depending on the availability of raw materials, the development of EU's high tech manufacturing industry including eco-technologies is at stake given the increased dependency of Europe on accessing raw materials outside of its borders (DG Enterprise and Industry - European Commission, 2012). According to the approach developed by the UK based Oxford Policy Management Institute and the International Council of Mining and Metals (ICMM), undertaking structural change in mining regions can be difficult for small or medium-sized communities, particularly those in peripheral regions. In cases such as these, there is a strong need for partnerships aimed at exploiting mineral wealth in order to ensure sustainable socio-economic development. Implementation of National and European policies in this field combined with the latest available mining technology as well as mineral partnership ventures might also have a very strong beneficial positive impact on regional development. Currently, the European mining industry represents one of the most modern and most innovative industrial sectors within the European Union. Most sustainable techniques used for exploration and extraction in Europe are supported by environmental protection and safety measures which are amongst the most stringent worldwide.

Source of energy

In countries with large coal resources, such as South Africa, India, and China, coal is an important source of energy contributing to economic growth. In countries with severe winters, such as Russia, Ukraine, Mongolia, and Kazakhstan, coal provides a lifeline for heating.

Tourism Industry

For tourism, the mining boom is producing economic benefits. The mining boom has directly or indirectly boosted national income that drives tourism demand. The mining boom has generated a range of benefits for promotion of Tourism:

Increased discretionary spending on leisure travel at home and abroad by increasing household incomes.

The rapid expansion in business travel leading to increased usage and yield for accommodation and aviation that are servicing the mining regions.

The demand for mining workers, which is creating significant problems for the tourism industry in attracting and retaining labour. As mining workers are now sourced from all over the country, this effect is not just restricted to areas adjacent to mines, but also to other cities and other regional areas.

Australia

The mining boom began in the mid-2000s leading to sharp increases in prices for mining-related commodities. It has, to date, delivered mixed impacts across Australian states and territories, and across industries. This is sometimes referred to as 'Dutch Disease', whereby a boom in an exporting industry, in this case mining, raises the terms of trade and the value of a currency. This is a follow-up report to Tourism Research Australia's (TRA) snapshot on the impact of the mining boom, released in November 2011. It showed that travel activity to the key 'mining' tourism regions (and for business more generally) had increased strongly and that accommodation and aviation industries in capital cities and in the mining areas were benefiting from the boom up to that time. Mining have Positive economic impacts for all industries in Australia. Strong growth in revenues and investment in the mining sector and for other industries aligned to mining and resources. Modelling undertaken for this report estimates that the mining boom has on average added 0.62 percent to total investment per year for each year in Australia. There has been a rapid expansion in job opportunities in mining and related sectors particularly for Fly-In/Fly-Out (FIFO) and Drive-In/Drive-Out (DIDO) workers in remote areas of the mining-rich states of Western Australia, Northern Territory and Queensland. Modelling for this report estimates that the boom has on average added 0.23 percent per year to Australian GDP, 1.07 percent to GSP in Western Australia, 0.71 percent to GSP in Queensland, and 0.66 percent per year to GSP in the Northern Territory. Considerable value-addition to the economy which has translated into strong income growth for the household sector and this has in turn led to an increase in household consumption of 0.47 percent per year at the national level.

Mining machinery

Mining promotes mining machinery. Mining machinery plays a very important position and role in economic construction, scientific and technological progress and social development. The development of mining and machinery is closely related to the human progress modern science and overall industrial level. Mining machinery serves for the energy, transportation and raw material base of industrial departments, the main task is to process coal, steel, nonferrous metals, chemicals, building materials and mining sectors, as well as provide advanced efficient technology and equipments for railway, highways, utilities and other large projects construction.

China

China mining machinery industry has experienced 30 Years of reform and opening up, the stages of introduction of digestion and absorption of foreign advanced technology and manufacturing to design harmony, self-designed, particularly in the implementation of the "Speed up the revitalisation of the state council on several opinions of the equipments manufacturing industry", has achieved two basically major changes. First, product development change from generic type to innovation type, second, the economic operation changes from extensive to efficiency. Since 2000, driven by the domestic basic industries and infrastructure development, mining machinery market demand, industrial output value and sales growth rate stabilised at 32 percent-40 percent. Market demand not only drive the growth of mining machinery; manufacturing economic, also promoting technological progress across the industry, and emerge a group of independent intellectual property rights of major new products. Keyfid Machinery, as manufacturer of mining machinery industry, dedicates to research and develop production of crusher equipments, sand making equipments, milling machine equipments and stone crushing production line, sand production line, development a hydraulic drive track-type mobile crushing plant, VM vertical mill machine, MTW European tech. grinding mill and etc. The success of these major new products development, has made a positive contribution to the national economy, which narrowing the gap with advanced countries, and enhance the ability to participate in international competition. China mining machinery industries are playing an increasing role in the international mining construction projects.

Sustainable economic development

There are many challenges and outstanding issues in Sustainable Mining. Some of the major challenges and important issues should be tackled on priority basis. Economic Development on a sustainable basis is a major challenge for any nation irrespective of its rate and scale of growth. Sustainable economic development is an approach which uses economic resources for development in a manner that allows the same resources to continue either indefinitely or for the given time period for future generations. Mining can not be sustainable without economic development and growth. Sustainable mining is possible with significant progress, development and growth of the economy of a country. On the other hand, mining contributes to economic development both directly and indirectly.[9] Hence, sustainable mining and sustainable economic development are inextricably blended.

In countries where mining and metals play a major role, there can be large discrepancies in the benefits that accrue at the national level and the benefits at the local level. Mining's economic footprint is significant. It puts pressure on local infrastructure and public services following the large scale in-migration that often occurs as mine development and operation proceeds. Targeted actions to build public sector capacity at the local level and to promote regional development initiatives are particularly important and necessary in this context, especially when mining is introduced into areas managed by small and poorly resourced local government authorities. Targeted actions should also focus on strengthening the local indirect economic benefits of mining, by integrating the sector more closely with the rest of the economy.[10] Mining, minerals and metals can, thus contribute to sustainable development.[11] Mining projects and the resulting mineral and metal products have a significant impact on human society and on the global eco-system. Mine projects are a source of livelihood and well-being for millions of people. For emerging nations they attract foreign direct investment, domestic investment, foreign earnings and government revenues. At the local level they provide an opportunity for employment, support for entrepreneurs to provide goods and services needed by mining activities and contribute to infrastructure and services in the local community. Often located in remote areas, mining can provide a unique means for stimulating local economic activity. Throughout history, mineral discoveries have catalyzed the development of sparsely populated and remote areas.

The discovery and development of mineral resources has provided an opportunity for addressing poverty reduction and helping countries in achieving the Millennium Development Goals (MDGs). Mining and minerals sector can bring about directly or indirectly, realisation of all the eight folds goals of MDGs such as; (i) Eradicate extreme poverty and hunger, (ii) Achieve universal primary education, (iii) Promote gender equality and empower women, (iv) Reduce child mortality, (v) Improve maternal health, (vi) Combat HIV/AIDS, malaria and other diseases, (vii) Ensure environmental sustainability and (viii) Develop a global partnership for development.

Importantly, while minerals and metals provide a material foundation for contemporary society, they are also a means for transforming current society into one marked by greater efficiencies, lower environmental stresses and more effective public services.

If managed responsibly and effectively, and in a context of good governance, mining and metals production can contribute meaningfully to an improvement in living standards, particularly in communities directly impacted by these operations. In the late 1990s, CEOs from some of the world's leading mining companies launched an effort to reposition the industry in terms of both performance and perception. Faced with a groundswell of public criticism, it commissioned the International Institute for Environment and Development (IIED) to undertake a global review of practices and develop an agenda for strengthening the industry's contribution to sustainable development. Some 50,000 people from around the world participated in the resulting process that came to be known as the Mining, Minerals and Sustainable Development (MMSD) project. Out of that review emerged an agenda for change to strengthen the contribution of mining, minerals and metals to sustainable development.[12]

More and more developing and emerging nations are finding that the role of mining and metals in their national economies can be a positive factor, that their dependency on mining and metals for economic strength is growing and that mining and metals can play a key role in addressing poverty reduction and other developmental issues. The mining and metals industry acts as an agent of change bring significant socio-environmental and economic change and promote sustainable development.[13]

Mining sector is better placed to comply major Sustainable Development Goals (SDGs) of United Nations. The Sustainable Development Goals officially came into force at the beginning of 2016. The 17 goals are; 1. End poverty in all its forms everywhere, Goal 2. End hunger, achieve food security and improved nutrition, and promote sustainable agriculture, Goal 3. Ensure healthy lives and promote well-being for all at all ages, Goal 4. Ensure inclusive and equitable quality education, and promote lifelong learning opportunities for all, Goal 5. Achieve gender equality and empower all women and girls, Goal 6. Ensure availability and sustainable management of water and sanitation for all, Goal 7. Ensure access to affordable, reliable, sustainable and modern energy for all, Goal 8. Promote sustained, inclusive and sustainable economic growth, full and productive employment and decent work for all, Goal 9. Build resilient infrastructure, promote inclusive and sustainable industrialisation, and foster innovation, Goal 10. Reduce inequality within and among countries, Goal 11. Make cities and human settlements inclusive, safe, resilient and sustainable, Goal 12. Ensure sustainable consumption and production patterns, Goal 13. Take urgent action to combat climate change and its impacts, Goal 14. Conserve and sustainably use the oceans, seas and marine resources for sustainable development, Goal 15. Protect, restore and promote sustainable use of terrestrial ecosystems, sustainably manage forests, combat desertification, and halt and reverse land degradation and halt biodiversity loss, Goal 16. Promote peaceful and inclusive societies for sustainable development, provide access to justice for all and build effective, accountable and inclusive institutions at all levels, Goal 17. Strengthen the means of implementation and revitalise the global partnership for sustainable development. A 2018 UN report had mentioned that mining has strong impacts on at least 11 of the 17 SDGs. It noted that 'mining is clearly relevant to SDG 1, 5, 6, 7, 8, 9, 12, 13, 15 and 16. At international level, governments of over 60 countries came together and formed an Intergovernmental Forum (IGF) on Mining, Minerals, Metals and Sustainable Development in 2002 to ensure that the mining sector focuses on sustainable development, negative impacts are minimised and benefits are shared by all. Contribution of mining on achieving important SDGs[14] are examined below.

End poverty (SDG 1)

Mining is a potential sector for substantial employment generation.

By providing employment to a larger section, it directly contributes to eradicate of poverty. Besides, mining is the largest revenue contributor to governments through royalties, taxes, and dividends thereby enabling the governments to spend more funds on poverty alleviation programmes. In order to create more positive impact the responsibility should be to facilitate equitable and inclusive access to employment opportunities; develop local supplier capacity; and strengthen local chains.[15]

End hunger, and achieve food and nutrition security through sustainable agriculture (SDG 2)

Mining companies usually invest a good amount on Water Infrastructure[16] and thus are responsible for agriculture. Moreover, mining companies provide road network for agricultural marketing by local farmers as well as other infrastructural support for sustainable agriculture.

Ensure healthy lives and promote well-being (SDG 3)

Large mining corporations/ companies are providing better health infrastructure and taking safety measures in and around mining areas.[17] Necessary health and precautionary measures as well as safety procedures become mandatory for any mining operations and or license to operate.[18] However, mining should adopt advance and harmless technologies to avoid any health hazards.

Ensure inclusive and equitable quality education (SDG 4)

Mines are mostly located in remote areas where the standard and level of education is very poor. As a part of their CSR, most of the miners and mining companies are, therefore, providing good educational infrastructure in their localities. However, there are still enough scope for inclusive education, more training and skill development programmes in their area of operation[19] and they should also concentrate on this.

Achieve gender equality and empower women and girls (SDG 5)

Mining companies should take a lead role in women empowerment[20] by ensuring equal pay for equal work, equal opportunities both outside and inside mining industries including recruitment, employment and promotion, equal representation of women in citizens' committees and stakeholder dialogues[21] etc. In Enterprise Social Commitment (ESC) programmes, mining companies should give priority to felt needs of local girls and women. Hence, mining sector can ensure gender equality and empower women and girls in various ways.

Ensure availability and sustainable management of water and sanitation (SDG 6)

Availability and sustainable management of water through efficient and scientific recycling and conservation methods should be attached utmost priority by the mining corporations.[22] They must adopt sanitation procedures and protocols for multifarious use of water. Besides using water for their industries, mining companies supply quality water to the surrounding habitations and water for agriculture as a part of their ESC policy.[23] However, many countries have their own water management policies and mining industries have to comply those regulations.

Ensure access to affordable and sustainable energy (SDG 7)

Mining industries can ensure access to affordable and sustainable energy by maintaining energy optimisation in entire mining cycle.[24] Energy sustainability is the hallmark of mining companies. They can provide access to affordable energy. Some countries have imposed energy stipulations and strict audit procedures for mining and other industries.[25] The mining companies have to adjust their energy demands and forge a balance between their energy demands and supply. They have to explore the possibilities of utilising renewable energy in their operations.[26]

Decent work and economic growth (SDG 8)

Mining sector has a significant contribution to economic growth by providing decent job/work opportunities, training and skill development to local people.[27] It also creates ancillary business opportunities in the locality. By creating excellent roads and communication networks[28] as well as other infrastructure facilities[29], mining companies can facilitate marketing, procurement etc. thereby making a positive change in the lives and living standards of the surrounding communities.

Infrastructure, Industrialisation and Innovation (SDG 9)

Normally, all mining companies build strong resilient infrastructure before and during their activities for use both by themselves and the local people, operate their industries comparatively for a longer period, must adopt innovative practices for their survival and therefore, best suited to achieve this important goal. Mining industries should take extra efforts to make their industries inclusive and sustainable industrialisation, and infuse innovative methods at all stages of their operation.

Sustainable consumption and production (SDG 12)

The natural and other resources are very much limited compared to their demand and hence, these calls for sustainable consumption and production of these resources recycling and reuse are utmost important.[30] Mining industries can make sustainable consumption and production of metals through intensive diversification, renewable technologies, waste minimisation and direct and supply chain management.[31]

Combat climate change (SDG 13)

The mining and metals industries are supplying and will continue to supply the materials required for a low carbon future. Besides, many mining companies are taking effective steps in reducing their own greenhouse gas (GHG) emissions.[32] Apart from their own emissions management, they are taking various cost saving energy efficiency initiatives. This will encourage other industries and sectors to take up similar drives. These measures along with Government regulations / policies on GHG emissions will go a long way in the transition to a low carbon economy and combating the climate change.[33]

Protect and promote life on land (SDG 15)

In compliance to Government stipulations, mining companies are taking effective steps to control adverse impacts on land, water and habitat.[34] Responsible mining practices by mining companies are the important yardstick for obtaining licence to operate. Responsible miners are taking adequate measures to protect and preserve ecology, environment and biodiversity.

Promote peaceful and inclusive societies (SDG 16)

Mining companies can promote peaceful and inclusive societies through involvement of local people in decision-making process during pre- and post-operation period[35], effective community engagement, systematic conflict resolution, upholding human rights[36] and respecting the rights of indigenous peoples.[37] Inclusion of all stakeholders through constant and continuous dialogues and implementation of various welfare measures for the marginalised groups can build trust with mining company and help in peaceful coexistence.

National Experiences

India is a significant producer and exporter of minerals. As per the Annual Report 2016-17 of the Ministry of Mines (MoM), India produces as many as 95 minerals. Mining is a vital contributor to the country's economy, forming a fundamental raw material for many important industries. The Indian mining industry is characterised by a large number of small operational mines along with a presence of few large-scale mining that contributes a major share to the overall production. From a macro view, India's domestic demand motivates and drives production growth of several minerals. Sustained growth in India's automotive sector has been driving demand for steel and aluminium, the power sector demands great proportions of fuel minerals, while construction industry is the driver for steel and aluminium as well as fuel minerals. The deposit of major minerals of mineral rich states of India is depicted in the Table 6.3. Highest deposit of coal is found in Jharkhand, while Karnataka has the country's lowest deposit of Chromite.[38]

Table 6.3: Major Mineral Production in States during 2016-17

(In lakh tonnes)

State	Bauxite	Coal	Chromite	Limestone	Manganese	Iron ore
1	2	3	4	5	6	7
Assam	-	4.44	-	15.03	-	-
Bihar			-	1.84	-	-
Andhra Pradesh	-	-	-	319.98	2.18	4.44
Gujarat	188.34	-	-	225.83	0.42	-
Haryana	-	-	-	71.25	-	-
Karnataka	-	-	0.01	270.27	2.20	242.89
Kerala	-	-	-	3.31	-	-
Maharashtra	16..64	294.71	-	108.56	5.42	14.69
Madhya Pradesh	5.64	981.07	-	319.85	5.63	10.38
Odisha	112.93	1257.93	29.34	42.96	5.01	899.95
Rajasthan	-	-	-	603.82	0.01	11.09
Tamil Nadu	0.07	-	-	220.39	-	-
Uttar Pradesh	-	144.01	-	24.43	-	-
West Bengal	-	242.15	-	-	-	-
Jharkhand	21.11	1083.52	-	10.24	-	193.98
Chhattisgarh	17.69	1253.34	-	283.21	-	275.94
Goa	-	-	-	-	-	70.96
Himachal Pradesh	-	-	-	98.95	-	-
J&K	-	0.01	-	7.12	-	-
Meghalaya	-	32.27	45.92	-	-	-
Telengana	-	541.67	-	222.97	0.10	-
India	227.05	5830.78	29.35	2487.83	8318.79	1719.48

Source: (i) Annual Report; Indian Bureau of Mines, 2016-17, Govt. of India, New Delhi (ii) Economic Survey; Govt. of Odisha ; 2017-18.

In India, Odisha state ranks the highest in recent years in terms of value of output of Minerals (34.3 percent) and its share has been increasing. Table 6.4 compares the major states of the country in 2017-18 in terms of percentage share of total value of mineral output in India. Table 6.5 illustrates the value of mineral production of major states and India from 2005-06 to 2016-17.

Table 6.4: Percentage Share of Value of Mineral Production by major States, 2017-18

Sl. No.	Name of the State	Percent to all India
1.	Odisha	34.3
2.	Rajsthan	18.8
3.	Chhattisgarh	15.7
4.	Karnataka	14.0
5.	Jharkhand	3.5
6.	Maharashtra	2.1
7.	Andhra Pradesh	1.9
8.	Assam	4.79
9.	Madhya Pradesh	3.7
10.	Gujarat	1.2
11.	Other States	2.9
	All India	**100**

Source: Economic Survey,Govt. of Odisha : 2018-19.

The value of mineral production of different states for six years i.e. during 2005-06 to 2007-08 and from 2014-15 to 2016-17 in percentage terms reveals an asymmetrical picture. The mineral production value shows an increasing trend in some states and comparatively decreasing trend for other states (Table 6.5).

Table 6.5: Value of Mineral Production of Major States

(In crores rupees)

State	2005-06	2006-07	2007-08	2014-15	2015-16	2016-17
1	*2*	*3*	*4*	*5*	*6*	*7*
Telengana	-	-	-	7080.07	8344.43	11720.34
				(3.34)	(4.00)	(5.42)
Assam	3088.24	3088.51	3156.20	9814.57	9260.66	9407.27
	(4.08)	(3.52)	(3.06)	(4.63)	(4.44)	(4.35)
Chhattisgarh	-	-	-	17569.71	13265.03	19082.44
				(8.29)	(6.36)	(8.83)
Gujarat	5346.13	5704.48	5703.75	10554.04	12058.43	10827.69
	(7.06)	(6.50)	(5.53)	(4.98)	(4.92)	(5.01)
Karnataka	2817.98	3619.15	6198.20	5782.47	4121.95	4930.39
	(3.72)	(4.12)	(6.01)	(2.73)	(1.98)	(2.28)

Maharashtra	3713.86	4045.27	4777.58	6316.65	5925.81	6029.89
	(4.91)	(4.61)	(4.63)	(2.98)	(2.84)	(2.79)
Madhya Pradesh	5900.08	6648.47	7937.91	12568.14	15624.52	13928.43
	(7.79)	(7.57)	(7.70)	(5.93)	(7.49)	(6.44)
Odisha	6622.00	7629.63	10627.05	26470.10	29528.57	29997.44
	(8.75)	(8.69)	(10.86)	(12.50)	(14.16)	(13.88)
Rajasthan	1432.70	2084.28	2470.44	22361.36	22689.53	23726.58
	(1.89)	(2.37)	(2.40)	(10.56)	(10.88)	(10.97)
Tamil Nadu	2347.20	2794.40	2961.15	5599.24	5321.99	6682.15
	(3.10)	(3.18)	(2.87)	(2.64)	(2.55)	(3.09)
Meghalaya	-	-	-	3550.43	2530.19	1912.83
				(1.67)	(1.23)	(0.88)
West Bengal	3174.56	3467.00	3072.02	4040.37	3528.52	6696.77
	(4.19)	(3.95)	(2.98)	(1.91)	(1.69)	(3.10)
Jharkhand	-	-	-	25645.19	24244.46	18531.05
				(12.11)	(11.63)	(8.57)
Offshore	-	-	-	49792.63	49186.84	46991.79
				(23.51)	(23.59)	(21.74)
All India	-	-	-	211712.22	208488.13	216188.01
				(100.00)	(100.00)	(100.00)

Notes: Figures in brackets are % share to all-India value.

Sources: (i) Regional Controller of Mines – Bhubaneswar, Government of India (ii) Directorate of Mines, Odisha (iii) Economic Survey, Government of Odisha; 2017-18.

Production of all minerals, fuel minerals like coal, metallic and non-metallic minerals and corresponding value in India during 2010-11 are presented in Table 6.6.

Table 6.6: Production and Value of Minerals Produced in India

Production: In lakh tonne
Value: In crores rupees

Sl. No.	Mineral	Prodn.	Value
I	*II*	*III*	*IV*
1	All Minerals	9875	108672
2	Fuel Minerals (coal)	5136	47480
3	Metallic Minerals	2242	40245
4	Non-metallic minerals	2497	20947

Source: Directorate of Mines, Government of Odisha.

The deposit, production of major minerals of India and Value of Mineral Production as discussed above boils down to the fact that mining sector plays a vital role in enriching the economy and driving the economic development of the country. The future for mining in India looks quite progressive but there are certain challenges that need immediate attention. The Indian mining sector is highly regulated with strong legal and regulatory mechanisms with the government enacting several acts, policies, rules at the central and state levels. Since mining sector has been considered as a polluting industry, causing environmental degradation, Government of India has now been quite progressive in establishing institutions and regulatory framework in order to counterbalance the negative externalities caused by mining activities.

Besides its contribution towards foreign exchange earnings, mining provides a major chunk of government of India's revenue earning. Mining contribution to economy differs from State to State depending on its mining area. Below are described mining status and an estimate of land under mining of the mineral rich states of India.[39]

Andhra Pradesh

In 1998, the undivided Andhra Pradesh had 1, 22,301 acres (49,514.57 ha) on lease to mining companies, many of them private, all of them belonging to the 1980s and 1990s. They included 16,523.24 acres to Singareni Collieries, over 5,000 acres for granites and other stones and similar areas for limestone, fireclay, graphite and other minerals. As many as 84,388 acres (69%) of it was private land, 29% common revenue and the rest forests. According to Fernandes, the total land under mining is bound to exceed 200,000 acres. The then AP Ministry of Mines informed the investors that it is the second largest store house of Mineral Resources in India, that it has 48 minerals, that it produces minerals worth Rs 4,857 crores which is 8 percent of India's production of Rs 59,509 crores. The sector contributes Rs 770 crores to the State exchequer and brings in Rs 634 crores in foreign exchange. It has identified the mining sector as one of the growth engines for the overall development of industry and infrastructure. Minerals awaiting exploration include 700 million tonnes of bauxite, diamonds over 50,000 sq. km, gold in 11 districts, many types of beach sand, clays, petroleum and natural gas. The State envisages mineral projects with the participation and investment from the private sector and considers it basic to a boost to industrial production. Recent spectacular discovery of large quantities of natural gas in the Krishna-Godavari Basin is expected to lead to long-term contracts with reasonable prices.

Jharkhand

Jharkhand accounts over a third of India's coal, around half of its mica reserves, 23 percent of iron ore, 34 percent of copper reserves and has other minerals like fireclay, manganese, uranium, kyanite and china clay. Because of this abundance, mining in Jharkhand began already in the nineteenth century. The Jharia coalmines belong to 1886, the Gurumahsini iron ore mines to 1911, Badampahar and Sulaipet to 1923, Noamundi to 1926, the mica mines in Hazaribagh and Koderma to the 1930s and the bauxite mines in Palamau and Lohardagga to the 1940s. In the late 1990s, Jharkhand had 398 working mines on 409,883.24 acres (165,944.63 ha) but the total with the mining companies was 515,124.59 acres (208,552.47 ha). Common with AP and Odisha is the slow disappearance of mica mines because of alternatives found for it. Many mica mines in Hazaribagh, Koderma and Giridih districts were closed down in the 1990s. Added to it is illegal mining, particularly of coal, much of it in the remote forest areas or on abandoned mines or those that have been declared unsafe by the Directorate of Mines. That accounts for the difference of more than one lakh acres between land under lease and the final total. Only 35.7 percent of land under mining is private and the rest revenue commons or forestland.[40]

West Bengal

Mining has been a major activity of West Bengal for two centuries. Its only major mineral is coal but it also has some minor metals. The State accounted for 25,908.54 million tonnes or 12.24 percent of India's coal reserves or the fourth highest in the country, coming after Jharkhand, Chhattisgarh and Odisha. The State had 114 coalmines in 1983 out of India's 498,[41] most of them are in Bardhaman, Malda and Purulia districts. Their number has declined since then but their size is bigger than in the past because of the shift to opencast mines at first and later globalisation and mechanisation that accompanies it. Apart from coal, the State has acquired land also for minor metals like dolomite. Gazette notifications 1947-1990 show that from the 1980s, there is a decline in the acquisitions for coal. The decline is true also of other development projects mainly because of labour unrest in the 1980s but the decline for mining seems to be because the minerals are nearly exhausted, not because of labour unrest. Because of non-availability of records, it is difficult to get all the information on land used for mining in the State. Gazette notifications show that 13,456.46 acres of private land and 1,440.92 acres of common land were acquired 1947-1990. Fernandes in 2006 indicated that Inquiries with various officials and other knowledgeable persons took us to around 30,000 acres used for mining in the 1980s and 1990s.

The Northeast States

The Northeast has 395 million tonnes of coal deposits apart from uranium, petroleum, limestone and other minerals, according to a news item[42]. Most coal is in Meghalaya and Assam. The petroleum deposits found by and large in Assam have been exploited since 1901. Since they are getting exhausted, new areas are being explored. No university in the region has a course in mining, most employees of mining companies are, therefore, from outside the region and there is resentment against them because of a feeling that they alienate the land of the local people but do not ensure any benefits to them. At the demand of the All Assam Students' Union (AASU) it has been decided to start this course at Mariani in Upper Assam.[43] For several years, the Uranium Corporation of India has been trying to mine uranium in Meghalaya but is meeting with resistance from the people. Tripura has gas deposits. The Gazette data show that only 105.47 acres have been acquired in Assam for mining till 2000 but mining leases in the State occupied 5,483.54 acres in 1991, of which 3,126.98 acres were on lease for 24 coalmines. According to the North Eastern Council website,[44] 6,397.3 acres were on lease in Dibrugarh district in 2004 for coal and more in Sibsagar but it does not give the exact area in the later case. It says that 4,940 acres are being explored in the Dibrugarh-Sibsagar belt and that coal is available in Tinsukia, Karbi Anglong and N. C. Hills but does not say whether exploration has started. So on excluding these areas, the land under lease for coal at may be 8,000 acres. In Assam too, it follows the pattern of fewer mines covering a bigger area. There are two limestone leases in Assam and the area used for it has remained unchanged since 1991. There are eight leases for crude oil, each of them of over 1,000 acres. The Nazira town has been built for this purpose, and hence, not less than 10,000 acres are on lease for petroleum. Thus, all the minerals including petroleum have used not less than 20,251 acres of land.

Goa

Much of Goa has a Pre-Cambrian rock formation. It has 23.55 MT of India's 176.48 MT (13.35%) manganese reserves and in situ deposits of 967.46 MT of iron ore of which 762.23 MT is recoverable, 362.56 MT probed, 200.26 MT probable and 199.41 MT possible according to IBM (1997B: p. 74-78). Its iron ore deposits are 7.27 percent of India's total of 12,745 MT[45] and are found in pink phylite that resembles the shales in the Singhbhum-Keonjhar-Bonai region of Jharkhand and Odisha.[46] Also small amounts of minerals like silica are mined. In 1992, mining gave some 8,500 direct jobs and a similar number in its allied sectors as indicated by Kamat. Mines are, thus, an important sector of the State's economy but there are signs that some of them are being abandoned and that mining is facing a crisis.

For three centuries Portugal ignored opportunities to develop a major iron ore export market in Goa though the fact that large mineral deposits existed was known at least from the seventeenth century. Only after Salazar assumed office and his regime badly in need of revenue and anxious to justify its claim to retain Goa, began to develop the mines. In 1941, a sample consignment of 1,000 tonnes was made to Belgium but regular exports to Japan began in 1947. Their speedy development after it was due to the concessions, low taxation on minerals and nominal import duty on mining machinery.[47] The decree of September 20, 1946 granted mining concessions in Goa and the concession was given in perpetuity. The Government's role was restricted to exercising supervision over the extraction of ores; the right to a concession was patrimonial under the Portuguese constitution. Such concessions were granted to around 600 private parties. After 1961, all new leases were brought under the Mines and Minerals (Regulation and Development) Act, 1957. About 80 leases were granted under it. The Goa government made several unsuccessful attempts to cancel the pre1961 concessions. Finally the Indian Parliament enacted the Goa, Daman and Diu Mineral Concessions (Abolition and Declaration of Mining Leases) Act, 1987 to abolish to treat them as leases under the corresponding Indian law, with retroactive effect from December 20, 1961. The owners challenged the Act in the Bombay High Court, Panaji Bench, and obtained a stay order. They could thus carry on mining operations but they have to go through the process required by law if their leases have to be renewed. The Government has laid down some conditions for their grant and renewal. Today over 11 percent of its land is on perpetual lease to private mining companies. Till 1989, the leases covered almost 500 Sq. km or 13.51 percent of Goa's landmass, some 350 sq. km of it on forestland. A vast area of 41,955 ha. or 11.33 percent of its land was under 526 leases for major minerals in late 1989, 268 of them on 22,645.11 ha. for iron ore and 244 on 16,316.06 ha. Ferro-manganese or Manganese. By 1994, the number of iron ore leases had declined to 220 and those of manganese to 176 covering 28,980.48 ha. but many applications were processed for new ones, 93 of them in 1997 alone as per report of the Department of Mines, 1998. In reply to Unstarred Question No. 578 of Shri Manohar Parrikar MLA, the then Chief Minister Mr Pratap Singh Rane told the State Assembly on 15th July 1998 that the State had 438 mining leases covering 28,349.63 ha. According to the Directorate of Mines and Industries, in 1998 Goa had 372.75 sq. km of land under iron ore, ferro-manganese and manganese leases but 253 leases covering 170.36 sq. km had been cancelled. Thus, at least 542 sq. km of land i.e. 14.64 percent of the State's landmass was under mining leases at some time or the other.[48] Not all the leases are worked simultaneously. In 1983, Goa had 153 (88 iron, 65 manganese) mines and 165 in 1985 on 26 percent of the leased area.[49] In 1985, it had 165 working mines covering 26 percent of the concessions. In 1994, it had 396 leases under operation, 220 of them iron ore and 176 manganese and the rest minor metals.[50] In 1992-93, Iron ore production was 294.4 lakh tonnes.[51] In 1997-98, it was 17,21,001 mt. Bauxite and manganese production seems to have declined more than that of iron ore, from 17,737 and 93,237 tonnes respectively in 1981-82 to 3,272 and 36,448 in 1991-92[52] and 12,110 and 26,867 in 1997-98. However, many mines have been abandoned and most lessees use the abandoned land as water reservoirs or fishponds or leave it fallows.[53]

Regional Scenario

In India, as observed in preceding section, the state of Odisha ranks the highest in terms of value of output of minerals, with an increasing trend over the years. Positively endowed with vast resources of varieties of minerals, Odisha occupies a prominent position in the mineral map of the country. It is, therefore, attempted to highlight the Odishan experiences on contribution of mining sector in shaping its economy and generating revenue for economic development. Odisha, being a mineral rich state in the country, abundant reserves of high grade Iron Ore, Bauxite, Chromite, Managanese Ore along with other minerals such as Coal, Limestone, Dolomite, Tin, Nickel, Vanadium, Lead, Graphite, Gold, Gemstones, Diamond and Decorative Stones and others are available in the State.[54] It is pertinent to indicate that out of 30 districts, 22 districts of Odisha posses minerals in its geographical boundary area. However, 11 districts possess major quantity of minerals in the State. A map of Odisha with visual representation of mineral rich districts is presented below. Districtwise percentage of mining area to geographical area in mining districts is shown in a map of Odisha.

An overview of mining sector in Odisha reveals that some minerals like Iron Ore, Limestone, manganese, graphite, dolomite, chromite and coal were known right from the beginning of first five year plan though their exploration and extraction were done in a moderate scale. Decades of planned development and extensive exploration of minerals in the State have brought to focus the large resources of these minerals and a few more minerals like bauxite, coal of thermal grade, rare earth minerals, magnetite and nickel Ore. Odisha has around 17 percent of the total mineral reserves of the country. The State has 95.26 percent of chromite of India[55] while 92.46 percent of the country's Nickel Ore are located and about 54.97

DISTRICTWISE PERCENTAGE OF MINING AREA TO GEOGRAPHICAL AREA

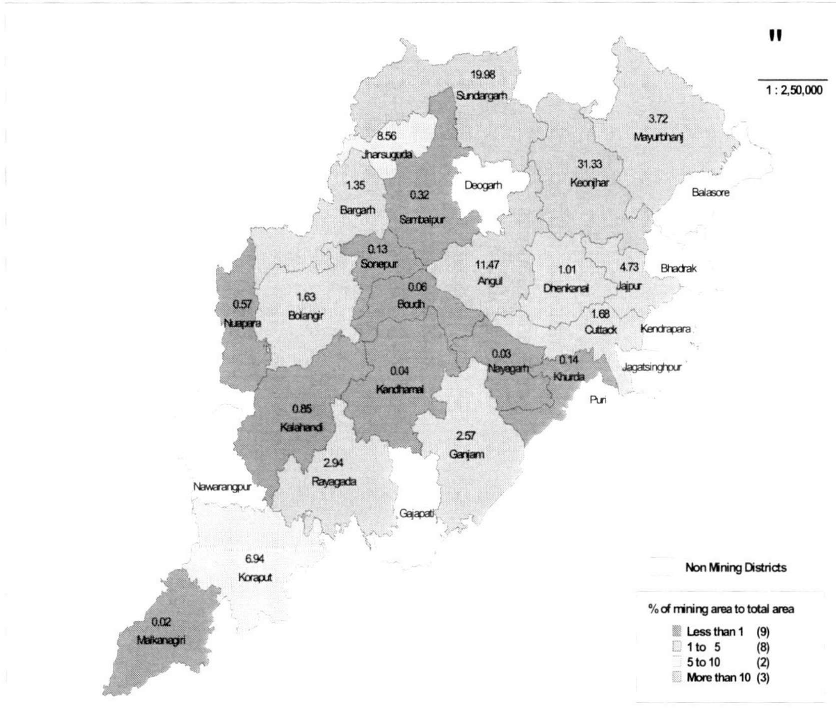

percent of the country's Bauxite reserves is also found in Odisha. Odisha has also got the country's 40.40 percent of manganese ore, 32.54 percent of Iron Ore, 24.90 percent of Fire Clay, 24.48 percent of Coal, 0.99 percent of limestone, 0.26 percent of asbestos, 21.90 percent of pyrophylite, 4.83 percent of quartlite and 0.02 percent of Tin. Odisha's share of major mineral deposits in India by the end of 2017-18 is depicted in Table 6.7. Highest deposit is Chromite (96%) followed by Nickel and lowest deposit is Coal (25%) and Iron ore (35%).

Table 6.7: Odisha's Share of Major Mineral Deposits in India

(by the end of 2017-18)

Sl. No.	Minerals/Ores	Percentage to all India
1.	Chromite	96
2.	Nickel	92
3.	Bauxite	53
4.	Manganese	45
5.	Iron Ore	35
6.	Coal	25

Source: (i) Economic Survey (Odisha) - 2018-19.

It may be noted that some mineral reserves of the State increased substantially from 2010-11 to 2017-18 as indicated in Table 6.8 whereas Bauxite resources increased from 1800.9 MTS to 2039.02 MTS, Coal reserves increased from 65,010.3 MTS to 77,2842.8 MTS, Iron Ore from 5008.3 to 7874.6 and manganese ore from 118.6 MTS to 217.8 MTS.

Table 6.8: Increase of major mineral resources between 2010-11 to 2017-18

(Million Tons (MTS)

Sl. No.	Mineral Resources	Mineral Resources during 2010-11	Mineral Resources during 2017-18	Percentage Increase
I	II	III	IV	V
1.	Bauxite	1800.9	2039.2	13
2.	Coal	65,010.3	77,284.8	19
3.	Iron Ore	5,008.3	7874.6	57
4.	Lime Stone	1000.7	2163.8	116
5.	Manganese Ore	118.6	217.8	84
6.	Nickel	4.4	19.2	341

Source: (i) Department of Steel & Mines, Government of Odisha, (ii) Economic Survey, Odisha – 2018-19.

Within the State, Coal Constitutes the lion's share (84.94%) of all mineral deposits followed by Iron Ore and Asbestos as on 2010-11 as detailed in Table 6.9.

Table 6.9: All Mineral Resources in Odisha

Sl. No.	Mineral Ores	Total Reserve (MTS) (Million Tons MTS)	Percentage of Total State Reserves of all Mineral
1.	Asbestos	0.0567	7.26343E-05
2.	Bauxite	1810.458	2.319249087
3.	Chinaclay	280.926	0.359874335
4.	Chromite	190.021	0.2434224
5.	Coal	66307.25	84.94150597
6.	Dolomite	673.045	0.862190121
7.	Fireclay	175.462	0.224771899
8.	Graphite	8.67465	0.011112478
9.	Iron Ore	5930.233	7.596800075
10.	Limestone	1782.987	2.284057941
11.	Lead and Zinc Ore (Basemetal)	4.98	0.006379524
12.	Manganese Ore	190.047	0.243455706
13.	Heavy Minerals	226.00	0.28951254
14.	Mica	0.10528	0.000134867
15.	Mineral Sand	222.082	0.284493468
16.	Nickel Ore	175.00	0.224180064
17.	Prophylite	12.292	0.015746408
18.	Quartz and Quartzite	70.114	0.089818063
19.	Tin	0.01549	1.98431E-05
20.	Vanadium Ore	2.500	0.003202572
	Total	**78062.24912**	**100**

Source: (i) Directorate of Geology, (ii) Activities of Steel & Mines Department: 2012-13, Government of Odisha, (iii) Economic Survey, Odisha – 2012-13.

There is a wide variation of mineral resources in different districts of the State. Out of 30 districts, mining area spreads in 22 districts of Odisha. Almost one-third of the minerals are confined to Keonjhar District alone. It is the most mined district of the State.[56] Keonjhar together with Sundargarh constitutes more than 50 percent of the State's Mineral Resources.[57] Abundant deposit of Bauxite is found in Koraput District, which has the distinction of being called as the Bauxite capital of India.[58] The districtwise availability of Mineral (Other than Coal) is shown in Table 6.10. Coal is mostly mined in Angul and Jharsuguda Districts.

Table 6.10: Mining Area Spread in different Districts of Odisha

Sl. No.	Name of the District	Percentage of Mining Area
1.	Angul	11.47
2.	Bargarh	1.35
3.	Bolangir	1.62
4.	Boudh	0.06
5.	Cuttack	1.67
6.	Dhenkanal	1.00
7.	Ganjam	2.57
8.	Jajpur	4.73
9.	Jharsuguda	8.56
10.	Kalahandi	0.84
11.	Kandhamal	0.04
12.	Keonjhar	31.33
13.	Khurda	0.13
14.	Koraput	6.94
15.	Malkangiri	0.02
16.	Mayurbhanj	3.72
17.	Nayagarh	0.02
18.	Nuapada	0.58
19.	Rayagada	2.94
20.	Sambalpur	0.31
21.	Sonepur	0.13
22.	Sundargarh	19.98
	ODISHA	100

Source: - Activities of Steel & Mines Department, Government of Odisha: 2012-13.

It has been observed that about 66 percent of coal has been extracted in Angul district and the rest from Jharsuguda, Sundergarh and Sambalpur districts. Iron ore extraction is mostly confined to Keonjhar district which accounts for 68 more than percent of total extraction, followed by Sundergarh (27 percent). Most bauxite mining takes place in Koraput district.

Koraput District has highest quantity of Bauxite reserve in the State. It produces more than 98 percent of the State's total Bauxite production.

A possible resource of 10.74 million tones of metallurgical grade bauxite has been estimated in Kusumdihi of Sundargarh district. Bauxite mines are mostly situated in Southern Odisha. Major companies extracting Bauxite in these areas are Nalco, Aluminium Company of America (Alcoa) L&T, Utkal Alumina, Balco, Phelps Dodge and Rpg Enterprises. Dolomite is abundantly available in Sundargarh District. Cement grade limestone is found in Garamura area of Nuapada district. Angul is the major coal producing district of the State. It has the highest quantity of coal reserves (more than 66 percent) followed by Jharsuguda, Sundargarh and Sambalpur Districts. Chromite is mined mostly in Jajpur, Dhenkanal and Keonjhar districts. More than 97 percent of chromite of the State is available in Jajpur District only. Keradapal and Tangarpara of Dhenkanal district have reportedly some chromite reserves. Coastal Puri and Jagatsinghpur districts are potential sources of heavy mineral in beach sand. Keonjhar district has highest quantity of Manganese reserves followed by Sundargarh district of the State. Iron ore is available in Sundargarh, Mayurbhanj and Keonjhar district. Dolomite is abundantly available in Sundargarh district. Cement grade limestone is found in Nuapada district.

Production of all minerals, fuel minerals like coal, metallic and non-metallic minerals and corresponding value in Odisha in two years for a comparative analysis is presented in Table 6.11.

Table 6.11: Production and Value of Minerals Produced in Odisha

Production: In lakh tonne
Value in crores of rupees

Mineral	Years, Production and Value			
	2010-11		2011-12	
	Prodn.	Value	Prodn.	Value
I	II	III	IV	V
All Minerals	2079	28287	1852	30204
Fuel Minerals (coal)	1111	5844	1052	5580
Metallic Minerals	909	23235	755	24418
Non-metallic minerals	59	236	46	207

Source: Directorate of Mines, Government of Odisha.

Growth of Mineral Sector in terms of total value of minerals produced in the state for the period from 2000-01 to 2016-17 displays an upward trend (Table 6.12).

Table 6.12: Value of Minerals Produced in Odisha

(Rs in crore)

Sl. No.	Year	Value of Minerals Produced in Odisha
1.	2000-01	2276
2.	2001-02	2910.47
3.	2002-03	3694.17
4.	2003-04	3877.75
5.	2004-05	6130.93
6.	2005-06	6604
7.	2006-07	7630
8.	2007-08	10627
9.	2008-09	15123
10.	2009-10	15317
11.	2010-11	28287
12.	2011-12	30204
13.	2014-15	54861.11
14.	2015-16	41535.57
15.	2016-17	41621.4

Source: (i) Economic Survey, Govt. of Odisha: 2012-13
(ii) Economic Survey, Govt. of Odisha: 2017-18.

Coal constitutes the major share in total production of major minerals. Table 6.13 illustrates mineral-wise and year-wise break-up of exploration/extraction of important minerals in Odisha (for three years). Rate of Annual extraction of mineral resources depicts the asymmetrical, varied and non-uniform pattern of exploration in major minerals.

Table 6.13: Rates of Exploration of Important Minerals

(In million tons)

Minerals/Ores	Total reserves in	Exploration during			Rate of Exploration to total reserves (%)		
	2011-12	2009-10	2010-11	2011-12	2009-10	2010-11	2011-12
1	2	3	4	5	6	7	8
Iron ore	4966.97	79.68	81.15	81.15	1.57	1.45	1.63
Chromite	162.29	3.41	4.30	3.79	2.00	2.59	2.33
Coal	71447.41	105.49	111.10	105.12	0.16	0.17	0.15
Bauxite	1795.81	4.88	4.86	5.05	0.27	0.27	0.28
Lime stone	997.62	2.71	3.81	3.14	0.27	0.38	0.31
Dolomite	327.01	1.32	1.49	1.11	0.40	0.45	0.34

Fire clay	175.46	0.06	0.1	0.00	0.04	0.06	0.00
Quartz, quart-zite and silica sand	70.11	0.14	0.032	0.07	0.20	0.04	0.10
Mineral sands	222.08	0.25	0.26	0.24	0.11	0.12	0.11
Graphite	4.33	0.02	0.02	0.03	0.58	0.46	0.69
Manganese ores	119.86	0.60	0.63	0.54	0.51	0.53	0.45

Sources: Directorate of Mines, Odisha, and Directorate of Geology, Odisha.

The rate of extraction of minerals to its total reserves in Odisha, is very minimal barring the case of Chromite and Iron Ore. Lack of adequate infrastructural facility, problems in availability of mining labour and to some extent, low rate of investment in mining sector of the state is responsible for this low rate of mineral exploitation. Regarding export of minerals, Iron Ore occupies the prominent place in the export basket of all minerals. It's share in total exports of mineral stood at 96.2 percent in 2011-12. However, the total exports of mineral and Ores from the state shows a fluctuating trend (Fig. 6.1). The quantity and value of export of minerals/ores from Odisha is also fluctuating over the years. An analysis of the export of different minerals shows that Iron Ore is the major export item among all minerals which increased from the years 2008-09 to 2010-11 though it decreased between 2007-08 to 2008-09 and again in 2011-12 (Table 6.14).

Figure 6.1: Total Exports of Minerals & Ores

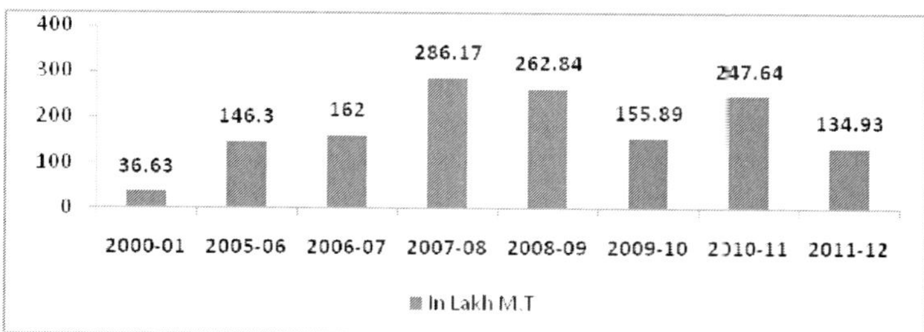

Table 6.14: Exports of Minerals/Ores from Odisha (Quantity in lakh tons, Value in crores of rupees)

Items	2007-08		2008-09		2009-10		2010-11		2011-12	
	Qty.	Value	Qty.	Value	Qty.	Value	Qty.	Value	Qty.	Value
I	II	III	IV	V	VI	VII	VIII	IX	X	XI
Chromite	9.07	1096.00	4.09	811.00	4.66	464.11	4.34	669.31	3.50	430.69
Iron Ore	167.00	5334.00	147.53	4017.24	148.68	4224.00	241.02	15155.22	129.67	3748.68
Mineral sand	2.00	49.40	1.08	26.65	2.55	72.32	2.25	63.81	1.56	52.42
Manganese	0.10	1.60	0.07	1.04	-	-	0.03	5.95	0	0
Others (thermal coal, Hard-coal and Ferrochrome)	108.00	1852.00	110.07	2232.45	-	-	-	-	-	-
Total	**286.17**	**8333.00**	**262.84**	**7814.64**	**155.89**	**4760.43**	**247.64**	**15894.29**	**134.73**	**4231.79**

Source: Directorate of Mines, Odisha.

In earlier years, however, both quantity and value of export of minerals shows an upward trend. An illustration of quantity and value of export of minerals earlier from 2000-01 and 2001-02 and later in 2015-16 and 2016-17 is given in Table 6.15.

Table 6.15: Export of Minerals/Ores from Odisha

(In lakh tones) (Rs In crores)

Sl. No.	Items	2000-01		2001-02		2015-16		2016-17	
		Qty	Value	Qty	Value	Qty	Value	Qty	Value
1.	Chromite	7.08	185.85	10.27	255.71	1.33	137.22	1.737	237.85
2.	Iron Ore	28.25	222.88	35.46	272.33	0.78	15.85	60.13	833.58
3.	Limestone	1.30	45.09	0.31	10.75	-	-	-	-
4.	Mineral Sand	-	-	-	-	0.03	0.00	2.0	189.97
5.	Manganese	-	-	-	-	4.25	5.64		
	Total	**36.63**	**453.82**	**46.04**	**538.79**	**6.58**	**158.72**	**63.86**	**1261.40**

Source: Government of Odisha, Economic Survey; 2005-06 and 2017-18.

A State's revenue and income, to a great extent, determines/ influences its economy and expenditure on economic development. Mining royalty is an important source of revenue for Odisha. The collection of mining revenue has steadily increased over past few years. The last revision of royalty has been done by the Central Government in the year 2009 though as per the provisions of MMDR Act, 1957, it was to be done in 2007 itself. This revision is, thus, delayed though frequent references have been made for revision of royalty rates and for imposition of Mineral Resources Rent Tax.[59] The overall trend in revenue receipts from the mining sector during 2000-01 to 2004-05 shows that it has moved upwards from Rs 360.41 crore in 2000-01 to Rs 670.51 crore in 2004-05 (Table 6.16). The revenue receipt, therefore, has nearly doubled during a period of five years. It has formed a lion's share of the total non-tax revenue of the state. It was 52.58 percent of the total non-tax revenue in 2000-01, which increased to 65.17 percent in 2004-05. It is indicative of the growing importance of the mining sector in the economy of Odisha.

Table 6.16: Revenue Receipt from Mining Sector and Total Non-tax Revenue of Odisha

(Rs In crore)

Year	Revenue receipts			Total own non-tax revenue of the State	Percentage of col. IV to col. V
	Cess on mining royalties	Mining royalties and other revenue from minerals	Total		
I	II	III	IV	V	VI
2000-01	0.09	360.32	360.41	685.5	52.58
2001-02	0.07	376.38	376.45	691.8	54.42
2002-03	-	443.53	443.53	961.2	46.14
2003-04	0.01	550.76	550.77	1094.5	50.32
2004-05	-	670.51	670.51	1028.9	65.17

Source: Government of Odisha, Economic Survey, 2005-06.

However, revenue collection by Government of Odisha from the year 2008-09 to 2017-18 also registered an upward trend barring two years i.e. 2014-15 and 2016-17 (Table 6.17). Collection of revenue substantially increased from Rs 1380.59 cr. in 2008-09 to Rs 6130.97 cr. during 2017-18.

Table 4.17: Year-wise Revenue Collection

(Rs in crores)

Year	Mining Revenue Collected
2008-09	1380.59
2009-10	2020.72
2010-11	3330.46
2011-12	4586.65
2013-14	5519.58
2014-15	5310.09
2015-16	5797.79
2016-17	4925.66
2017-18	6130.97

Source: (i) Activities of Steel & Mines Department, Odisha, 2012-13 (ii) Directorate of Mines, Odisha (iii) Economic Survey; 2018-19, Govt. of Odisha.

Thus, revenue collection from minerals increased steadily over the years and this is a pointer to the fact that mining sector is a potential source of revenue to the state and substantially supports the states' economic development.

One of the important indicators of economic development of a nation is the level of employment and the efficiency and sustainability of a sector in Country's economy to a large extent, based, inter-alia, on employment it generates. Mining and quarrying provides employment to different sections including tribal, indigenous, marginalised groups and both skilled and unskilled people. Table 6.18, 6.19 and 6.20 respectively reports the number of workers directly engaged in various mining activities in different years from 2000-01 to 2003-04, 2004-05 to 2011-12 and 2014-15 to 2017-18.

Table 6.18: Number of workers directly employed in major mineral activities

(2000-01 to 2003-04)

Sl. No.	Mineral ore/ district	2000-01	2001-02	2002-03	2003-04
I	II	III	IV	V	VI
1.	Bauxite (Koraput, Sundergarh)	603	458	724	792
2.	Chromite (Dhenkanal, Jajpur, Keonjhar)	6743	5624	4786	4424
3.	Coal (Angul, Jharsuguda	19583	17069	17657	17318
4.	Dolomite and limestone (Baragarh, Bolangir, Nuapada	4868	3302	3245	3516
5.	Iron ore (Jajpur, Keonjhar)	13255	11983	10523	11418
6.	Manganese ore (Keonjhar, Sundergarh, Rayagarh)	4081	3437	3660	3892
7.	Others	3804	3262	3572	2383
	Total	**52,937**	**45,135**	**44,167**	**43,743**

Source: Govt. of Odisha, Economic Survey; 2005-06.

The whole purpose of displaying the workers engagement in different periods in three different tables is to make a comparative analysis of erratic nature of employment in mining sector over a period of time. It shows a declining trend in some years and then upward trend and vice versa. This is also shown in figure 6.2 up to 2011-12.

Table 6.19: Number of workers directly employed in major mineral activities

(2004-05 to 2011-12)

Mineral Ores/ District	2004-05	2005-06	2006-07	2007-08	2008-09	2009-10	2010-11	2011-12
I	**II**	**III**	**IV**	**V**	**VI**	**VII**	**VIII**	**IX**
Bauxite (Koraput, Sundergarh)	775	839	866	678	634	664	564	932
Chromite (Dhenkanal, Jajpur, Keonjhar)	6607	8236	8452	9816	6528	7826	7571	7053

Coal (Angul, Jharsuguda, Sundergarh, Sambalpur)	17624	14500	13985	12747	13467	13875	15389	16330
Dolomite and lime stone (Bargarh, Bolangir, Koraput, Sundergarh)	1626	2378	1822	1843	2206	2312	2441	496
Iron ore (Jajpur, Keonjhar, Mayurbhanj, Sundergarh)	19592	20782	16677	18912	16838	14679	20071	17257
Manganese ore (Keonjhar, Sundergarh, Rayagada, Bolangir)	1513	1505	2612	2655	2294	2538	4069	2641
Others (china clay, quartz & quantity, graphite, mineral sand, gem stone, pyrophilite etc.)	2100	7524	2962	2525	2200	1811	1772	3530
Total	**49837**	**55764**	**47376**	**49176**	**44167**	**43705**	**51877**	**48239**

Source: Directorate of Mines, Odisha.

Figure 6.2: Workers directly employed in major mineral activities in Odisha, 2000-01 to 2011-12

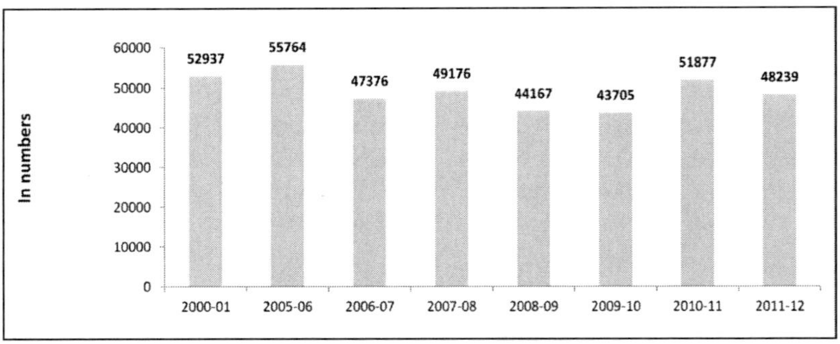

The highest numbers of people are employed in the coal and iron ore mining activities. The total number of people employed in the mining sector in Odisha was 52,937 in 2001-02. Thereafter, it has shown a declining trend up to 2003-04. The employment situation in this sector improved in 2004-05 and the figure reached 49,794 with an increase of 13.83 percent over its previous year because of expansion of mining activities in the State. Again it increased to a total of 55,940 in the year 2017-18.

Table 6.20: Number of workers directly employed in major mineral activities

(2014-15 to 2017-18)

Mineral Ores/District	2014-15	2015-16	2016-17	2017-18
I	II	III	IV	V
Bauxite (Koraput, Sundergarh)	674	896	1215	1350
Chromite (Dhenkanal, Jajpur, Keonjhar)	5328	5974	5974	6074
Coal (Angul, Jharsuguda, Sundergarh, Sambalpur)	19700	19800	20800	22200
Dolomite and lime stone (Bargarh, Bolangir, Koraput, Sundergarh)	367	450	450	550
Iron ore (Jajpur, Keonjhar, Mayurbhanj, Sundergarh)	17611	18100	18600	19600
Manganese ore (Keonjhar, Sundergarh, Rayagada, Bolangir)	2359	2415	2915	3915
Others (china clay, quartz & quarzite, graphite, mineral sand, gem stone, pyrophilite etc.)	1331	1451	2451	2251
Total	**47370**	**49086**	**52405**	**55940**

Source: (i) Directorate of Mines, Odisha (ii) Govt. of Odisha, Economic Survey; 2018-19.

The sector has been increasingly employing labour-saving and capital-intensive production techniques and technology over the years, for which total employment in the sector has displayed an unimpressive figure. This sector has become more mechanised. However, employment, which touched 48,239 by the end of 2011-12, decreased by 7.01 percent over 2010-11. Further, nearly 70 percent of employees are engaged in the iron ore and coal sub-sectors, as may be seen from figure 6.3 which depicts the composition of employment by mineral type in 2011-12.

Figure 6.3: Composition of Employment by Mineral Type, 2011-12

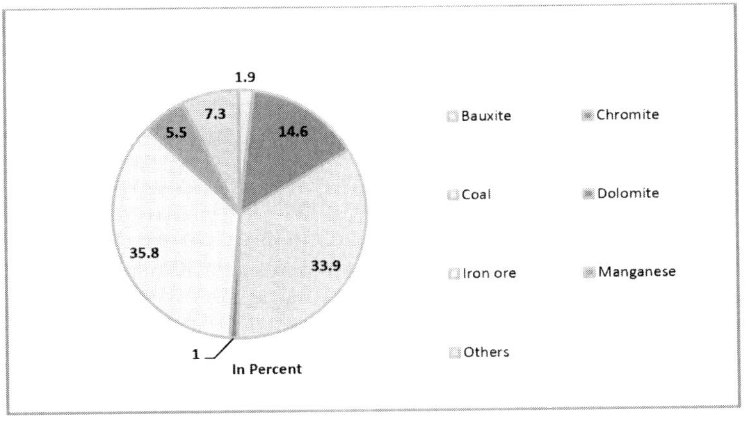

The trend of employment in Mining Sector is fluctuating though the overall trend over the years has shown a more or less an increasing trend since 2015-16. The fact, therefore, remains that mining sector generates substantial employment in the state.

Thus, in terms of production, foreign exchange earnings through exports, generation of revenue and employment, mining sector greatly contributes to economic development of the state. Now, let us discuss regarding investment, mining leases granted and MOUs signed with different companies by the Government of Odisha, the factors which influences the economy of the state.

Data about investment in mining by April 2006 shows that Odisha had the highest amount (Rs 19,588 crore), almost double of Jharkhand. By March 2009, Odisha had bettered it, inching close to the Rs 1,50,000 crore mark. If we compare the average size of mining projects for Odisha and its close rival Jharkhand, in April 2006, it can be seen that Odisha's Rs 653 crore per project is not only the highest, but also more than double the Jharkhand size (Table 6.21). By March 2009, Odisha had leapfrogged to an average size of Rs 2,264 crore, almost three times that of the closest rival, Rajasthan. So, when it comes to mining companies, Odisha can easily edge out Jharkhand very much in sync with its 2006 top rank in the nation.

Table 6.21: Investment in Mining

(Amount in Rs crore and Size in number)

State	Nos: April 2006	Amt: April 2006	Average Size	Nos: March 2009	Amt: March 2009	Average Size
Andhra Pradesh	57	10,327	181	67	49,607	740
Assam	9	3,359	373	11	4,108	373
Bihar	7	152	22	8	536	67
Chhattisgarh	50	5,867	117	60	10,966	183
Gujarat	22	8,360	380	45	11,952	266
Jharkhand	25	7,913	317	48	13,837	288
Karnataka	25	2,384	95	25	12,133	485
Madhya Pradesh	28	1,552	55	33	7,579	230
Maharashtra	19	3,388	178	17	5,741	338
Odisha	30	19,588	653	65	1,47,153	2,264
Rajasthan	16	5,802	363	21	16,965	808
Tamil Nadu	15	8,571	571	16	10,456	654
Uttar Pradesh				4	987	247
Uttarakhand				1	19	19

West Bengal	17	2,713	160	21	4,287	204
All India	408	1,11,015	272	519	3,42,588	660
BIMARU	126	21,286	169	175	50,889	291
Non-BIMARU	164	39,102	238	202	98,284	487
ALL	**290**	**60,388**	**208**	**377**	**1,49,173**	**396**

Source: CMIE Cap Ex: June 2006 and June 2009.

Regarding mining leases, total leases both in numbers and area as well as working leases from 1990-91 to 2011-12 shows a fluctuating trend. Table 6.22 depicts year-wise lease area of mining in Odisha. Total number of mining leases in the state declined from 629 (124.22 hectares) in 1990.91 to 597 in 2009-10 and in 2010-11 and 2011-12 it remained constant (600 nos.). Coal and Iron Ore are prominent minerals in leased out areas during 2011-12.

Table 6. 22: Number of Mining Leases, 1990-91 to 2011-12

	Total Leases		**Working Leases**	
	Nos.	**Area in '000 hectares**	**Nos.**	**Area in '000 hectares**
1	2	3	4	5
1990-91	629	124.22	399	93.36
2000-01	613	112.47	335	82.96
2003-04	607	101.95	339	73.91
2004-05	594	99.08	331	72.2
2005-06	571	86.42	335	59.08
2006-07	602	97.02	370	74.44
2007-08	586	94.7	374	68.15
2008-09	596	96.626	376	74.024
2009-10	597	95.957	330	70.520
2010-11	600	96..570	78	26.622
2011-12	600	98.438	128	64.457

Source: Directorate of Mines, Odisha.

However, during the year 2012-13, only 2 (Two) mining leases (ML) covering 361.827 ha. have been granted by the Government of Odisha. Besides, 1 (one) prospecting licence (PL) covering 676.254 ha. has also been granted by the State Government. Emphasis has been given on grant of mining leases for mineral based industries for value addition, preferably in the State so as to generate more employment and promote industrialisation in the State. Status of mining leases for different circles of the State is depicted in Table 6.23.

Table 6.23: Mining Lease in the State

CIRCLE	TL	W	TD	S	NW (TD+S)	L	D	NF	TW
Baripada	44	6	15	21	36		1		1
Berhampur	2	1						1	
Bhawanipatna	76	2	32	28	60	1	2	10	1
Bolangir	74	1	25	17	42	1	15	15	
Cuttack	13		2	9	11		1	1	
Jajpur Road	21	13	7	1	8				
Joda	101	32		66	66	1	1	1	
Keonjhar	21	2	6	6	12	1	3		3
Koira	72	26	8	34	42		2		2
Koraput	34	6	6	21	27			1	
Phulbani	5		2		2		2	1	
Rourkela	63	10	24	19	43	2	5	3	
Sambalpur	49	18	3	16	19		8	3	1
Talcher	25	14	7	2	9			2	
Total	**600**	**131**	**137**	**240**	**377**	**6**	**40**	**38**	**8**

(TL: Total Leases; W: Working; TD: Temporary Discontinued; S: Suspended; NW: Non Working (TD+S); L: Lapsed; D: Determined; NF: RML not filed; TW: Temporary working for dispatch)

Source: Activities of Steel & Mines Department, Government of Odisha, 2012-13.

To regulate the mineral trade, grant of trading licences has been ensured by the Government of Odisha as provided in the Odisha Minerals (Prevention of Theft Smuggling and Illegal Mining and Regulation of Possession, Storage, Trading and Transportation) Rules, 2007. The circle-wise breakup of the trading licence issued is furnished in Table 6.24.

Table 6.24: Circlewise Trading Licences Granted

Circle Name	Total	Valid Licence	Suspended	Cancelled / Expired
Baripada	181	162	1	18
Berhampur	214	200	7	7
Bhawanipatna	9	1	0	7
Bolangir	20	10	2	8
Cuttack	297	195	20	82
Jajpur Road	355	239	9	107
Joda	1105	676	178	248
Keonjhar	232	163	27	42
Koira	622	376	41	204
Koraput	27	25	0	2
Phulbani	0	0	0	0

Source: Activities of Steel & Mines Department, Government of Odisha, 2012-13.

As far as MOUs signed with Companies are concerned, the Government of Odisha has been adopting a neo-liberal approach in its development policy towards backwardness, poverty and under-development. The state was in the forefront in implementing privatisation in electricity and neo-liberal reforms in various other sectors. Mining and Industrialisation are the prime movers of its development plank and government is constantly encouraging and inviting both domestic and foreign capital in these sectors in tandem with Globalisation and the policy and approach of Government of India to attract foreign direct investment in post-reform phase through 'policies of deregulation'. A word 'Provincial Darwinison' has been coined to imply a process in which State/Provincial governments of India with federal governance structure have been competing with each other to attract foreign and domestic capital by suitably amending their policies and programmes. Odisha with rich and abundant mineral base naturally attracts the foreign direct investment. Compared to its mineral deposits, the level of production/utilisation of these mineral resources is very low and hence, the State Government has strongly adopted the policy of attracting investment in mining sector. With liberalisation of the Indian Economy, the vast mineral reserves of the State have also attracted the dominant multinational companies/corporations to establish plants, industries in the State. A number of MoUs have been signed by the State Government with the reputed companies to establish mineral based industries. Prominent among these area are south Korean Company, POSCO, Vedanta Alumina Ltd., TATA Steel, Utkal Alumina Ltd., Jindal, Neelachal Ispat Nigam Ltd., Sterlite Iron and Steel, Essar, Tisco, Konark MCT Coke Ltd., Visa and others as indicated in Table 6.25.

Table 6.25: MoUs signed with Companies for mineral based industries in Odisha

Sl. No.	MoUs signed with Companies for mineral
1.	Arcelor Mittal India Ltd
2.	Sterlite Iron & Steel Co Ltd
3.	POSCO India Ltd
4.	Jindal Steel & Power Ltd,
5.	SSL Energy Ltd.
6.	Uttam Galva Steels Ltd.
7.	Nilachal I Spot Nigam Ltd.
8.	Konark MCT Coke Ltd
9.	ESSAR Steel Odisha Ltd.
10.	Bhusan Power & Steel Ltd.

11.	Aditya Aluminium Ltd.
12.	Tata Steel Limited
13.	Welspun Power & Steel Ltd.
14.	Vedanta Aluminium Ltd.
15.	Jindal India Thermal Power

Governments of Odisha have signed MOUs with a number of Mineral based industries. It is reported that the State level single-window Clearance Authority recently cleared many private Project proposals to set up iron ore based plants.

The fact follows from the regional scenario is that mining has a significant contribution for the economy and economic development particularly for the state of Odisha. Apart from generating huge revenue thereby enabling the government to fund its development programmes, mining is a potentially high source of job creation and provides employment to a large number of unemployed people. Odisha, a relatively backward and under developed State is the early initiator of economic reforms in various sectors to improve its growth performances.

Negative Economic Impacts

In spite of many positive outcomes of mining on economy as discussed above with experiences from regional to global level, Mining has distinct direct and indirect negative economic impacts. Mine closure cause adverse impact on the economy and create barriers to economic restructuring. Mining, as well as the cessation of mining where it has become uneconomic, can also be a cause of poverty or adversely affect the living conditions of the poor and other vulnerable groups. Large losses by state-owned mining industries, especially coal mining, have been a significant barrier to economic restructuring and recovery, especially in Eastern Europe and the former Soviet Union. Closure of noneconomic mines has added to poverty, especially in mono-industry communities and mineral-dependent regions. In addition to the loss of jobs among the local population, essential public goods and services originally provided by the mining company—transport and water, for example, cease to be delivered, with particularly harmful effects on the poor and other vulnerable groups. Therefore, stringent mine closure regulations should be framed and implemented. E.S. van Druten and M.C. Bekker of Graduate School of Technology Management, University of Pretoria, South Africa have developed an inclusive model for successful mine closures.[60] According to them, the complete and successful closure of a mine in the South African environment remains problematic. Closure remains an elusive mining phase due to unclear roles and responsibilities, inaccurate closure management data, and the inability of stakeholders to conceptualise and integrate closure information into business processes. Currently there are no proven methods available to address the complexities encountered during a closure application, and to date no mine has been granted full closure. Many mines are left abandoned and un-rehabilitated, resulting in risk to investors, surrounding communities, and regulators. Various requirements and aspects were identified and listed according to four categories: namely, primary, secondary, process drivers, and information sources. These aspects should be considered in the development of an integrated mine closure model. Integrated mine closure planning (IMCP) requires closure planning and execution to start prior to the commencement of physical mining activities and continue until final closure of a mining operation.

Mining often leads to macroeconomic mismanagement and corruption. Not all countries with substantial natural resources provide an attractive framework for foreign direct investment to leverage existing opportunities for generating fiscal income; nor are mining companies, in particular, when state owned, always managed with appropriate financial and operational effectiveness and efficiency. According to Evans, "Mineral-dependent economies are often more prone to governance and corruption issues than non-mineral-dependent economies. Also, negative consequences of macroeconomic mismanagement may be harsher in the context of mining than in a non-mining economy: the presence of mining will inflate wages and keep the exchange rate strong, which can prevent other sectors, for example, agriculture, from being internationally competitive and thus from realising the opportunity for export-driven growth. The costs of macroeconomic mismanagement are high, given the often substantial fiscal revenue from mining, particularly when considering that these natural resources are non-renewable. There is a pervasive blindness to the fact that our limited resources are being permanently destroyed for the sake of short-term gain that current forms of extraction, and the trend for ever-increasing extraction and consumption of mineral products is totally unsustainable". In India, some companies are exploiting India's resources, to an extent, oppressing the people and exercise foreign financial control. Most of them, as observed by B.K. Mishra, are foreign companies/ corporations and financial institutions adopting a foreign system of relating and valuing and they have a single aim: maximising profit. The company, as it evolved in the US under the 'robber barons' such as Rockefeller and J.P. Morgan, exists purely to make a profit. Effects of its actions on people are considered irrelevant. It represents a totally hierarchical, unaccountable form of power, with no long-term safeguards. One significant contrast in the Indian context is the ancient example of the Arthashastra, which emphasized the creation of wealth through mining, but insisted that the State should own and control all important mines and not allow them to get into private hands.

Conclusion

Economic development and for that matter sustainable development is an outstanding challenge for nations across the globe. Mining contributes immensely to economic development for both developed, developing and under developed countries in a variety of ways. It has substantial contribution to India's economy. As a major mineral State in the national front, Odisha occupies a prominent place in terms of mineral deposits. Odisha's share in India is highest deposit in Chromite (96%), Nickel (92%), Bauxite (53%), Coal (25%) and Iron ore (35%). Mining royalty is an important source of revenue with a major share in the total non-tax revenue of the state. The collection of mining revenue has been increasing over the years. Mining and quarrying generates employment for different sections of people including tribal groups. Thus, mining sector has a vital role in shaping the economy of a developing state like Odisha.

After having discussed the issues of mining and economy in general and and experiences of some of the mineral rich countries as well as India and Odisha region in particular in preceding paragraphs, it would be pertinent to look at the environmental and social issues, the two other outstanding and key challenges of mining sector.

End Notes & References

1. Dash (Dr). L.N., *"Economics of Mining in Odisha"*, Odisha Review, Government of Odisha.
2. International Council of Mining and Metals (ICMM), *"Mining's contribution to sustainable development; an overview"*, June, 2012.
3. Dash. (Dr). L.N., *Op. Cit.*
4. Panigrahi, H.S., *"Responsible & Sustainable Mining vis-à-vis SDGs."*
5. To illustrate the mining's global contribution on economic development, cases of some mineral rich countries have been cited.
6. ICMM, *Op. Cit.*
7. Generation of indirect employment by mining sector in the country of Brazil.
8. http://www.opml.co.uk/paper/role-mining-national-economic development. *"The role of mining in national economic development"*.
9. ICMM, Raw Materials Group, Oxford Policy Management, 2014. *"The Role of Mining in national economies"*, 2nd edition.

10. Inter-sectoral linkage is an important parameter for sustainable development.
11. ICMM, 2012, *Op. Cit.*
12. United Nations, *"Transforming our world: the 2030, Agenda for Sustainable Development"*, 2015.
13. IIED, Mining, Minerals and Sustainable Development (MMSD).
14. Out of 17SDGs, contribution of mining in achieving 13 SDGs i.e. SDG-1, 2, 3, 4, 5, 6, 7, 8, 9, 12, 13, 15 and 16 are examined.
15. Panigrahi, H. Sekhar; *'Responsible & Sustainable Mining vis-à-vis SDGs'*
16. CCSI, *"Leveraging Mining Investments in Water Infrastructure for Broad Economic Development: Models, Opportunities, and Challenges"*, 2014. http://ccsi.columbia.edu/files/2014/05/CCSI-Policy-Paper-Leveraging-Mining-Related-Water-Infrastructure-forDevelopment-March-2014.
17. World Bank blogs, *Better Health in Mines and Mining Communities: A Shared Responsibility*, 2015, http//blogs.worldbank.org/health/better-health-mines-and-mining-communities-shared-responsibility.
18. World Health Organisation, WHO factsheet, http://www.who.int/mediacentre/factsheets/2003/fs134/en
19. MMG, "*Indigenous and host communities employment and training*". http//www.mmg.com/en/Careers/Vocational-pathways/Indigenous-and-host-communities-employment-and-training.aspx
20. PwC, "*Mining for talent 2014: A review of women on boards in the mining industry",* 2014. http//www.pwc.co.uk/mining/publications/mining-for-talent-2014.jhtml
21. Oxfam, Women, *"Communities and Mining: The gender impacts of mining and the role of gender impact assessment",* 2009.
22. WBCSD, "*Water and Sustainable Development Program. Water: Facts and trends"*. http//www.unwater.org/downloads/Water_facts_and_trends.pdf
23. CCSI, *"Leveraging Mining Investments in Water Infrastructure for Broad Economic Development: Models, Opportunities, and Challenges",* 2014. http://ccsi.columbia.edu/files/2014/05/CCSI-Policy-Paper-Leveraging-Mining-Related-Water-Infrastructure-forDevelopment-March-20141.
24. World Mining Congress, *"Production Energy Optimisation in Mining"*, 2013.
25. Natural Resources Canada, "*Green Mining Initiative",* http://www.nrcan.gc.ca/mining-materials/green-mining/8178
26. Carbon War Room, *"Sunshine for Mines: Implementing Renewable Energy for Off-grid Operations"*, 2014. http://carbonwarroom.com/what-we-do/research-publications/sunshine-mines-implementing-renewable-energy-grid-operations 5
27. ICMM, "*The role of mining in national economies"*, 2012, https://www.icmm.com/document/4440
28. CCSI, *"Leveraging Mining Demand for Internet and Telecommunications Infrastructure for Broad Economic Development: Models, Opportunities, and Challenges",* 2014. http://ccsi.columbia.edu/files/2014/05/CCI-Policy-Paper-Leveraging-Mining-Related-ICTInfrastructure-for-Development-June-2014.pdf
29. CCSI, *"A Framework to Approach Shared Use of Mining-Related Infrastructure",* 2014. http://ccsi.columbia.edu/files/2014/05/Case-Study_Mozambique-March-2014.pdf

30. ICMM, *"Maximizing Value: Guidance on Implementing Materials Stewardship in the Minerals and Metals Value Chain"*, 2006.
31. ICMM, *"Minerals and Metals Management, 2020: Status report"*, 2015.
32. ICMM, "A *global approach to collaboration: Annual Review"*, 2016.
33. IPCC, *"Climate Change 2014 Synthesis Report: Summary for Policy-makers"*, 2014.
34. IFC, *"Performance Standard 6: Biodiversity Conservation and Sustainable Management of Living Natural Resources"*, 2012.
35. Oxfam, *"Community Consent Index 2015: Oil, gas, and mining company public positions on Free, Prior, and Informed Consent"*, 2015.
36. ICMM, *"Human rights in the mining and metals industry: Integrating human rights due diligence into corporate risk management processes"*, 2012.
37. (i) ICMM, *"Indigenous Peoples and Mining"*, 2013. http://commdev.org/wp-content/uploads/2015/06/ICMM-Indigenous-Peoples-andMining-Position-Statement.

    (ii) United Nations Declaration on the Rights of Indigenous Peoples, 2007. 38. Economic Survey, Government of Odisha: 2012-13.
38. Indian Bureau of Mines (IBM), Annual Report, 2016-17.
39. Collected from different reports of Government of Odisha.
40. Ekka and Asif; 2000.
41. CMIE Report, 1986, p. 93.
42. The Sentinel, October 12, 2007.
43. The Assam Tribune, October 14, 2000.
44. Extracted, www.nerdatabank.nic.in
45. Indian Bureau of mines (IBM) report 1991, p. 5 and p. 19.
46. Indian Bureau of mines (IBM) report, 1997, p. 59 and p. 63.
47. Fernandes and Naik, 2001.
48. Department of Mines, Government of Goa, 1998.
49. CMIE, *Op.Cit.* 1986.
50. IBM report, 1997.
51. IBM report, 1994.
52. *Ibid.*
53. Fernandes and Naik, *Op.Cit.*
54. Activities of Steel and Mines Department, Government of Odisha:–2012-13.
55. *Ibid.*
56. Palash, Mining watch, blogspot.Com/2007/12, December 24, 2007.
57. Information collected from Steel and Mines Department, Government of Odisha.
58. Palash, *Op.Cit.*
59. Activities of Steel and Mines Department, *Op.Cit.*, p. 18.
60. E.S. van Druten; M.C. Bekker, Towards an inclusive model to address unsuccessful mine closures in South Africa, Graduate School of Technology Management, University of Pretoria, South Africa.

Annexures

Annexure 6.1: Top 20 mining countries in terms of production value

Sl. No.	Rank and country	2010 Production value (current US$ million)	2010 Production value %	2000 Production value (current US$ million)	2000 Production value %	Change in production value 2000-2010 %	2010 Production value as % of 2010 GDP	2000 Production value as % of 2000 GDP	2010 Mineral export contribution %
1	2	3	4	5	6	7	8	9	10
1.	Australia	71,955	15.6	16,444	14.7	337.8	7.8	3.9	40.3
2.	China	69,281	15.0	10,576	9.4	555.1	1.2	0.9	1.5
3.	Brazil	47,027	10.2	7,754	6.9	506.5	2.3	1.2	19.0
4.	Chile	31,275	6.8	10,452	9.3	199.2	14.7	13.9	65.9
5.	Russian Federation	28,680	6.2	10,776	9.6	166.1	1.9	4.1	6.6
6.	South Africa	27,116	5.9	12,694	11.3	113.6	7.5	9.6	37.4
7.	India	26,042	5.6	2,930	2.6	788.8	1.5	0.6	17.9
8.	United States	22,957	5.0	11,253	10.0	104.0	0.2	0.1	6.2
9.	Peru	18,832	4.1	4,682	4.2	302.2	12.0	8.8	62.7
10.	Canada	13,984	3.0	7,853	7.0	78.1	0.9	1.1	11.9
11.	Indonesia	12,225	2.6	4,948	4.4	147.1	1.7	3.0	10.6
12.	Ukraine	9,283	2.0	1,807	1.6	413.6	6.7	5.8	8.2
13.	Mexico	8,361	1.8	2,426	2.2	244.6	0.8	0.4	4.9
14.	Kazakhstan	7,248	1.6	2,390	2.1	203.3	4.9	13.1	13.1

15.	Iran, Islamic rep.	4,387	0.9	802	0.7	446.8	1.3	0.8	3.7
16.	Philippines	4,221	0.9	397	0.4	964.1	2.1	0.5	6.8
17.	Sweden	3,974	0.9	1,058	0.9	275.5	0.9	0.4	5.3
18.	Ghana	3,964	0.9	1,015	0.9	290.4	12.7	20.4	25.4
19.	Zambia	3,850	0.8	616	0.5	524.7	23.8	19.0	83.6
20.	Papua New Guinea	3,166	0.7	1,338	1.2	136.5	33.4	38.0	54.0
Total top - 20		417, 867							
As % of the World Production		88%							

Source: World Bank, UNCTAD stat. and Raw Materials Group.

Annexure 6.2: Reliance on export of metallic minerals

Sl. No.	Rank and country	GDP/capita (PPP at current prices), 2009, US$ %	Mineral export contri-bution 1996 %	Mineral export contri-bution 2005 %	Mineral contri-bution 2010 %	Change in mineral contribution 1996-2010 (Percentage points)
1	2	3	4	5	6	7
1.	Botswana	$13,384	58.7	86.5	83.7	25
2.	Zambia	$1,430	79.4	64.0	83.6	4
3.	Dem. Rep. of the Congo	$319	72.4	70.2	78.3	6
4.	Mongolia	$3,522	60.3	70.1	77.6	17
5.	Suriname	-	68.0	64.3	75.4	7
6.	French Poly-nesia	-	69.2	55.3	67.1	-2
7.	Chile	$14,311	47.7	56.5	65.9	18
8.	Guinea	$1,048	77.1	84.0	65.2	-12
9.	Peru	$8,629	48.3	57.9	62.7	14
10.	Mauritania	$1,929	36.1	49.3	60.4	24
11.	Northern Mari-ana Islands	–	3.3	4.5	58.9	56
12.	Mozambique	$855	6.1	66.9	57.0	51
13.	Mali	$1,186	8.5	37.2	54.8k	46
14.	Sierra Leone	$808	30.6	58.2	54.3	24
15.	Papua New Guinea	$2,281	24.5	39.2	54.0	30
16.	Namibia	$6,410	36.2	41.2	53.4	17
17.	Nauru	–	73.1	25.2	50.8	-22
18.	Armenia	$5,279	23.9	39.8	50.6	27
19.	Jamaica	$7,633	49.7	68.5	49.6	0
20.	Cuba	–	15.1	39.2	47.7	33

Source: OPM, 2011.

Annexure 6.3: Mining contribution index (MCI) ratings of countries

Sl. No.	MCI range	No. of Countries	Name of the Country	MCI ratings
I	II	III	IV	V
1	100 to 90	1	Zambia	97.7
		2	Northern Mariana Islands	97.6
		3	Papua New Guinea	95.5
		4	Montenegro	95.3
		5	Mauritania	95.3
		6	French Polynesia	94.8
		7	Mali	94.2
		8	Iceland	93.6
		9	Mongolia	93.3
		10	Congo (Dem Rep)	93.2
		11	Guyana	93.1
		12	Chile	92.1
		13	Congo. Rep	91.5
		14	Somalia	91.3
		15	Laos	91.1
		16	Burkina Faso	90.2
2	Below 90 to 80	1	Cuba	89.4
		2	Georgia	89.2
		3	Liberia	89.2
		4	Bolivia	88.0
		5	Peru	88.0
		6	Australia	87.9
		7	New Caledonia	87.5
		8	United Arab Emirates	86.6
		9	Nambia	86.5
		10	Suriname	85.9
		11	Ghana	84.9
		12	Gambia, The	84.7
		13	Bahrain	82.5
		14	Tanzania	82.3
		15	South Africa	81.2
		16	Malawi	80.9
		17	Armenia	80.8

3	Below 80 to 70	1	Brazil	79.6
		2	Bhutan	79.0
		3	Benin	77.4
		4	Madagascar	77.1
		5	Togo	76.2
		6	Cyprus	74.1
		7	Dominica	74.1
		8	Lebanon	74.1
		9	Uzbekistan	73.2
		10	Bulgaria	70.5
4	Below 70 to 60	1	Philippines	69.9
		2	Hong Kong SAR, China	69.3
		3	Macao SAR, China	69.1
		4	Senegal	67.6
		5	Greenland	67.5
		6	Canada	67.1
		7	Indonesia	66.4
		8	Ukraine	66.3
		9	Oman	65.8
		10	Guinea	65.3
		11	Serbia	64.6
		12	Gabon	64.2
		13	Honduras	63.9
		14	Djibouti	63.7
		15	Dominican Republic	63.7
		16	Botswana	61.9
		17	Guatemala	60.7
5	Below 60 to 50	1	Nicaragua	59.7
		2	Colombia	59.4
		3	Ethiopia	59.1
		4	Zimbabwe	58.7
		5	Syrian Arab Republic	58.5
		6	Switzerland	58.0
		7	Morocco	57.7
		8	Kyrgyz Republic	56.9

		9	Albania	56.5
		10	Kenya	56.1
		11	Niger	55.9
		12	El Salvador	55.9
		13	Aruba	55.7
		14	Mexico	55.5
		15	Myanmar	55.4
		16	Egypt, Arab Rep.	55.4
		17	Turkey	55.4
		18	Turks and Caicos Islands	55.2
		19	Sweden	54.9
		20	Iran, Islamic Rep.	54.7
		21	Kazakhstan	54.0
		22	Bermuda	53.3
		23	Trinidad and Tobago	52.4
		24	Belgium	52.1
		25	United Kingdom	52.1
		26	Cambodia	51.9
		27	Finland	51.9
		28	Jamaica	51.7
		29	Sierra Leone	51.3
		30	Argentina	51.0
		31	Italy	50.5
		32	India	50.3
6	Below 50 to 40	1	Thailand	49.9
		2	United States	49.8
		3	Croatia	49.8
		4	St. Vincent and the Grenadines	49.8
		5	Tonga	49.5
		6	Sudan	49.4
		7	Mozambique	49.3
		8	St. Lucia	48.8
		9	Luxembourg	48.6
		10	Netherlands	47.9
		11	Algeria	47.8
		12	Russian Federation	47.6
		13	Maldives	47.4

		14	Moldova	47.4
		15	Greece	47.2
		16	Poland	45.9
		17	Latvia	45.5
		18	Central African Republic	45.5
		19	Tajikistan	45.5
		20	Haiti	45.3
		21	Israel	45.3
		22	Rwanda	44.5
		23	Portugal	44.4
		24	Vietnam	44.0
		25	Bosnia and Herzegovina	43.6
		26	Mauritius	43.4
		27	New Zealand	43.2
		28	Lesotho	43.2
		29	Fiji	43.0
		30	Estonia	42.7
		31	Singapore	42.7
		32	Pakistan	42.0
		33	Jordan	41.7
		34	Afghanistan	40.6
		35	France	40.3
7	Below 40 to 30	1	Slovak Republic	39.9
		2	Malta	39.4
		3	Burundi	38.9
		4	St. Kitts and Nevis	38.9
		5	Denmark	38.2
		6	Norway	37.8
		7	Turkmenistan	37.5
		8	Bahamas, The	37.0
		9	Czech Republic	36.3
		10	Japan	36.2
		11	Cape Verde	35.4
		12	Sri Lanka	35.4
		13	Barbodos	35.1
		14	Korea, Dem. Rep	35.1
		15	Ecuador	34.5

		16	Equatorial Guinea	34.2
		17	Germany	33.6
		18	Venezuela, RB	33.6
		19	Eritrea	33.5
		20	Malaysia	33.4
		21	Spain	33.4
		22	American Samoa	32.3
		23	Korea, Rep.	32.1
		24	Austria	32.1
		25	China	32.0
		26	Cote d' Ivoire	31.3
		27	Slovenia	31.1
		28	Uganda	30.7
		29	Libya	30.4
		30	Tunisia	30.3
		31	Ireland	30.0
		32	Macedonia, FYR	30.0
8	Below 30 to 20	1	Iraq	29.5
		2	Cayman Ialands	29.0
		3	Nigeria	28.6
		4	Nepal	28.3
		5	Qatar	28.3
		6	Marshall islands	28.1
		7	Romania	27.8
		8	Cameroon	27.7
		9	Guinea-Bissau	27.6
		10	Tuvalu	27.4
		11	Belarus	25.7
		12	Kuwait	25.5
		13	Kuwait	25.2
		14	Hungary	25.2
		15	Uruguay	24.6
		16	Vanuatu	24.3
		17	Bangladesh	24.1
		18	Comoros	24.1
		19	Samoa	24.1
		20	Lithuania	23.8

		21	Saudi Arabia	23.1
		22	Solomon Islands	22.9
		23	Yemen, Rep.	22.9
		24	Seychelles	22.2
		25	Antigua and Barbuda	21.5
		26	Brunei Darussalam	20.3
9	Below 20 to 10	1	Swaziland	19.8
		2	Costa Rica	18.9
		3	Panama	18.9
		4	Paraguay	18.9
		5	Chad	18.4
		6	Gibraltar	18.2
		7	Angola	17.2
		8	Sao Tome and Principle	15.6
		9	Mayotte	14.9
		10	Timor-Leste	14.6
		11	Belize	14.6
		12	Grenada	14.6
		13	Palau	13.4
		14	Azerbaijan	13.3
		15	Channel Islands	12.3
		16	Isle of Man	12.3
		17	Kosovo	12.3
		18	Liechtenstein	12.3
		19	Monaco	12.3
		20	Puerto Rico	12.3
		21	San Marino	12.3
		22	Virgin Islands (U.S)	12.3
		23	West Bank and Gaza	12.3
		24	Kiribati	11.1
		25	Micronesia, Fed, Sts.	10.6
10	Below 10	1	Andorra	8.5
		2	Guam	4.7

Source: Web extract, Mining's contribution to sustainable development, data by country, October- 2012.

Chapter – 7

Mining and Environment

Mining activities and minerals affect favourably to the economy, contributes to economic growth and development but have adverse impact on the ecology and environment. Mining, in some form or the other, exist since beginning of human civilisation and is, no doubt, necessary for sustenance and subsistence of human beings. Mining has also some positive contribution on the environment though adverse impacts outweigh the favourable effects. It is imperative to highlight both the adverse and positive impacts of mining and put a concluding note at the end.

Adverse impacts

Mining operation/ extraction of minerals have negative impacts on ecology and environment. It is deleterious and harmful to the area, people and all other living organisms far beyond the mines. It has adverse impact on natural resources viz. land, water, air etc. Mining affects landscape pollutes groundwater, surface water, air and noise, causes soil erosion, dampens plant and animal life.

AIR

Mining operation at different stages and method of extraction directly affect the air quality. It uses Energy and CO_2 and emits of nitrogen oxide and sulphur oxide as unrefined materials goes to the surface and mineral deposits are exposed from the mine site. Dust spreading is a common feature during digging activities and blasting. Besides, road dust due to vehicle movement and fine dust from engines and explosives create air pollution. During extraction of the minerals, generally the soil is removed, vegetation is also removed and in the process the soil is directly exposed to air and weather which ultimately gets in contact with toxic materials. Spreading of smoke during fires and deforestation activities also contaminates air. Toxic smoke erupts from primitive mercury-gold extraction creates health hazards.

WATER

Mining has potentially strong effects both on ground and surface water in a vast area. It affects agronomy, nature, wildlife and forest. Change in river water both quality and quantity happens due to leaching of chemicals from tailings, removal of overburden layers into groundwater, surface run-off water and surface water and leaching of production chemicals like mercury into ground and surface water. Occurrence of acid mine drainage spreads acidity and toxicity in surrounding areas. Increase in river sedimentation due to soil erosion cause siltation and water pollution. River water quantity is also affected because of erosion caused due to deforestation, digging, increased run-off etc. Erosion shortens life time of dam reservoirs and may destroy fishing grounds. Sometimes, there is shortage and limited access and pollution is caused by waste water from the habitants/miners. According to Mannylumanao[1], "the sources of water pollution as a result of irresponsible mining can include processing plants, pond's tailing, mining underground irresponsibly, improper waste-disposal areas, active or abandoned surface. The sediments are most likely from the increasing soil erosion that can cause siltation or the smothering of stream beds. Siltification caused by mining can affect irrigation distribution, swimming, fisheries, domestic water supplies, and other that uses water from streams. Irresponsible mining results in unnaturally high concentrations of chemicals such as arsenic, mercury and sulfuric acid over a wide area surrounding the mine".

LAND

Mining, without responsible practice, can be a cause of destruction of landscape due to open pit mining, waste piles, piles of tailings and creation of artificial lakes causing insect growth and unnatural geomorphology. There will be a loss of habitat for mankind, animal and plant. There will also be a loss of potential touristic quality and devastated landscape will not attract tourists. "Mining can cause physical destruction to its surrounding land by creating landscape holes such as open pits and piles of waste rock. Such breakdowns can contribute to the retrogression of the area's flora and fauna. There will be also a huge possibility that many of the surface features that were present before the mining activities started cannot be replaced after the process has ended if not rehabilitated. Landscape declination caused by the movements of the ground on the surface of the earth as a result of collapsing overlaying sheets can cause damage to roads and buildings in the vicinity land".[2] A huge mass of forest and agricultural land are being leased out for mining thereby severely affecting not only the mankind but also flora and fauna.

Noise

Apart from air, water and land, mining also cause noise pollution. Large mining projects put machines/machineries in different stages of mining operation. Running of engines of these machines create high sound. Explosives used for blasting make unbearable noise. Besides, heavy vehicles ply during day and night also create noise.

Mining, thus, causes environmental hazards and many negative ecological impacts. Mining operations, whether smaller or large-scale, are inherently disruptive to the environment, producing enormous quantities of waste that can have deleterious impacts for decades. The environmental deterioration caused by mining occurs mainly as a result of inappropriate and wasteful working practices and rehabilitation measures. Mining has a number of common stages or activities, each of which has potentially-adverse impacts on the natural environment, society and cultural heritage, the health and safety of mine workers, and communities based in close proximity to operations. As indicated by Noronha, the social and environmental impacts are more pervasive in regions where operations are newly established or are closing down. Both large and small-scale mining, not only adversely affect the environment and many other critical sectors associated with it. Damage caused through water pollution, water quantity, tailing management, noise, dust, and land disturbance are not only responsible for Environmental degradation but also create social issues by adversely affecting the health sanitation and livelihood of the marginalised and vulnerable groups. Let us now discuss the environmental impact of mining in the global context, national scenario and experiences of our case study region, Odisha in the sections that follows.

Globally, throughout eras mining provided most of the materials accumulated today in the techno sphere, understood herein as the total of infrastructures including buildings, machines, tools, etc., that support the world population. Recent estimates, according to Zalasiewicz and others during 2016, suggest that the techno sphere mass might be of about 30 trillion (1012) tonnes, while, for comparison, human biomass is about five orders of magnitude lower. This means that mining and ore milling activities produced a very high volume of materials using procedures often very disruptive to the environment and, for long time, without evaluation of impacts. In very recent years, the development of Environmental Impact Assessment (EIA) procedures aiming at identifying the potential effects and damages caused by developments in the environment and society, as observed by Jain and others, are helping to foresee and estimate gains, costs, losses, and consequences of such projects from various perspectives. Variations of EIA have been developed such as the Strategic Environmental Impact Assessment (SEIA) to help evaluating and planning beyond the narrow scope of one project and even the Radiological Environmental Impact Assessment (REIA) for projects involving radioactive materials, such as an uranium mine or a nuclear facility. Various authors[3], reports of IAEA[4], NEA[5] and OECD[6], described the environmental impact of mining projects of European and other countries. According to their analysis, 'in the past, in European countries, it was common to see mines and cities and their activities entangled in a small territory. This has been the case, of the Ruhr basin in Germany, Loire in France, and West Midlands (Birmingham) in UK, to cite just a few very old mining regions. In recent years, the regulations and concerns with environmental resources lead to mitigation of pollution and remediation of the environmental impact. One good example is the Erzgebirge (Ore Mountains) in Saxony, Germany, near the border with Czech Republic, where first silver and associated base metals and later uranium was extracted to supply the URSS nuclear programme. With Germany reunification, in 1991 started a large environmental remediation programme that involved clean-up of contaminated areas, radioactive waste confinement, installation of mine water treatment plants, and rebuilding the landscape and villages. The state-owned former mining company Wismut GmbH has been implementing this work. Remediation was needed because the exposure of large human communities to radiation and radioactive waste is not acceptable according to current radiation protection standards i.e., IAEA, 2014. Similar scenarios of exposure to harmful contaminants (uranium or other metals) exist in many regions including west Europe where, due to dense occupation of territory and miners' settlements near the former mines, almost every mine has villages or cities nearby. In some regions, the cost of environmental remediation of mining legacies has been high and paid by public funds because the former mining companies do not exist any longer. For example, the environmental remediation cost of the Saxony areas in Germany amounts already to 65 billion Euros, the cost of remediation abandoned mine sites in Portugal is of 139 million Euros, and the project to relocate and confine the uranium mining waste in Colorado (MOAB project) is budgeted at one billion dollars. The value of a healthy environment and control of soil, air, and water quality justifies the cost of these environmental remediation works and recuperation of contaminated areas. With the move of mining activities to other continents and remote areas, some unforeseen exposures did occur. For example, uranium mining in Niger started in the Sahara desert where no settlements existed and mining companies built the mining infrastructure from scratch. Because it was the desert, there was no population critical group to protect and, in the absence of water in the surface, mining impact was foreseen as minimal. However, in a short time, the access to water from the wells drilled by the mine and the opportunity to recycle scrap materials (metal, timber, plastic) attracted people. Rapidly a city grew around the mine site, counting today 80,000 inhabitants. Agriculture, using wastewater from the mines developed also to supply vegetables and fruits to the miner's families, and this way a large potentially exposed critical group was born in the desert. All these artisanal and agricultural activities created pathways for the transfer of radioactive contaminants and radiation exposure of population, requiring measures to reduce exposure to radioactivity. Mining outside Africa, for example, in Indonesia, China, and Peru, has also developed in situations of elevated risk to humans and the environment, with large mines surrounded by cities where conventional safety and prevention of contamination and exposure to toxic metals was not ensured. Remediation of mining environmental impacts on these ecosystems might be nearly impossible at the present'. Carvalho observed that, mining impacts always existed and we still have mines opened in Roman times with acid mine drainage, waste piles, and environmental contamination. Large-scale mining grew during the nineteenth and twentieth centuries with impacts neglected and hidden costs. Impacts on miners' health due to occupational exposure to toxic contaminants (lung cancer, silicosis, lead poisoning, etc.) were not accounted for, for a long time. Until very recently, environmental impacts on nature and on populations at large were not duly considered either.

In India, The Energy Research Institute (TERI, 1997) undertaken a study in around 520 sq. km of Goa. It was revealed that out of 3,350 ha. under mining and associated activities, 1,300 ha. were under dumps and it resulted in widespread environmental degradation. There was evidence of air and water pollution in much of the mining area and the surrounding villages, especially a high concentration of suspended particulate matter. Wastewater and runoffs from dumps containing both dissolved and suspended solids pollute the rivers, streams and other water sources throughout the year but adequate control measures are lacking. In Singhbhum district of Jharkhand, it was reported that the coal towns of Megahaluburu, Kiriburu, Gua and Noamundi with a population of some

20,000 persons lack drainage and dump their effluents in the surrounding villages and sewage in the rivers and rivulets, thus polluting them.[7] Also studies in Chhattisgarh show the same trend. The environmental degradation is an ongoing process and has an impact on people. The dust, noise and water pollution force the residents to leave the area when their health deteriorates or land becomes infertile. The number of such indirect DPs is substantial but no way has been found of counting them. Besides, legally they move out of their own volition so they are not considered DPs though the project forces them to move out.[8] For example, in the Talcher coalmines in Odisha, many residents told that they were planning to move out because they could not live amid the dust and the noise of constant explosions.[9] The second process resulting in environmental degradation is the transition to destructive dependence. Communities that have for centuries used the forest, water and land resources in a sustainable manner, change their culture because of impoverishment. For sheer survival, they make a transition from constructive to destructive dependence on the source and destroy or overuse it. For example, a majority of the tribal and other forest dwellers have started cutting trees for sale as firewood. Fishing communities have given up traditions that restricted fishing in certain seasons or controlled the extent of fishing.[10] Jharkhand is one of the richest areas in the whole country, rich in minerals with huge reserves of coal, iron ore, mica, bauxite and limestone and considerable reserves of copper, chromite, asbestos, kyanite, china clay, fire clay, steatite, uranium, manganese, dolomite, tungsten, gold, etc. The distribution of these minerals is mostly localised. Coal is found in the gondwana rocks of the Damodar basin. The well known coalfields of the area are Raniganj, Jharia, east and west Bokaro, Ramgarh, and south and north Karanpura. If Damodar valley is the repository of coal resources, Singhbhum is a veritable museum of non-fuel minerals with huge deposits iron ore and sizeable deposits of copper, uranium, chromite, asbestos, kyanite, china clay, manganese, etc. The Kolhan series of iron ore in Singhbhum is one of the richest iron belts in the world. The Singhbhum copper-uranium belt contains the largest deposits of copper and uranium, in the country today. The world's largest deposit of high grade kyanite occurs at Lapsa Buru in the Singhbhum shear zone. Mica is an important mineral for whose deposits the Jharkhand area leads all countries in the world. The opening of coal mining in Dhanbad area during the second half of the nineteenth century and the establishment of the Tata Iron and Steel Company in Jamshedpur in Singhbhum district in 1907 marked the beginning of the large-scale exploitation of mineral and other industrial resources in this area. Coal, however, is the biggest mining industry of Jharkhand. The large-scale mining and allied activities going on in the Jharkhand region have caused severe damage to the land resources of the area. Vast areas of rich forests and agricultural lands belonging to the indigenous people have laid waste because of haphazard mining. Underground mining operations, especially of coal, have created unsafe surface conditions in many areas warranting diversion of roads, railway lines, etc. and the shifting of a number of townships. This large-scale mining in the region have adversely affected ground water table in many areas with the result that the yield of water from the wells of adjoining villages has drastically reduced. Further, effluents discharged from mine sites have seriously polluted the streams and underground waters of the area. Mining and allied operations are causing serious air pollution in the region. Open cast quarries, coal washeries, thermal power plants, coke-oven plants, cement factories, fertiliser plant etc, contribute to serious air pollution. The large-scale exploitation of the natural resources of the region through the development of mines, industries, etc. has adversely affected the indigenous people of Jharkhand. They are systematically and methodically being dispossessed of the ownership of their means of production, of the products of their labour and of the very means of human existence. They are dispossessed of their political autonomy and their communities broken up. Similar is the situation with other States. In the name of 'development' for national interest, the Jharkhand areas are witnessing not development but the misery of its people and of its natural wealth through a process colonialist and capitalist exploitation. The brutality inherent in the process of industrialisation- the plundering of its mineral wealth, and the decimation of its forests which provided much of the livelihood for its people, has not only reduced the majority of its inhabitants to destitution but has also brought the area to the brink of an ecological disaster.

Minerals are considered national resources and sub soil rights over lands belong to the state and not to the individual owners. The powers of eminent domain, which are enjoyed by the state while acquisition of lands for mining, positions local communities in totally helpless situations. Although apparently, there are complex procedures for acquisition of lands and for obtaining mining leases for exploration, prospecting and extraction, in reality, mining projects are sanctioned more with the view of encouraging and catering to the interests of the mining industries rather than to protect the interests of the local communities and the rest of the natural resources which exist in a mining area apart from the minerals, as reflected in the New Mineral Policy of 1993. The shifting focus of India's economic reforms are reflected in the series of recommendations to the Mineral Policy as well as to push for amendments in other Acts which are related to mining activities like the Labour laws, the Coal Nationalisation

Act, The Fifth Schedule, the Land Acquisition Act, the Forest Conservation Act, the Environment Protection Act, the PESA and others, shows an increasing disrespect of the state towards the rights of communities. As communities have no access to project information, they have to rely on the revenue officials or on the local politicians, to inform them of the extent of land proposed for occupation. In many cases lands and forests and common properties are acquired far more than sanctioned for the projects. This blatant violation was exposed in one of the mining projects of the Indian Rayon's and Industries in the scheduled area of Andhra Pradesh where the area occupied by the company was greater than the area actually sanctioned as per the government order. It was only with the information gathered from government orders and circulars, which were unearthed by the local movement by the Adivasi women who physically stopped the company from entering their land that prevented further encroachment by the company.

Coming to the case of Odisha region, another major mining state, has the dubious distinction of clearing the maximum amount of forest land for mining in the country. Of the total forest land cleared for mining in India, Odisha accounts for 17 per cent. The state's water resources are as stressed, contrary to the belief that Odisha is water- surplus. The state's hilly terrains, with their natural springs, are being destroyed by mining, contends the CSE report. Odisha's second largest river, the Brahmani, is one of the 10 most polluted rivers in India – "due to the large-scale mining operations on its banks". The state's six million strong tribal population has borne the brunt of these environmental impacts, says CSE, which has displaced about 500,000 people (mostly tribals) in the state. The horrors of land acquisition are experienced by mining affected communities in a similar manner or perhaps more severe in nature, than that of dams and other big projects. When mining leases are given, the immediate threat to local communities is displacement and land alienation, which are inevitable features especially with regard to large-scale mining. The abrupt disruption and dislocations of livelihoods by massive dispossession and displacement of population in the mining sector in the back drop of an agrarian economy has accentuated poverty and marginalisation of the State. The area under tenancy in Odisha is higher than the national average. As many as 85 percent of ownership holdings were marginal with less than a hectare owned by them during 2003 and hence, Odisha's agriculture is 'small holders agriculture'. The large and medium farmers account for barely one percent in the state owning around 11 percent total areas as indicated in Table no. 7.1.

Table 7. 1 : Ownership Holdings in Odisha and All India

State	Year	Percentage of house holds						Percentage of Area owned					
		Mar-ginal	Small	Semi medium	Medium	Large	All	Marginal	Small	Semi medium	Medium	Large	All
Odisha	2003	85.50	9.70	3.70	0.90	0.10	100	41.52	27.06	19.72	9.98	1.78	100
	1992	75.15	14.42	7.34	2.40	0.12	100	26.37	27.16	25.99	18.08	2.40	100
	1982	66.06	20.84	9.31	3.42	0.37	100	19.88	29.73	25.04	19.50	5.84	100
All India	2003	79.60	10.80	6.00	3.00	0.60	100	23.05	20.38	21.98	23.08	11.55	100
	1992	71.88	13.42	9.28	4.54	0.88	100	16.93	18.59	24.58	26.07	13.83	100
	1982	66.64	14.70	10.78	6.45	1.42	100	12.22	16.49	23.58	29.83	18.07	100

Sources: NSS Report No. 491: Household Ownership Holdings in India, 2003.

There has been a significant increase in the area under tenancy between 1981-82 to 2002-03 in the State. It increased from 9.48 percent to 13.15 percent whereas at the all India level, it recorded a declining trend as evident from Table 7. 2.

Table 7.2: Incidence of Tenancy in Rural Odisha and India, 1981-82 to 2002-03

State	Year	EOH	MH	ELIH	NONLIH	OWA	LIA	NON-LIA
1	2	3	4	5	6	7	8	9
Odisha	1981-82	74.27	24.06	1.49	0.18	83.97	9.92	6.12
	1991-92	70.46	17.69	1.70	10.15	84.41	9.48	6.11
	2002-03	77.60	12.00	8.40	2.00	83.78	13.15	3.07
All-India	1981-82	80.58	16.24	2.37	0.81	91.08	7.18	1.74
	1991-92	81.98	8.87	3.85	5.30	87.91	8.28	3.81
	2002-03	88.60	6.99	3.06	1.35	92.57	6.60	0.84

Sources: (i) Report on Landholdings (2) 27th Round 1982, NSS Report No. 331 (ii) Report on Some Aspects of Operational Holdings; 48th Round 1991-92, Report No. 407 (iii) Report on Some Aspects of Operational Holdings, 2002-03, 59th Round, NSS Report No. 492.

(Notes: EOH- Entirely Owned Holdings; MH = Mixed Holdings: ELIH= Entirely Leased-in Holdings; NONLIH= Neither Owned Nor Leased-in Holdings; OWA = Owned Areas; LIA = Leased-in Area; NONLIA = Neither Owned Nor Leased-in Area.)

District-wise (erstwhile undivided) analysis of area leased into total operational area from 1980-81 to 2000-01 reveals that it increased substantially from 1.27 percent to 7.52 percent as indicated in Table 7.3:

Table 7.3: Area Leased-in to Total Operational Area in district-wise from 1980-81 to 2000-01 (in percent)

District	1980-81	1985-86	1990-91	1995-96	2000-01
Balasore	2.53	1.23	2.21	15.02	19.31
Bolangir	0.22	0.32	0.04	0.78	0.39
Cuttack	2.58	4.35	0.53	13.95	23.10
Dhenkanal	1.19	3.22	0.43	6.10	8.17
Ganjam	3.15	1.87	0.12	8.59	9.87
Kalahandi	0.17	0.65	0.22	0.63	0.57
Keonjhar	0.64	0.20	0.33	1.88	3.73
Koraput	0.47	0.21	NA	1.90	1.67
Mayurbhanj	1.17	1.93	0.52	4.06	5.80
Phulbani	0.73	0.43	0.04	1.93	2.15
Puri	1.74	1.23	0.09	7.74	11.84

Sambalpur	0.90	1.25	0.75	3.12	5.33
Sundergarh	0.33	0.23	0.25	1.41	1.63
Odisha	1.27	1.45	0.45	5.60	7.52

Sources: Report on Agricultural Censuses of Odisha, Various Years.

It may be noted that Cuttack district has registered highest percentage of area leased into total operational area (20.5 percent) followed by Balasore district (around 17 percent) and Puri district (10 percent) and all these districts belong to coastal areas of the State and are not mineral rich districts. Mineral reserves are more in northern and southern districts of the State. There has been glaring regional imbalance/ Spatial unevenness in the levels of development of the coastal districts and the northern and southern districts and in later case; there is a greater concentration of scheduled castes, scheduled tribes and also greater concentration of Poverty, malnutrition and illiteracy. According to Banikantha Mishra,[11] to delve deeper into the relative importance of Agriculture, Industry, Mining, Quarrying and Manufacturing, we analysed the change in per capita district domestic product for four sectors like Agriculture, Industry, Mining and Manufacturing between 1993-94 and 2001-02. Agriculture has suffered in all districts. That mining (including quarrying) shows higher increases in all no mining districts implies that the mining is quietly spreading across uncharted areas of the state. That mining districts have the best show in registered manufacturing is puzzling, since, in Odisha, value addition and industrial use of minerals take place outside the minerally rich districts. Observing the preponderance of the mining sector in Odisha, it is imperative to analyse the sector in more detail. Overall, during 1993-94 and 2003-04, the rate of mineral exploitation in Odisha has increased – on the average per year – by 10.3 percent, while its value has gone up by 12.8 percent. Export of minerals in quantity terms has gone up by 15.7 percent, higher than the increase in exploitation rate, implying that a smaller percentage of the exploited minerals are used now to meet domestic needs. The value of exports has also increased. Despite all this, this capital-intensive sector has been ruthless on employment, reducing the number of workers by an average 4 percent annually. We discover some chilling truths when we focus on the most important minerals, namely, iron ore, chromite, coal, and bauxite. The average annual rate of exploitation (as a percentage of reserves at the beginning of the year) looks relatively low for all the four minerals. But when we ferret out the rate at which the quantity exploited has increased annually, the figures are quite high, from 7.3 percent for bauxite to 15.9 percent for iron ore. If this growth rate in exploitation is maintained, the state's existing iron ore and chromite reserves would get exhausted in 20 years and coal and bauxite reserves in 50 years. When we look at the employment-generation by these activities, we find an across-the-board fall in the number of workers. The sole exception is the number of workers engaged in bauxite mining, which has gone up by around 6 percent per year. Bauxite accounts only for 1 percent of the total mineral workers employed, whereas coal, iron ore, and chromite, the big employers in the mining sector, employ anywhere between 13 percent and 35 percent each of the total mineral workers. The fall in the number of workers per lakh tonne employed highlights the increasing mechanisation in this sector. The value of goods exported, brings to the fore the growing importance of mining and the dwindling importance of agriculture. Whereas exports of agriculture and forest products from the state fell by an average of around 20 percent a year during 1993-94 to 2003-04, mineral and metallurgical export registered a surge. That Odisha has been pushing its mining sector in a big way is well documented by analysing its forest resources. The total forest area went up in mining districts while that in KBK and industrial districts it fell (posssibly because they are also getting industrialised), but the non-industrial districts exhibited improvement. In fact, if we look at forest area diverted for non-forest use, it went up from 789 hectares at the end of 1993-94 to 28,769 hectares as at the end of 2003-04, an average annual increase at the rate of 43 percent. Mining accounted for one-third, irrigation one-fourth, transmission lines one-ninth, and industries one-twelfth of diversion of the 28,769 hectares. According to figures released in early 2010 by the Ministry of Environment and Forests, Odisha has one of the highest rates of diversion of forest land. In fact, the state leads the nation in diversion of forest land for mining during the last three years.[12] Against this background, one is left wondering whether the case filed in the Odisha High Court by the Society of Retired Forest Officers, challenging the automatic rights of tribals on forest land due to its concern about interests of the forests and ecological security has also raised concerns about diversion of forest land for use in mining.[13] Besides, out of the total land of 79,339 hectares allotted for mining leases as on 31 December 2005, more than 50 percent is forest area. Fortunately, as per the Directorate of Mines, only a quarter of that forest had been officially allowed for diversion as of the end of 2005. Looking at specific minerals, it is found that iron ore accounts for the highest percentage of land allocated towards mining by the end of 2005. B. Mishra further maintained that even in the Short Run, 'We Would Be All Dead'.[14] Resource exhaustion is also a big threat. As we have seen above, if the current growth in the exploitation amount is maintained, bauxite and coal reserves would last for around 50 years, but iron ore a paltry 20 years! Some were, however, more optimistic and estimated that, if all the existing MOUs were converted into projects, the bauxite reserve would last around 50 years, coal 100 years, and iron ore 25 years. Some of the above worries, especially about iron ore, may sound unfounded. More recent data would, however, put things in perspective. The 27.33 percent growth in Odisha's iron ore exploitation during 2002-03-2007-08 (which can be derived from figures given in various issues of *Economic Survey of Odisha*) is much higher than the 16.65% growth in India's production during 2003-08, which was itself the second highest in the world, next only to China.[15] Some other iron ore figures for India and Odisha may be equally of concern. Export of iron ore by Odisha during 2002-03-2007-08 went up by 29.11 percent, more than the rate of increase in its exploitation. Among the countries of the world, production in 2007 as a percentage of the reserve base is the highest for India at 5.17 percent, conspicuously above the second highest figure of 3.06% for Brazil, while, in exports as a percentage of the reserve-base, it stands second, marginally behind Brazil, thanks to India's low exports-to-production ratio. Now, even as per the Odisha government data, as of 3 November 2005, MOUs had been signed for standard and mega steel projects for production of 21.94 and 36.10 MTPA, at an investments of Rs 28,570.18 crore and Rs 1,08,586.17cr., respectively. To appreciate the magnitude of this expansion, one merely needs to reckon that the 58.04 MTPA total additional capacity envisaged in these MOUs was around 1.26 times the state's 2004-05 iron ore exploitation of 46 million tonnes (MT) and 78% of its 2007-08 figure of 74.50 MT. Even industrialists like Naveen Jindal echo the earlier concern and believe that the "13-billion tonne haematite ore available in the country would be exhausted in 20 years, going by the 10% growth of the steel industry" (Nagaraj; 2006). Concern by the Indian Bureau of Mines does not differ too much. "Again, if 43 steel projects who have signed MOUs with state government will start their production, then nearly 90 MT additional iron ore will be raised every year, and this will rise to 133 MT of iron ore raising. So also, it is expected that, if the iron ore raising continues at this rate, then in the next 25 to 30 years, the resources will be exhausted" (*New Indian Express* 2006). Others are still more "optimistic" about iron-ore and believe that, if "exploitation of iron ore continues at this rate, given the iron ore mining capacity and assuming that no new mine is explored, Odisha's iron ore reserve will be completely depleted after 40 years".[16] The sops that governments dish out to corporations to attract their investment do not end with low priced raw material and cheap and excess land. But, these "concessions" do not always escape scrutiny.[17] The Odisha High Court had pulled up Odisha's IDCOL board for failing to protect the state's interest by giving away Rs 20,000 crore in incentives to

Jindal for exploration of Tangarapada mines. At a macro level, the RBI has sharply criticised Odisha's SEZ policy for offering "to exempt all taxes, levies, stamp duties, registration charges, and electricity duty for 20 years for the industrial units coming up in the zones".[18] "Odisha is minerally endowed State. So, it is imperative to look at the values of its mineral production, along with that of the other states. Odisha has the highest average annual growth of the value of production from 1994-95 to 2007-08. Besides, as per *Indian Minerals Yearbook 2005*, published by the Indian Bureau of Mines, Odisha is one of two states in our sample that had leased area per mine in 2003-04 in excess of 150 hectares. So, it is worthwhile to analyse how Odisha fares in this sphere vis-à-vis the whole country. Though Odisha's 2003-04 exploitation rate is less than its share in India's reserves, this share has drastically fallen in bauxite. This clearly suggests that Odisha is exploiting bauxite at a much faster rate than it should be doing. A comparative analysis of the value of minerals produced in Odisha and India lead to a similar conclusion. From 1994-95 to 2006-07, the value of non-metallic minerals (which include graphite, limestone, etc) produced in the state has increased by eight percent per year, as against a slight decline in the country. But, in fuel minerals (coal, petroleum, etc) and metallic minerals (bauxite, chromite, iron ore, etc), Odisha is well ahead, leading to a hugely increased share in the country's production. That is why the overall average- annual-change for Odisha during this period is around a high 16 percent, almost four times as much as the all-India figure. That also explains why Odisha's share in the total value of minerals produced in India went up from 4.2 percent to 15.7 percent during this period. Policymakers in the state seem to have confused between the state's static and dynamic comparative advantage in mines. Even if mining were to be allowed, the purely economic logic of overexploitation of mineral resources and the flurry of private (domestic and foreign) investment in mining and mineral-based industries in Odisha in the post-liberalisation era is the unlocking of the state's natural (and, hence, static) "comparative advantage" by removal of the freight equalisation subsidy. The resultant market forces have made the prices of minerals of Odisha supply-determined, which has driven prices to zero in a notional sense, resulting in manipulative high supplies at some positive price. Thus, the focus has been on maximising the revenues in the short run (for a few years) rather than maximising the present value of "all" expected future revenues. A principle of preservation and containment of social and environmental costs must not allow an output-to-reserve (proven) ratio of more than some "sustainable" benchmark, so that the mineral resources last for a very long time. Then, Odisha can bargain for a price that would be closer to world market prices and reflect the true economic, social, and environmental cost of the state's mineral resources. This will then be the truly "market clearing" price by eliminating whatever artificial "excess demand" there is. If the static comparative advantage is thus taken care of and not thrown to the wind (as some oil-rich African countries like Nigeria did), this can – though need not necessarily – be turned into a "dynamic comparative advantage" through proper public investment policies to address pressing social deprivation issues. If, on the other hand, the cost principle, followed in the case of mineral resources (non-renewable) of Odisha, enjoying a natural comparative advantage, is based on its static aspect and hence is short-sighted, rather than its true cost based on the principle of preservation and social ethics, then, given the environment of "poverty and deprivation" in the state, the oft-quoted adage of "poverty amidst plenty" could well be destructive."[19] The new wave of mining projects planned for Odisha, starting with Utkal, Vedanta and Posco, represent huge injections of Foreign Direct Investment. Since the new National Mineral Policy brought in 1993, the share of mining ventures that can be owned by foreign companies has risen from 26 percent to 74 percent, and lobbyists are trying to get this raised to 100 percent. This is one way that profits from these projects are streamlined now for taking out of India. Foreign Banks which have invested in Vedanta and other joint ventures are becoming right now the effective owners of some of Odisha's richest mineral deposits, with a view to rapid extraction, and minimal checks on the consequences for the environment and local people. It is well known that the main profits from FDI really go abroad,[20] and in mining, the foreign companies are actually gaining control of India's most precious resources: its mineral deposits and flowing water. From bauxite's natural placing in the earth it ensures the land's fertility. Damage from mining the bauxite cappings on south Odisha's greatest mountains is irreversible, both for nature and for indigenous communities. As Kishan Pattnayak,[21] put this: 'the first step of mining is displacement. In Odisha, there is not a single successful story of rehabilitation by the government of displaced families, who otherwise have been surviving on the natural resources of the area by living peacefully since hundreds of years on their own land.' Odisha state pollution Control Board has attempted a holistic appraisal of the all round eco-degradation brought about by Odisha's mining sector. The large mines and the beneficiation plants/washeries, waste dumps and effluents discharged increased pollution in and around the mines. The mines mostly being located in forest areas are responsible for increased deforestations. For opencast mines, besides air and water pollutions, it involves excavations of land, loss of soil, degradation/disfiguring of surface area and deforestation.

Valuable topsoil and usable land is lost forever, which cannot be fully recovered, thus overall affecting the land use pattern. Natural drainage system and ground water table is disturbed which not only affects the inhabitants in close proximity but also people living in faraway places. In Keonjhar, the findings of a study show that the land use pattern surrounding the mines area are changing rapidly. The major sufferers of this change are forest and agricultural land. Apart from the above two, another important finding of the study is the sharp increase in wasteland which is putting negative implication on livelihood of tribal. Apart from environmental degradation, this resource extractive development process has put a negative impact on the tribals inhabiting surroundings of mining. The local economies that used to depend on agriculture, forest produces and traditional occupations like fishery, handicrafts etc are on the verge of collapse or extinction.

A study by Sills and others in the Iron ore mining district of Keonjhar, Odisha shows that both the forest stock and forest diversity proxied by 'Percent forest around village' and 'percent of forest species collected' respectively, are significantly correlated with the distance to iron ore mines, with correlation coefficients of 0.25 and 0.68, respectively. These correlation coefficients indicate that both the quantity and quality of forest increase with distance to iron ore mines. Results on Non Timber Forest Product (NTFP) shows that, households farther from mines obtain more benefits from forests compared to those living closer to the mines.[22] The Barsua and Kolta iron ore mines and their allied activities have destroyed prime forests and wildlife corridors in the Bonai hills of Sundargarh district.[23] It is apprehended that open-cast mining would lead to the disappearance of the streams and pollute the major drinking water sources for the tribal people.[24]

Positive Impacts

Despite the adverse impacts of mining elaborated in foregoing sections, mining has some positive contributions on environment as discussed below.

Contribution towards a low carbon economy

The importance and imperative of low-carbon economic development has been widely acknowledged globally in recent years. Reducing the environmental impacts of mining activities is recognised as a key component of low-carbon economic development. ICMM in a series of publications has described that the mining and metals industry is well placed to contribute to the resolution of the climate change challenge. Many mining and metals companies are measuring, managing and reducing their own greenhouse gas (GHG) emissions. More significantly, the industry is also supplying the materials needed to build a low carbon future. Lowering GHG emissions is a considerable challenge for many industry sectors. In the future, factors such as rapid population growth, high levels of industrialisation and continued economic development in a number of regions will make this all the more challenging. Such future trends are expected to increase aggregate and per capita consumption levels of minerals and metals as increasing population and development translates into rising demand for goods, infrastructure and housing. Minerals and metals are integral inputs to human development and the advancement of society. The industry already plays a critical role in supplying the inputs for current technologies and production practices. Similarly, it will provide vital inputs for fully realising GHG emissions abatement opportunities across all sectors of the economy in the transition to a low carbon future. In addition to managing their own emissions, the mining and metals industry also makes a broader contribution in providing the minerals and metals required for the transition to a low carbon economy. Many abatement options have the potential for reducing global GHG emissions, including reducing consumption, low carbon energy technologies and improvements in transportation systems and building infrastructure. Energy efficiency initiatives in the mining and metals industry are already driven by cost savings opportunities and security of supply considerations. Emissions management is another driver. Well-designed policies will encourage further energy efficiency improvements across all sectors of the economy in the transition to a low carbon economy.

Promotes Green Mining

Joshua Kirkey defines green mining as "technologies, best practices and mine processes that are implemented as a means to reduce the environmental impacts associated with the extraction and processing of greenhouse gases, used water and minerals".[25] In short, the development of innovative green mining technologies aims to improve the mining sector's economic and environmental performance simultaneously. Mining experts like Ying Wang, Yuxuan Lei and Shuyi Wang[26] of China University of Mining and Technology defined a green mining system as a synergetic operation system composed of a mining subsystem and an environmental governance subsystem. Through conducting a case study of China's coal enterprises, using the 2019 data collected from Chinese coal mining enterprises based on the constructed indicator system, they identified the mining subsystem as the first production stage and the environmental governance subsystem as the second production stage. To evaluate green mining efficiency, an entire green mining efficiency indicator system was constructed by analyzing the main inputs and outputs of the two subsystems. Data Envelopment Analysis (DEA) model was used to assess green mining efficiency in terms of mining efficiency and environmental governance efficiency. According to this empirical study's results, the exported parameters information from the two-stage combined DEA model supports green mining efficiency improvement as quantitative evidence. Specifically, the authorities should formulate regulations and establish standards to encourage the application of green mining technologies and simultaneously curb the use of old mining methodologies at the coal enterprise level. The government should also foster green mining technology innovations through cooperation mechanisms by forging efficient operation frameworks organised by government sectors, commercial corporations, and coal enterprises.[27]

Conclusion

Mining activities should not pose a potential threat to the health and livelihood of the poor and vulnerable groups who have little mobility or means of alleviating negative impacts. The harmful impacts of mining should be addressed through development of new technology that uses fewer chemicals during extraction and processing. That apart, adequate and sufficient rehabilitation provisions should have been guaranteed for the rehabilitation and relocation of the displaced families including indigenous communities. Moreover, mining companies should always endeavour to create a win-win situation for all stakeholders. However, it is important to clarify that, in spite of experiencing its share of environmental and health related problems that adversely impact human quality of life, small-scale mining plays a pivotal role in alleviating poverty in the developing world and contributes significantly to national revenues and foreign exchange earnings. These important socio-economic contributions make small-scale mining an indispensable economic activity. Countries should, therefore, take some steps to obtain the greatest benefits from mining. They need to monitor the social and environmental impacts of communities and regions affected by mining activities and in particular, by mine closure. Countries can establish a sound mineral regulation and licencing system for commercial-scale mining. This involves creating "a level playing field," with ease of entry and responsible exit, sound fiscal policy and avoidance of subsidised state-owned mining enterprises (SOE), or, if they already exist, the privatisation of SOEs. They can ensure sound macro-economic policies so that mineral-rich countries benefit from the developmental impact that mining can have, rather than seeing non-mining sectors obstructed and opportunities wasted and attract private-sector investment through appropriate laws and regulations. Questions of ownership, land and water use, socio-environmental standards, and responsibilities need to be reliably clarified and appropriately implemented and monitored. Moreover, it should encourage planning for mine closure, already upon the first investment by a mining company, by supporting the build up of local administrative and management capacity and by designing and implementing appropriate regulation and oversight for mine closure. Besides, countries can mitigate health, environmental, and socio-cultural risks for the population engaged in mining through communication, information and education, as well as through agreements with the project authorities.

End Notes & References

1. Mannylumanao, *"The environmental effects of irresponsible mining industry."*
2. *Ibid.*
3. (i) John Wiley & Sons Ltd. and the Association of Applied Biologists, publishers, *"Food and Energy Security"*, 2017, (ii) F. P. Carvalho, *"Mining and Sustainable Development."*
4. International Atomic Energy Agency, (IAEA), 2005.
5. National Environment Agency, (NEA), 2014.
6. Organisation for Economic Cooperation and Devlopment (OECD), 2014.
7. George, 2002; p. 17.
8. Ganguly, Thukral, 1999.
9. Fernandes and Raj, 1992.
10. Bharali, 2007.
11. Banikantha Mishra, p. 16.
12. Dharitri, Odia daily, 2010.
13. New Indian Express 2008.
14. Mishra, B.K.- Op.Cit.
15. Ericsson, Magnus and Anton Löf; 2009.
16. Das, A. 2005: p. 4680.
17. Economic & Political Weekly May 15, 2010, Vol. xlv, No. 2061.
18. New Indian Express 2009.
19. Mishra, B.K.
20. Woodward, 2001.
21. Kishan Pattnayak in social worker in Orissa.
22. Naik, 1993.
23. Vagholikar, Moghe and Dutta, 2003.
24. Das, 2005.

25. Joshua Kirkey was the Communications Advisor for Natural Resources, Canada (NRC), 2014.
26. Ying Wang, Yuxuan Lei and Shuyi Wang, "*Green Mining Efficiency and Improvement Countermeasures for China's Coal Mining Industry*", School of Management, China University of Mining and Technology, Beijing, China.
27. *Ibid.*

Chapter – 8

Mining Induced Displacement and Rehabilitation

Mining plays a crucial role in economic development of a nation but with a social cost, the cost being mining induced never ending social problems. Several authors have commented on the potentially-adverse social impacts of mining, which include displacement of local people from ancestral lands, marginalisation, and oppression of people belonging to lower economic classes. Mining, however, has multifarious adverse impacts on the people and their socio-cultural set up. The negative impacts are in the form of dispossession of land, involuntary displacement, unsatisfactory resettlement and rehabilitation (R&R), impoverishment of Project Displaced Families, health hazards, woes of women and indigenous tribal communities etc. though it has some favourable effects on the society and social structure. Mining-induced displacement and resettlement (MIDR) is an important issue and successful resettlement and rehabilitation of displaced people is a major challenge in Sustainable, Inclusive and Responsible Mining. MIDR, that was already a problem has become a major risk from the point of view of social sustainability. An effort, therefore, has been made to look at MIDR in the past and future trends internationally, nationally and regionally. Global scenario of MIDR, India's experiences on land and displacement due to mining projects, MIDR situation in regional perspective with an experience of a mineral rich state as well as other outstanding social issues have been analyzed in succeeding paragraphs of this section.

Dispossession of Land

The land, much of which are used for mining is Common Property Resources (CPRs) while Indian land laws going back to the colonial age recognise only individual ownership. The CPR dependants, who have lived on that land for centuries before this law was enacted, are considered encroachers. Many of them are thus evicted and are not counted among the displaced. The Land Acquisition Act of India, before last amendment, was draconian and obsolete and gives over-riding powers to the State to encroach onto people's lands for any 'public propose' including mining. In most mining situations, communities become aware of projects only at the time of eviction when the bull-dozers are brought in many times with police presence. The public hearings are organised in a biased and almost discreet manner where the local administration, the statutory bodies for clearing the projects and the mining companies are in close collusion to ensure that there are no effective objections raised by the public. Most mining companies also operate on a deemed consent basis so that even without the mandatory clearances, mining operations are expanded beyond lease areas and periods. Landlessness takes the form of downward mobility from large to medium landholders, from medium or small to marginal and small and marginal farmers to landless. It has a caste-tribe as well as a gender dimension. In Odisha, for example, among the mining displaced families, 16.7 percent of the tribal and 13 percent of the Dalit DPs became landless against 3.6 percent of the general castes. In AP, among all the DPs including those of mining, landlessness rose from 10.9 to 36.5 percent and in Assam from 15.56 to 24.38 percent.[1] The average area cultivated declined, for example in Assam from 3.04 to 1.45 acres per family. Also support mechanisms such as ponds, wells, poultry, cattle and diary animals that supplement the farm income decline. According to Fernadeses and Bharali, Joblessness takes two forms viz. lower access to work and downward occupational mobility. In AP, the proportion of cultivators among the DPs/ PAPs declined from 83.72 to percent 41.61 percent. Besides, 45 percent of the cultivators became landless agricultural labourers or daily wage earners. In West Bengal, access to work declined from 91.02 percent to 53.18 percent and in Assam from 77.27 percent to 56.41 percent. 50 percent of the cultivators became daily wage earners or domestic or other unskilled workers. In Odisha, access to work among the mining DPs declined from 89.3 percent to 57.5 percent. The shift was much greater among the tribal DPs because very few of them own *patta* land, so Coal India Limited (CIL) did not give them a job. In Odisha and Chhattisgarh, much of the mining land is CPRs. In Odisha, according to Fernandes and Asif, as of 1997, they are 30 percent of the land used for coal even in the non-tribal Talcher area of Angul district where the proportion of the CPRs is low. They are nearly 60 percent of the land on lease for non-coal mines in the tribal areas. In AP 30 percent of the 1,22,301 acres used by mines are CPRs. Most coalmines of West Bengal are in its tribal areas, so the CPR proportion is high. Its exact extent could not be got but it is clear that very little of it is private. For example, the Gazette notifications show that out of 13,456.46 acres of private land notified 1947-1990, only 2,900.38 acres are of 1981 to 1900. The rest was acquired 1947 to 1980. The remaining 27,000 of the 30,000 acres used for mining are CPRs.

Land under Mining

The extent of mining land used in the some mining States is shown in Table 6.1 which gives the type of land used in the States from which information is available, most of it of the 1980s and 1990s and some from the 1970s. Since mining began long before it, the area under the control of mining companies is much bigger than this, probably thrice as much. It also means that much of the land under their control belongs to the abandoned mines category. But as the Goa experience shows, the lessees are not ready to acknowledge it. Instead, fertile paddy land has been acquired in the State in recent years on both sides of the Konkan railway. If all its plans go through, Goa may acquire 7.2 percent of its landmass for new projects when it is possible to use the abandoned land for new industries. Besides, two-thirds of the land used in most States is tribal commons or forests. Rehabilitation of mining DPs is lower than that of the others. One sees no reason why abandoned land should not be used to rehabilitate their DPs. As the Table 8.1 shows, among six States, the highest area of land has been used for mining in Jharkhand (2,08,552.47 ha.) and lowest in Assam (11,394.75 ha).

Table 8.1: An Estimate of Land under Mining in the 1990s in some States (ha.)

Sl. No.	State	Private	Common	Forest	Others	Total
(1)	(2)	(3)	(4)	(5)	(6)	(7)
1.	Andhra	22019.43	14359.11	990.28		37368.82
2.	Assam	42.70	3196.17*		8155.88	11394.75
3.	Goa	18066	18818	17316		54,200
4.	Jharkhand	74562.35	63296.03	70694.09		208552.47
5.	Odisha	27663.38	67465.09	70918.87		166047.34

6.	W. Bengal	10121.56	1440.32*			11562.38
7.	Total	155810.32	168575.32	159919.24	8155.88	492460.76
8.	%	31.17	34.47	32.69	1.67	100.00

Source: Lok Sabha Unstarred Question No. 5410, dated 15.05.2007.

Courtesy: Table from Doc Centre, Indian Social Institute, New Delhi.

*What is not Tenancy land is Govt. Land, CPRs.

It is difficult to get data on the land used by each mining company but Table 6.1 gives data on that owned by Coal India Ltd (CIL) in 2007. Most information is only after 1970 or even the 1980s. For what precedes the 1980s one has to depend on estimates.[2] The size of the mines today is bigger than in the past. For example, coal is India's most important mineral catering to a third of its energy needs. The size of its coalmines has grown from an average of 150 acres in the 1960s to 800 acres in the 1980s[3] and to some 1,500 acres today because the last three decades have witnessed a shift from underground to opencast mines in order to exploit lower quality coal. Open cast mines require more land and displace more persons but create fewer jobs than underground mines do. Also the bigger size of other mines causes higher displacement. For example, the Freeport mine in Indonesia caused 15,000 DPs. They were 20-30,000 in the Ghana Tarkwan gold mine and 37,000 in those of southern Africa according to Downing.

The Geological Survey of India estimates that on 1st January 2000, India's coal reserves down to a depth of 1,200 meters were 2,11,593.61 million tonnes. Of this total 10,778.58 million tonnes are proven reserves, 10,894.15 indicated reserves and 4,235.81 are inferred reserves.[4] Its exploitation results in much land loss and displacement. Besides, mechanisation of the mines during the last two decades reduces employment. Till the mid-1980s, the T. N. Singh Formula 1967 had stipulated that industries and mines give a job each to the families they displaced. Since they began to mechanise their operations from the mid-1980s, this Formula was abandoned in 1986.[5] It is worthwhile to see its impact, among others, in the jobs given by CIL. It gave a job each to 11,901 (36.34%) of the 32,751 families displaced in 1981-1985.[6] But till 1992, immediately after mechanisation in the Upper Karanpura Valley of Jharkhand the first 5 of the 25 mines that were to have 1,00,000 DPs, gave a job each to 638 (10.18%) of the 6,265 families displaced. With traditional transport, the NALCO mines activated in the late 1980s in Koraput district, Odisha would have created 10,000 jobs and rehabilitated the 50,000 DP/PAPs of the Upper Kolab dam and 6,000 of the NALCO Plant in the same district. Their income would have created more indirect jobs. But the mines were fully mechanised and created some 300 skilled and semi-skilled jobs that went to outsiders since its predominantly tribal PAPs lacked the skills they required.[7]

Table 8.2 presents the total land acquired by eight Coal companies of India. It also depicts the total tenancy land acquired as well as tenancy land in possession.

Table 8.2: Company-wise Status of Land Acquisition by CIL and Its Subsidiaries in India, June 2007, (Area in Hectares)

Sl. No.	Company	Total Land Acquired *	Total Tenancy Land Acquired	Tenancy Land in Possession
1	*2*	*3*	*4*	*5*
1.	Eastern Coalfields Limited (ECL)	13093	12022	9145
2.	Bharat Coking Coal Limited (BCCL)	4280	3857	1937
3.	Central Coalfields Limited (CCL)	35736	12758	4883
4.	South Eastern Coalfields Limited (SECL)	20538	12812	11873
5.	Western Coalfields Limited (WCL)	19785	17089	14330
6.	North Coalfields Limited (NCL)	15948	5612	5275
7.	Mahanadi Coalfields Limited (MCL)	19965	9180	4252
8.	North Eastern Coalfields (NES)	25041	41.47	41.47
	Total	**154386**	**73371.47**	**51736.47**

Source: Lok Sabha Unstarred Question No. 5410, dated 15.05.2007.

Courtesy: Table from Doc Centre, Indian Social Institute, New Delhi.

* What is not Tenancy land is Govt. land and CPRs

It is revealed from the table that the total land acquired by eight coal companies in India as on June, 2007 is 154386 Ha. Of which CCL has acquired highest area (35736 Ha).

In Karnataka, MID is substantial and DPs were not rehabilitated properly. In Karnataka, Supreme Court of India stipulated norms for Karnataka Mines and said no mining in State till reclamation and Rehabilitation measures are implemented. The Supreme Court said, it would not allow resumption of mining of iron and ore in Karnataka unless there is a statutory compliance and full implementation of reclamation and rehabilitation (R&R) measures. "You don't press for resumption of mining activities unless we are sure that there is a statutory compliance. No mining activities will resume unless reclamation and rehabilitation (R&R) measures are fully implemented. All statutory compliance should be there", a special forest bench headed by Justice Aftab Alam observed. The bench, which accepted all recommendations of the apex court-appointed Central Empowered Committee (CEC) asked the authorities in Karnataka to fully cooperate in the implementation of the suggestions. The report said there was no need for any orders for resumption of mining activities till the recommendations are implemented.[8] Mathew Areeparampil have discussed the displacement due to mining in Jharkhand.[9] According to him, for a proper understanding of the problem of displacement of people due to mining we have to look at it from the wider perspective of dispossession that results from such mining activities. Displacement becomes a problem because of the dispossession that results from it. Land acquisition for mining is a major problem in UP. UP farmers are opposing vehemently and fighting tooth and nail against land acquisition.[10]

Now let us examine the land acquisition for mining projects in Odisha, the state that is rich in mineral deposits. Extraction of minerals through mining and quarrying gathered momentum just after independence in the State. There are around 700 mines distributed throughout the State. Exploitation of minerals from the mines requires large-scale land acquisition. The mines are mostly located in the tribal areas where the tribal people have been living in what is called 'encroached Government land' for generations together, having no patta or legal evidence of individual possession of land. Further, most of the mines, barring a few are operated by private parties on a lease basis. They manage to take over the land by satisfying its owners (Who is either partly affected or totally displaced due to the mining operation) with some compensation or other. Since such transactions are never legally registered, it is difficult to arrive at any reliable database on the extent of displacement caused by mining. In mining operations in Odisha, the maximum displacement can be found in the coal mining sector. As the coal mining projects usually carry out open cast mining or strip mining, they bring about involuntary displacement of the people from their original homes. Mining of iron ores also, in the tribal areas in north Odisha, has displaced many people. But the extent of this displacement has neither been fully documented, nor properly assessed.

Between 1951 to 1995, according to one estimate[11] 410137.24 Acres of land had been acquired for mining projects which constitutes 17.36 percent of total land acquired for development projects as detailed in Table 8.3.

Table 8.3: Land Acquisition for Mining Projects in Odisha between 1951 to 1995

(In Acres)

Sl. No.	Types of land	Land Acquired for mining	Percentage to total	Total Land Acquired for all dev. Projects	Percentage to total
1	2	3	4	5	6
1.	Private	68328.86	16.66	987651.85	41.81
2.	Public	166638.76	40.63	661090.94	27.99
3.	Forest	175169.62	42.71	713449.24	30.20
	Total	**410137.24**	**17.36**	**2362192.03**	**100.00**

Source : (i) Development induced displacement & Rehabilitation in Odisha 1951 to 1995 by Dr.Walter Fernandes & MD. Asif of Indian Social Institute, New Delhi, 1998 (ii) "Basachyutanka Durdasha", Indian Social Foundation, Gaunli Bichar, Bhubaneswar.

Displacement

Displacement, in common parlance, means moving from usual location and generally refers to putting or being out of place. Displacement may be voluntary and willing or involuntary and forcible. A person forced or removed from his or her own home or country is called a displaced person. However, in the present context, for the purpose of the study, displacement may be referred to as involuntary or forced migration of population from a particular area/location for mining activities. Mining causes massive displacement of people settled in the mining and adjacent areas. Involuntary displacement of population and forced relocation is a profound social disruption and the most negative externalities of development. It entails a huge socio-psychological and economic stress on PAPs.

Globally, no systematic international survey has been made of the scale of Mining Induced Displacement and Rehabilitation (MIDR). The little information available points to high displacement in the past and a rising trend for the future as rich mineral deposits are found in areas with relatively high density of politically powerless populations, low land prices because of the backwardness of the area, poor definition of land tenure and open-cast mining. In every country including India, most persons affected are tribal and other indigenous peoples. Higher displacement will also result in greater impoverishment of the displaced. Even the World Bank acknowledged this problem when it said in its Group's policy on involuntary resettlement: "Bank experience indicates that involuntary resettlement under development projects, if unmitigated, often gives rise to severe economic, social and environmental risks: productive systems are dismantled; people face impoverishment when their productive assets or income sources are lost; people are relocated to environments where their productive skills may be less applicable and the competition for resources greater; community institutions and social networks are weakened; kin groups are dispersed; and cultural identity, traditional authority, and the potential for mutual help are diminished or lost."[12]

The mining affected persons face problems similar to those of displaced persons (DP) or deprived of sustenance without physical relocation (PAP) by other types of projects but its environmental impact seems to be greater than that of other projects. 'Mining, it has been realised, affects environmental and social change no matter where it occurs. Mining related disruptions can impact the physical environment and or local communities' as observed by Miranda. Displacement specialists, e.g., M. Cernea speaks of its impact as loss of physical income earning assets such as homes, productive land, jobs and subsistence resources and non-physical assets like community support, cultural sites, social systems, identity and mutual help mechanisms. Food insecurity, poor access to health care and education and social disruption are among its impacts. Pandey is of the opinion that the tribal and indigenous peoples particularly women suffer more than others do. It also means that, the social impact of mining goes far beyond loss of land. According to Downing, "Failure to mitigate or avoid these risks may generate "new poverty," as opposed to the "old poverty" that peoples suffered before displacement. Certain groups, especially indigenous peoples, the elderly and women have been found to be more vulnerable to displacement-induced impoverishment risks". Its greater negative impact on women known already a century ago motivated the International Labour Organisation to evolve the Convention concerning the Employment of Women on Underground in Mines of all kinds and the hazard of child labour resulted in the Convention concerning the Minimum Age for Underground Work. These and other measures and policies emanated from the realisation of the need to prevent further impoverishment of the already poor in the mining areas.

In African major mineral countries, Ghana for example, mining communities are compelled by law to vacate the land or are restricted to safe zones where the land is required for mining purposes. There is invariably conflict between mining companies and their host communities. The main problem is with enumeration of properties and rate negotiations. Furthermore, there is also difficulty in acquiring suitable alternative land for resettlement of communities. The Government of Ghana has put in place certain measures to mitigate the adverse effects of mining in order to ensure peaceful coexistence between the mining company and its host community. Minerals and Mining law provides the regulatory and legal framework of mining in Ghana. It promotes as much as possible, peaceful co-existence between mining companies, their host communities and empowers regulatory agencies. The law is periodically amended to address new challenges.

In Tanzania, mining has had adverse socio-cultural impacts. Apart from displacement, three were cases of unemployment, child labour, accidents and theft. An influx of foreign mining companies has made it more difficult for locals to secure land. In the 1980s, the Tanzanian government amended mineral policies for the sole purpose of creating a favourable investment climate for foreign mining companies. As a result, several small-scale miners and farmers have lost their mine sites, agricultural and grazing lands. The long-term implications of such displacement include accelerated food insecurity to landless classes, increased poverty and intensified environmental degradation. Besides, displacement has already caused conflicts between the local people and the mine operators. There have since been additional social conflicts between small-scale miners and the large-scale foreign mining companies, as the small-scale miners have begun to find that areas previously open to prospecting and mining of gold is now under the control of a private foreign company. Mihayo, for example has described the nature of such disputes that have occurred at the different mines. Profound conflict among mineral stakeholders suggests that there is a weak or inadequate enforcement of natural resources policies in Tanzania. Thus, case studies of Ghana and Tanzania present a diluted picture on impact and policies of displacement.

In India, mining Sector plays a vital role at the national level both in terms of its social and economic context, as discussed earlier. But, the India's Environment Report[13] paints horrific picture of the devastation that has been wrought by mining in the country. The statistics are shocking; (i) Between 1950 and 1991, mining displaced about 2.6 million people. Not even 25 percent of these displaced have been rehabilitated. (ii) For every one percent that mining contributes to India's GDP, it displaces 3-4 times more people than all the development projects put together. In India, where population increases already place unsustainable pressures on fragile environments, it is estimated that over the last four decades some twenty million people have been displaced, while only approximately one quarter of that number have been rehabilitated.[14] Many have simply moved to find new lives in the growing urban centres or have moved into more marginal areas and thereby increased pressure on fragile environments and accelerated the destruction of forest cover. Many have also been impoverished as a result of these processes, and have unquestioningly accepted the inevitability of the resettlement process, and the authority of

Government to exercise its right of eminent domain. This development is intrinsic to liberalisation formalised by the economic policy of July 1991. It allows up to 74 percent foreign investment in the exploration and mining of diamonds and precious stones and 100 percent in the automatic route for processing of minerals and metallurgy. Efforts are being made to dilute the environmental laws to suit the needs of the private investors. According to Kalshian, because of the adverse environmental impacts, the World Bank that provides technical assistance to member countries, commissioned in 2004 a review of extractive industries but rejected the report presumably because it suggested, among others, that the Bank move out of this sector since it disrupts the life of the indigenous peoples. Besides, since most minerals are in the tribal areas, the Centre has been trying to dilute the Fifth Schedule because of the Supreme Court judgement in Samata vs the State of Andhra Pradesh.[15] It concerned land allocation to Birla Periclase in a tribal area. In a majority judgement, the court held that in AP the person includes the State and property includes prospecting for mining licences. It declared all transfers of prospecting licenses to private companies void. Since it could go against the economic policy, the Ministry of Mines, GoI, in its letter to the Committee of Secretaries,[16] said that if this judgement is followed, large tracts of land with bauxite in AP would never be exploited. It could also apply to other States and hamper mineral exploitation of the whole country. Based on the advice of the Attorney General, it suggested amendments to the Fifth Schedule under Article 368 so as not to affect the basic structure of the Constitution. Experiences of different major mining states and the extent of MIDR as well as rehabilitation approaches are not uniform. Hence, general policy parameters, Mining Acts/Rules were promulgated by Government of India to regulate mining activities and streamline the process of rehabilitation of displaced people. According to an estimate, [17] during the period from 1951 to 1990 out of total displacement of 1,85,00000 populations by different development activities 21,00,000 people were displaced by 1210 mining projects constituting 11.3 percent of total displacement. However, experience indicates that until the nineties of the last century, R&R was not being considered seriously and in the result, it was almost the last chapter in any project history. This is the reason for which collection of data on D.Ps and P.A.Ps (project Affected Persons) and its tabulation was neither systematic nor dependable. Hence the actual number of D.Ps and P.A.Ps may be higher than the estimates.[18]

Thus, lack of revisions of database is a handicap and therefore, the estimate of the numbers displaced in India ranges from 3 to 30 million.[19] Odisha, Chhotanagpur in southern Bihar and Chhatisgarh in Madhya

Pradesh (undivided) bear the brunt of development induced displacements in India. 52.48 percent of all mines in India are concentrated in the four states of the Eastern region. According to Walter Fernandes, displacement due to mines in these regions was 11,50,000 persons out of which 76.09 percent were tribals. Fernandes has described the mining induced Displacement and Rehabilitation (MIDR) of different States of India. It is more difficult to get data on the DP/PAPs due to mining projects than on land used for mining. Even the total of the DP/PAPs in India is not available, much worse of their category. However, Table 8.4 depicts a conservative estimate of mining induced displacement (MID) in eight states. The table shows that mining induced displacement (MD) is highest in West Bengal (4,18,061) followed by Jharkhand (4,02,882). MID is lowest in Kerala (78) followed by Gujarat (4128) and Goa (4740). In Goa, many must have moved out of the area when mining activity was at its peak in the 1980s. Others from within and outside Goa migrated to these areas. So mining settlements grew, most of them temporary or semi-temporary. One does not know the number they displaced. The limited information in Assam shows that Bensali Coal in Sibsagar district had 5 DPs and 10 PAPs on 2.91 acres from the data of Deputy Commissioner, Sibsagar 1987. The partial data on much of the land gives the same density of 1.7 DPs and 3.4 PAPs. Thus, the 8,000 acres used for mining would have caused 13,680 DPs and 27,520 PAPs or a total of 41,200. It is an underestimate but in the absence of other data one has to limit oneself to it. A study in Meghalaya shows that coal mining has displaced several thousand but it is difficult to come to an exact figure since much of the land is community owned. So it is mined through an informal understanding with the community or individual owners.

Table 8.4: Mining-Induced Displacement (MID) in some States

Sl. No.	State	Years	Tribals	Dalits	Others	Com-bined	Total
1.	Andhra	1980-95	-	-	-	100541	100541
2.	Assam	1980-2000	-	-	-	41200	41200
3.	Goa	1980s	-	-	-	4740	4740
4.	Gujarat	1980-2000	-	-	-	4128	4128
5.	Jharkhand	1980-95	83543	63352	255987	-	402882
6.	Odisha	1960-95	150000	45000	105000	-	300000
7.	Kerala	1990s	-	-	-	78	78
8.	W. Bengal	1960-2000	2459	99244	316358	-	418061
	Total		**236002**	**207596**	**329449**	**462266**	**1571630**

Source: Lok Sabha Unstarred Question No. 5410, dated 15.05.2007. Courtesy: Table from Doc. Centre, Indian Social Institute, New Delhi (what is not Tenancy land is Govt. land and CPRs)

In other States too, limited data on displacement and deprivation could be got only from the 1980s. According to Fernandes, the initial figure of 2.55 million (1951-1900) arrived at from secondary sources is an underestimate. Table 8.4 based on primary data-based studies gives a little over 1.5 million mining DP/PAPs in the 1980s and 1990s in a total of 60 million DP/PAPs of all the projects between 1947-2000.[20] It does not include Maharashtra, MP and Chhattisgarh that have high deposits. Many others are up to 1995. When these States are studied and data on the others updated, one can expect the number between 1947-2000 to be five million.[21]

Resettlement and Rehabilitation (R&R)

Resettlement and Rehabilitation (R&R) is an essential part of development Projects. The people who are affected by implementation of mining projects or any development project for that matter, who have lost their home and hearth, land and assets with socio-economic hardships to the benefit of others must be compensated with monetary awards for the expropriated assets and must be rehabilitated to enable them at least to restore their previous standard of living and reconstruct their socio-economic system to the extent possible. R&R activities involve a number of stakeholders. The most important stakeholder is the Government who initiated the process of displacement. The others are the donor agencies, DPs/PAPs, hosts of the beneficiaries, civil societies and others. Government decides the compensation entitlements and mode of resettlements, DPs, PAPs try to get maximum benefits and even fight over the choice of rehabilitation, there is clash or confrontation between the hosts and DPs over access to and pressure on the common property/ resources. Conflict resolution between various stakeholders is a major problem in R&R. Conflicts arise over different issues of rehabilitation. Compensation money, process of relocation and resettlement, extension of infrastructure facilities so on and so forth. Hence, conflict resolution between various stake holders in R&R process is a primary problem for the agencies responsible for its implementation. Absence of proper coordination between different stakeholders, between different activities/ sub-activities and between different units and sub-units in resettlement process is a vital issue. Involuntary displacement of population and forced relocation is a profound social disruption and the most negative externalities of development. It entails a huge socio-psychological and economic stress on PAPs. Hence, a systematic and participatory approach needs to be followed in R&R. Processes involved in R&R are broadly, Data Collection and socio-economic base line/bench mark survey, Identification of Project

Affected Area (PAA), Enumeration of Project Affected Persons (PAP), Land Acquisition (LA) and Land Alienation, Compensation, Evaluation and Resettlement.

Like displacement, reliable and complete data is not also available in case of rehabilitation in mining projects. Studies made by authors like Fernandes. Murickan, Asif, Naik et al. indicate rehabilitation in mining projects of some states. The studies also show that rehabilitation is weak by all the projects but it is lower by mines than by others. Odisha has resettled 37.27 percent of all its DPs. AP 28.82 percent, Kerala 13.8 percent 1951-1995, Goa 40.78 percent as of 1965-1995, West Bengal nine percent and Assam the DPs of some 10 projects as on 1947-2000 (Table 8.5).

Table 8.5: An estimate of Rehabilitation in mining projects of some States

Sl. No.	State	Percentage of Rehabilitation
(1)	(2)	(3)
1.	Odisha	37.27
2.	Andhra Pradesh	28.82
3.	Kerala	13.8
4.	Goa	40.78
5.	West Bengal	9

Source: Data taken from study reports of Fernandes and Naik, (2001), Fernandes and Asif (1997) Fernandes, (2001), Murickan et al (2003).

At the all India level, however, the displacement of persons due to mining Projects is 21,00,000 between 1951 to 1990 against which 5,25,000 persons constituting around 25 percent were rehabilitated according to an estimate by Fernades (1991). In Odisha rehabilitation of displacement due to mining projects is slightly better standing at 60 percent.

An alternative to resettlement is jobs. As stated above, till the 1980s, CIL used to give some jobs but their number has declined. Even one job per family is not an alternative because after land alienation the remaining family members do not have resources for self-employment. Possibly because of fewer jobs, CIL changed its norms for giving jobs. In what later became Mahanadi Coal Limited in Odisha, in the 1980s the norm was a job given for three acres of land acquired.[22] In parts of Jharkhand it was changed to two acres if the person was a matriculate. Such jobs went by and large to young men from the dominant castes even in tribal majority districts because they alone had *patta* land and had access to high school education. These and other cases point to the neglect of the DPs. The World Bank is reported to have felt that CIL's inability to deal with the DPs was making it less attractive for capital markets since providing inheritable employment to displaced families resulted in a sustainable economic liability but its failure to rehabilitate the DPs was a social liability. So the "World Bank required a distinct $US 63 M Coal Sector Environmental and Social Mitigation Project (CSESMP)—known as the Coal India Rehabilitation Project (CIRP)—as a condition for CIL's obtaining a $US 500 M loan. A large part of the CSESMP was a pilot project that helped CIL implements appropriate policies for environmental and social mitigation of MIDR and test the effectiveness of these policies on 25 of its 469 coalmines".[23] Also the CIL rehabilitation policy (CIL; 1994) seems to be its outcome. What is said of CIL is equally true of other mines. In fact, CIL has given more jobs to its DPs than most others have done. It has also resettled some till 1985. However, after liberalisation, profit has attained greater importance than in the past and the social component has got weak in the economy as a whole. The loss of jobs is only one of its signs. Its other face is the demand from private companies for more mining land particularly in the tribal areas of Jharkhand, Odisha and Chhattisgarh. Thus, there seems to be more displacement in Middle India than in the last some years, much of it tribal but their rehabilitation schemes are weak.

The R&R of Population displaced by mines and protection of the environment is an important, but often neglected, aspect of development strategies in both rural and urban area. There has until recently been a largely residual approach to dealing with such populations; pay them the minimum cash compensation and after a period of transitional help, leave them to their own destinies. A coalition of forces is now aligned in often complex ways to ensure that the minimalist, residualist and welfarist approaches are challenged and the rights and entitlements of project-affected persons (PAPs) are protected and defended. The latter are then at the 'coal face' of negotiations about future sustainable livelihood, and primary stakeholders in debates about the direction of the development effort. Recognition that resettlement and rehabilitation (R&R) process need to be seen as development projects in their own right is slowly emerging. The implications will need to be forged.[24] In contrast to public sector mining, the impact of private mining operations on surrounding populations has been uniformly negative. This is because private mining companies undertook unscientific mining practices and while they displaced the surrounding populations, often consisting of large number of tribal groups, from their subsistence activities they did little to aid them in acquiring skills or the new disciplinary requirements of an industry.[25]

Resettlement is more difficult, more expensive, and more time consuming than hitherto acknowledged. Large numbers of people, usually very poor, have to be moved and their former incomes re-established and, there they are below the poverty line, hopefully improved in a short period of time to avoid destitution. Weak implementing agencies, lack of funds and of detailed information on affected people, in addition to the lack of suitable areas to resettle displaced people, have complicated the task. By bringing R&R into more central focus, a shift from an ad hoc reactionary and welfarist approach to a planned, proactive and participatory one can be more successfully negotiated.[26] In most pre-mining situations in rural and tribal areas, land is a joint ownership of brothers and sons. The labour is jointly distributed and invested and the output is jointly enjoyed. The rehabilitation procedures followed so far in mining projects has proved that families are neutralised as jobs (if given) are extended only to one of the sons or brothers whose wages or income from the mines becomes his individual income and not shared with the rest of the joint family. This has thrown entire families into a state of poverty and lack of livelihoods, forcing them to migrate and break up social relationships. Monetary compensations are always wasted away in a short period leaving the families impoverished. Rehabilitation has never addressed the need for women's livelihoods and land rights as it is only the family as a unit and the men in the family who are considered to be compensated for their loss of lands. Projects which have been set up decades ago, like the NALCO in Damanjodi in Odisha have yet to complete their rehabilitation activities properly. As there was no policy on rehabilitation, this had been implemented from individual project to project basis where the nature of rehabilitation depended on the voluntary interest of the company and not on any set guidelines or rules.[27]

As regards mining induced Displacement and Rehabilitation in Odisha, as many as 1,00,000 persons were displaced due to mining projects in Odisha during the period from 1951 to 1995 of which 60,000 persons were rehabilitated and hence 40 percent of displaced persons were not rehabilitated. The details are at Table 8.6.

Table 8.6: Persons displaced and Rehabilitated due to mining projects in Odisha (1951 to 1995)

Sl. No.	Category of persons	Number
1	*2*	*3*
1.	Persons displaced	1,00,000
2.	Persons rehabilitated	60,000
3.	Percentage of rehabilitation	60.00
4.	Persons not rehabilitated	40,000
5.	Percentage of people not rehabilitated	40.00

Source: Development induced displacement & Rehabilitation in Odisha 1951 to 1995 by Dr. Walter Fernandes & MD. Asif of Indian Social Institute, New Delhi, 1998.

There is massive displacement in Coal Mining Projects. The extent of land acquired for Coal mining projects in Odisha by 1996 both Government and Private is around 2427 hectares in 79 villages and 3143 families were displaced as detailed below in Table 6.7. There are around 700 mines distributed throughout the State. Exploitation of minerals from the mines require large-scale land acquisition. The mines are mostly located in the tribal areas. Further, most of the mines, barring a few are operated by private parties on a lease basis. They manage to take over the land by satisfying its owners (Who is either partly affected or totally displaced due to the mining operation) with some compensation or other. Since such transactions are never legally registered, it is difficult to arrive at any reliable registered database on the extent of displacement caused by mining. In mining operations in Odisha, the maximum displacement can be found in the coal mining sector. Mining of iron ores in the tribal areas in north Odisha has also displaced many people.[28] Coal mining operations all over the State have caused major displacement.[29]

Table 8.7: Land acquired for coal mining projects in Odisha

Sl. No.	Project	No. of villages	Extent of land acquired both government and private in hectares	No. of families displaced
1	*2*	*3*	*4*	*5*
1.	Talcher area	61	1498	1790
2.	Ib valley area	18	929	1353
Total		**79**	**2427**	**3143**

Source: (a) Development of projects and induced Displacement: Wr. A.K. Dalua (b) Depriving the under privileged for development, Balaji Pandey.

As evident from the Table 8.7, for Talcher Coal Mining Project and lb Valley Projects, 1790 and 1353 families were displaced respectively. Regarding resettlement, according to Balaji Pandey, in one village consisting of 22 families has already been affected by the MCL in Talcher; no one has as yet been resettled in any rehabilitation colony. Some of displaced families have made their own arrangements for rehabilitation in the nearby village while the rest have been given MCL staff quarters temporarily. Fifty-six jobs have been given to members of these outstee families (MCL, Talcher, 1994). In the lb Valley area, 110 displaced families from the two affected villages are resettled in the colonies. A total of 199 jobs have been provided to the members of these displaced families.[30] Displacement and rehabilitation of two other Coal mining projects is presented in Table 8.8, out of 631 displaced families, only 11 percent were rehabilitated in case of Utkal B1 Coal Mining and only 12 percent were rehabilitated in case of Utkal Coal Ltd.

Table 8.8: Displacement and Rehabilitation in other Coal Mining projects

Sl. No	Project	No. of families displaced	No. of families not Rehabilitated	Percentage of families rehabilitated
1	*2*	*3*	*4*	*5*
1.	Utkal B1 Coal Mining	631	560 (89%)	11
2.	Utkal Coal Ltd.	143	132 (88%)	12

Source: The Samaj, Odia Daily, dated 15.12.2014.

In Raijharan Coal Mining Project, 1074 families were affected out of which 622 families are to be displaced as on August. 2013, already 59 displaced families were resettled rehabilitated. The details of displacement and rehabilitation of Raijharan Coal Mining Project in Angul district in Chhendipada Tahasil is indicated in Table 8.9.

Table 8.9: Displacement and Rehabilitation of Raijharan Coal Mining Project in Angul District (As on August, 2013)

Sl. No.	Description Activities/ Events	Number
1	*2*	*3*
1	Land acquired Govt. Private Total	 31.030 454.480 485.510
2	Families affected	1074
3.	Families already displaced	59

4.	Families to be displaced	563
5.	Total displacement of families	622
6.	Families Rehabilitated/Resettled	59*
7	Families provided Jobs	11
8.	Families opted for cash in lien of job	28

Source:- Information collected from industrial Development Corporatio, Odisha, Bhubaneswar. *out of 59 families, 34 families resettled in R&R colonies and 25 families opted for self relocation.

In Bauxite Mining Projects of Odisha, the Displacement and Rehabilitation scenario is not also encouraging. Alumina (Aluminum oxide, A1203) is produced from Bauxite by refining of alumna's bonding with Iron, Silica, and about 40 other mineral elements found in substantial quantities in Bauxite.[31] India's first major aluminium company was INDAL (Indian Aluminium). It was set up as a subsidiary of ALCAN (Aluminium Canada), a principal supplier to the UK arms industry as well as to the US, and originally a subsidiary of ALCOA (AL. CO. of America). Alcan's set-up was designed in part to facilitate US control of aluminium cartels centred in Europe. Since the 1990s at least, the aluminium cartel has been directly controlled by the US administration. The foundation-stone for the Hirakund dam was laid by Odisha's last British Governor, and then again by Nehru in 1948. By 1959, it was supplying Alcan/ Indal's Hirakund smelter, whose bauxite was mined and refined in present-day Jharkhand to the north.

The first aluminium company was set up in the state by Indalco, which was sponsored by Alcan, the Canadian aluminium company. The cheap hydropower of Hirakud propelled it to set up its plant there. It brought bauxite from mines in Bihar. It was only in the early 1970s that the Government authorities got an inkling of Odisha's huge bauxite reserve. Just after the completion of identification of the bauxite mines in 1980, the public sector undertaking Nalco was set up in Odisha; to the extent that aluminium Pechiney of France provided the technical know-how and $400 million loan to this venture, it may be taken as the first glimpse of foreign interest in Odisha's bauxite. The proposed Balco at Gandhamardan-to be built with technical support from the US, Hungary, and erstwhile Russia-had to be shelved due to opposition from local people. From the early 1990s, the beginning of the liberalisation regime-most projects that were proposed were primarily private. Two of these projects came into the limelight. L&T had joined hands with the American company Alcoa to set up an export-oriented plant at Kodingmali that was to purchase machinery from Switzerland-based Alusuisse. The firm Alusuisse was also to extend technical support to Utkal Alumina to be set up in Baflimali by a Joint Venture of Indalco, Tata, Hydro Aluminium of Norway, and possibly Alcan of Canada. Among the purely Indian firms evincing interest in this sector have been Hindalco and Sterlite. Then came NALCO (National AL.CO), 'the pride of Odisha', one of India's biggest profit-making PSUs, set up in 1980. An article in EPW in 1981 by Rajagopalam et. al. gave a range of economic arguments against setting up NALCO, in particular the low price for bauxite enforced by external pressures (i.e. the aluminium cartel), plus excessive consumption of electricity, water, etc., and excessive pollution. Damanjodi refinery and the Upper Kolab dam displaced over 50,000 people, Adivasis the majority. The rail link between Koraput and Rayagada, built to facilitate NALCO and future aluminium firms, had negative effects on the forest and interior Adivasi villages over a wide area. And NALCO's smelter at Angul has inflicted serious pollution on thousands of people, killing all the fish in a long stretch of the Nandira and Brahmani rivers, as well as killing people and damaging extensive areas of cultivated land when its toxic waste fly-ash ponds have flooded. Tribal villages all around NALCO's bauxite mine on the long summit of Panchpat Mali have suffered pollution and lost their land's former fertility, while over 200 deaths in work accidents at the Damanjodi plant nearby have gone largely unreported and uncompensated, as have deaths from fluoridises around the Angul smelter, and deaths from industrial pollution among workers in all these plants. So NALCO's high profits come at a huge human and environmental cost that has never been 'properly calculated'. Part of the subsidies and 'externalities' of aluminium production are examined below. When NALCO's example was emulated by BALCO (Bharat AL.CO), wanting to mine the summit of Gandhamardan, a movement arose to prevent this, representing an alliance of local Adivasis, Dalits, Hindu activists and many others. They came together to save the Gandhamardan Mountain range, which has an exceptional wealth of forest cover, as well as mythical and medieval temple heritage. Balco built a colony for several thousand workers, which is now derelict. The planned bauxite mine on the ridge was finally declared illegal on environmental grounds in 1987, after hundreds of protestors had endured police beatings and arrest, and women had stopped the passage of mining vehicles by laying their children in front of them and to maintain the blockade, saying: 'What future do they have, if you destroy our Mountain?' There are certain facts about Bauxite worth mentioning here. There are only seven countries in the world who extract Bauxite more than Odisha does. Around 10 percent of total Bauxite produced in the world is extracted from mines of Odisha. Quality of Bauxite in Odisha is superior than most of the countries. There is 45 to 48 percent of alumina (Al_2O_3)

in it and content of unrequited silica is only 2 to 3 percent. Many reputed international and national companies are extracting Bauxite from Odisha mines. Indian Aluminum Company ltd. (INDALCO) established under the aegis of Aluminum Company of Canada (ALCAN) during 1938 in Kolkata was the first company to extract Bauxite from Odisha. National Aluminum Company (NALCO), a public sector enterprise of India established in 1981 with technical knowhow from Aluminum Pachiney'of France started extracting Bauxite from Panchpatmali Bauxite mining of Odisha since 1986. Besides, Bharat Aluminium Company (BALCO), a public sector enterprise of India established with technical know-how from US, USSR and Hungary got permission to produce Bauxite from Odisha. Other Private companies such as Aluminium Company of America (ALCOA), LARSON and TUBRO of India established with technical know- how and machineries from ALU Suisse of Switzerland, Hydro Aluminum of Norway, TATA Company, Karnataka Cables Ltd. established with Joint ownership of Phelps Dodge of America and R.P. Goenka enterprise of India and with technical know-how from Kaizer Engineers, Kaizer Aluminum Renold international of America and ALUTERVE of Hungery, Hindalco-industries Ltd and sterlite industries Ltd. of India are extracting/received permission to extract Bauxite from Orissa.[32] For extracting Bauxite at Damanjodi, NALCO acquired 3444 hectares of land and affected 15423 persons in Koraput. Out of 15423 affected persons, 2368 persons in 610 families in 15 villages were fully affected and 736 persons in 178 families in four villages were partially affected.[33] Two of India's six working smelters, and one of its six refineries, are located in Odisha, as well as a string of big dams to give electricity and water to these factories. The present plans for intensive mining of Odisha's bauxite alongside a complex of big dams, aluminium factories and rail links actually date back to the 1920s.

Mining proponents commonly err in understating the numbers of people to be displaced. That there are such huge and seemingly irreconcilable differences of opinion about the numbers of people likely to be impacted or displaced suggests that the proponents are not being realistic in jointly agreeing on the precise numbers involved. The main difference of opinion between the project sponsors (Hindalco & Alcan) and the impacted people and their advocates, concern how many people may be impacted. Most of the people are vulnerable ethnic minorities, Adivasis or other indigenous peoples. Estimates of the people negatively affected by the Utkal project range from 750 (NorskHydro's estimate), to 3500 (Utkal's estimate) to 60,000 (Norwatch, 1998). The variation in these numbers may be derived from differing interpretations of how broadly to cast the net in terms of being 'affected' by the project, whether, for instance the figure should be limited to those whom will be physically displaced or adversely affected in an indirect manner (such as decreased access to public land, for example). Utkal states that there are 1750 "property holding" families, and 1900 "affected" families. But many potentially impacted people may not hold property. Few Adivasis own paper titles to the lands they have used for centuries. The 1750 families estimate neglects key categories of people, such as: (i) Families working as traditional labourers or sharecroppers on land that they do not own; (ii) Families without paper titles for land they have peacefully used for years (usucapion rights); (iii) Families who depend on public land and who have 'use rights' under the Fifth Schedule of the constitution of India, and under Odisha Law. Utkal states that they have documents showing support from elected representatives from all but one of the "official project-affected" villages. Impacted groups have not been provided with any evidence of this statement. Alcan recently admitted that these documents hold no legal weight, since they are the result of "spontaneous meetings". While the Baphlimali Hills area is to be allocated to mining, assessments by various groups reveal that the requirements of 2,610 hectares of land, including 1,000 hectares of cultivable land for the factory/wastage dump alone, will cripple the livelihoods of most settlements in the area. In fact, many villages stand to lose 75 percent of cultivable land and will not even be considered displaced, rendering the people virtually landless. Adivasis have been warned that few jobs will be open to them, mainly as construction labourers or menials. One member of the "fully displaced" family category may receive one permanent job.[34] Hindalco's plan for mining Kodinga Mali (on the SW rim of Kashipur),with a refinery at Kansariguda village, and an expansion of the Hirakud smelter. Larsen and Toubro's plans for mining Siji Mali and/or Kuturu Mali (N and NE of Kashipur) from Sunger (where a movement was active against its plans in the '90s), plus a refinery near Kalyansingpur, in a joint venture with Dubai Al.Co.[35] The Sterlite/Vedanta's project near Lanjigarh is to mine the northwest ridge of the Niyamgiris. Construction of their refinery is under way, after evicting Adivasis who had agreed to accept compensation. These evictions took place early in 2004. Adivasis in the Lanjigarh area including those who refused compensation, whose villages stand just outside the boundary wall, live now under a shadow of tremendous pressure from the company people, contractors and police; as well as a climate of fear, after one leader of their Movement was run over on the newly metalled Lanjigarh-Doikal road on 27th March 2005. The legality of the Vedanta project, and its impact on the exceptional Forest which covers the lease area, has been questioned on numerous fronts by the Central Empowered Committee of the Supreme Court, who inspected it in Dec. 2004 and June

2005.[36] The refinery is being constructed by an Australia-based company called Worley, through foreign finance. The Indian company Vedanta Alumina is owned via Sterlite Industries by Vedanta Resources, which was launched as a UK company on the London Stock Exchange in December 2003, and invested in by several leading European banks. This investment is also funding a huge new smelter complex at Korba, and plans for another smelter near Hirakud. Khandual Mali near Karlapat (west of Lanjigarh) was under threat, initially from a local company, Sulakshmi Mines and Minerals, aiming to make money through selling the mining rights. An attempted sale to BHP Billiton (the world's biggest mining company) was declared illegal in 2004, after BHP had already announced its intentions to set up a bauxite mine and refinery in Kalahandi. This Mountain is the one referred to by Fox in the 1920s. Both environmentally and socially, it is an area of exceptional importance and sensitivity, and the villagers living on the mountain have made clear their resistance to mining it. Other projects being planned include Gandhamardan again, whose mining rights have been purchased by a Canadian Company called Continental Resources; Mountains in the Kuttia Kondh territory, whose bauxite was recently surveyed by Jimpex; and Deo Mali, Koraput's (and Odisha's) highest Mountain, which Nalco has designed on. In Chattisgarh, Balco-Sterlite-Vedanta is setting up a bauxite mine at Bodai-Daldali, on a mountain near Kawardha, bordering MP and Kanha National Park, where Baiga tribals have already been displaced.[37] Displacement and rehabilitation of two other mining projects is presented in Table 8.10.

Table 8.10: Displacement and Rehabilitation in Mining projects

Sl. No.	Project	No. of families displaced	No. of families Rehabilitated
1	*2*	*3*	*4*
1	Lanjigarh Alumina Refinery	121	121
2	Utkal Alumina, Rayagada	183	183

Source: The Samaj, Odia Daily, dated 15.12.2014.

The British started large-scale coal and iron-ore mining in north Odisha. But the escalating displacement since Independence, at least 3 million indigenous cultivators thrown off their land in Odisha alone, by mines, dams and metal factories' is unprecedented. Plans for Odisha's rapid industrialisation also involve a huge increase in coal and iron-ore extraction. POSCO's steel project is just one out of 37 current plans for steel plants. 'Kalinganagar' (Jajpur District), where recent protests by Adivasis unwilling to be displaced were met with ferocious police suppression, involves plans for a vast industrial park. And every day, several thousand trucks carry iron-ore from north Odisha to Paradeep for export. The northern districts of Keonjhar, Mayurbhanj, Sundargarh, Jharsaguda and Jajpur have already faced mass displacement and deforestation, as well as pollution, from iron and coal mining.[38]

As regards cost benefit of Bauxite and Aluminium, since the Second World War, a key strategic aim of the US has been to acquire a large and regular supply of bauxite/aluminium as cheaply as possible from third world countries. The IBA (International Bauxite Association), a producers' cartel of bauxite-exporting countries, managed to double the price they were getting for bauxite in the early 1970s, under the leadership of Michael Manley in Jamaica, whose Government faced a campaign of severe destabilisation by CIA operations as a result[39], and this success has never been repeated. As the al.cos themselves put this, getting a cheap source of bauxite is their 'starting point of value creation'. As it is refined, smelted, and fabricated, each stage of production ups the profit margin. But the highest profits are made further along the line, in the sophisticated alloys and Metal Matrix Composites demanded by Defence. This 'strategic importance' is a key reason Abdul Kalam gives for expanding India's aluminium industry.[40] Aluminium's history in the third world needs to be understood in this context. Al.cos comes into a region advertising huge economic benefits in employment and for boosting the local economy. As Anderson says frankly in 1951, and subsequent analysis confirms,[41] these benefits do not materialize. This is clear from examining the extent of subsidy in materials alone; but also from the economic cost which Govt. institutions are beginning to attach to pollution.

Government of Odisha had for the first time implemented a policy for rehabilitation of displaced persons/ families in case of Mining projects in the State during 1998[42] and then during 2015, promulgated Odisha Mineral Exploration Policy, 2015 vide resolution No. 6065 dated 04.07.2015. On the issues of land acquisition and displacement by mining, the limited data show that land acquisition has been high, particularly of the tribal and other CPRs. A large number of their dependants were not counted till recently. So their exact number was not known. But it is clear from the studies that their numbers are high and that they feel the negative impacts of displacement and environmental degradation more than the others do. Women feel it more in every community particularly because the number of jobs has declined and they go mostly to men from the dominant castes. So the subalterns in general and women in particular are ignored. Poverty is bound to grow if corrective measures are not taken.[43]

Impacts of MIDR

Studies and experience of involuntary resettlement and rehabilitation by mining during the last some decades show the near impossibility of restoring the DP/PAPs to their former economic and social condition not because of financial constraints alone. They have failed because a majority of them are tribal-indigenous people and most mines are in the administratively neglected "backward" areas whose inhabitants have very little exposure to the formal economy into which they are pushed. But the backwardness of these areas reduces the cost of land and makes it economically viable for the project. A second reason of this failure is the assumption that compensation suffices to rehabilitate a displaced economy and community. The problem goes far beyond compensation to the failure to realise that development itself may cause poverty in a traditional society. Pandey observed that "The local inhabitants, due to the introduction of these projects in their areas, gradually lose their control over the resources they had once access to on the one hand, and, they are shifted to a pattern of an economy with which they are not acquainted on the other. While their resources are exploited intensively by the project authorities, they are not able to enjoy their benefits because the process of development is heavily loaded against them".

Indigenous Communities

Mining projects frequently are located in remote areas where indigenous communities are members of a distinct cultural group, often a minority within a community of minorities. Here mining activities have a negative impact on the livelihood of indigenous people, especially with regard to issues concerning land tenure, often causing socio-cultural conflicts within and among communities. The Scheduled Tribes are mostly affected and worst sufferers in mining projects of the State. They lost their home and hearth, livelihood and means of living permanently. Homelessness accentuated their pauperisation. They suffer from the loss of habitat for mankind, animal and plant, Loss of hunting grounds, agricultural / forestry production facilities and Loss of social equilibrium due to influx of new inhabitants. The lure of new opportunities can create in-migration that may cause new tensions in the community between existing residents and newcomers. These poor people have no or little mobility or means of alleviating negative impacts.

Local non-mining population

Mining has various negative impacts on non-mining sectors. Large mining operations can inadvertently have an adverse effect on the ability of the local non-mining population to achieve and sustain economic self-sufficiency. The ability of this population to earn a living can be threatened or impaired by the mine's use of natural resources, such as land and water, on which the poor depend. Besides, the higher incomes of mine workers can lead to rising local prices, with the poor left behind; at the same time, the poor and non-mining population may have only limited access to services provided by the mine. In remote areas, the demands of mining operations on infrastructure services, for example, may put those services beyond the reach of the poor, either because their prices have become prohibitively high or because of simple usage limits. Overall, mining also can drive regional price levels to a point that leaves the poor unable to afford basic goods and services.

Socio-cultural conflicts

The establishment of mining activities in otherwise rural areas or in the "wilderness" are potentially disruptive to the socio-cultural surrounding. Mining operations most often bring in its wake large scale migrant workers who live with or without their families and within disrupted social contexts. With different socio-cultural milieu, local people may find it difficult to adjust and accommodate those people. The traditional villages became mining slums and displaced camps with growing socio-cultural conflicts. Food security can be threatened or compromised by mining-related factors such as loss of agricultural land; shortage of water supply which can be affected when the demands of mining operations divert excessive amounts of water from the local supply, Piling of mineral and stone waste; noise; dust and land disturbance often associated with mining activities. Each of these, therefore, presents a potential threat to the health and livelihood of the poor and vulnerable groups who have little mobility or means of alleviating negative impacts. Such environmental damage can be caused by small-scale mining as well as by large-scale mining, if no appropriate precautions are taken or deemed affordable. The case of Mahanadi Coal Fields where many destitute women were found, badly implemented resettlement programmes, destroyed agriculture lands, homestead kitchen gardens, forest resources, fishing and livestock breeding, which were the main livelihood and food security sources for women. Water is one main problem expressed by women in all the mine sites. Mining has contaminated water bodies by dumping mine tailings, depleted ground water affecting agriculture and domestic water needs.

Health hazards

The harsh living conditions for miners in small-scale mining as well as in large-scale mining, along with the lack of information and education about prevention, often can contribute to a high prevalence of human immunodeficiency virus (HIV) and other communicable diseases among miners and their families. Also, work-related injuries and health risks—lung cancer, for example; reduce the miners' life expectancy and often put families in particularly precarious situations. There are Health and Safety issues due to mining operation. Acceleration of mining activities may lead to number of death at mine and on road, number of injuries at mine and on road as well as increase in Illness due to respiratory problems, Cancer, HIV etc. Small-Scale and large-scale mines employ outsiders and engage migrant labours. This situation can encourage a high prevalence of HIV/AIDS and other communicable diseases in and around mining communities. Mining wastes contaminate and pollute both surface and ground water. Women and children suffer from serious health problems because of polluted water and air like skin diseases, asthma, respiratory ailments, silicosis, tuberculosis, filariasis, hydrocele, cancer, and reproductive health problems, physical and mental disabilities. Malaria has increased in every place where there is mining due to stagnant mine pits. Mental illness and stress among women increased due to loss of livelihood, increase in sexual abuse and domestic violence.

Impoverishment

Lack of proper rehabilitation and resettlement of displaced families result in impoverishment. Most farmers have become daily wage earners, their income has declined by more than half, and over 50 percent have gone below the poverty line and pulled children out of school in order to turn them into child labourers. Many have taken to crime or prostitution. Drunkenness as a coping mechanism has increased so has wife beating. Respiratory and malnutrition based diseases like tuberculosis are prevalent. Many DPs fill the slums but are evicted from them in order to keep the city beautiful. Such impoverishment is seen in landlessness as well as in joblessness that are major two of the impoverishment risks. Then follow other risks like homelessness, food insecurity and health risks.[44] The situation is expected to deteriorate with liberalisation to which mechanisation is intrinsic. The International Labour Organisation estimates that 12 million jobs were lost in India during the first six years of liberalisation. It includes the loss of jobs in mining because of mechanisation mentioned above. Besides, most skilled jobs go to outsiders since very few land losers have the skills they require. The data given on Talcher and Jharkhand show that the tribals and Dalits suffer the impact of joblessness and downward occupational mobility more than the others do. Another recent major change is separation of the industrial area from the mining area. An effort was made during the last few decades to have the industry or thermal plant close to the mines. It gave the displaced persons a few jobs, most of them exploitative, meant for sheer survival. Even that possibility disappears with the separation of the two and with mechanisation.

Identity crisis

The PAPs/DPs face identity crisis. After the loss of land and other assets, according to Balaji Pandey, the transition to a new life begins, which involves a variety of problems for the displaced persons. If they are resettled near the original village in the same environment and with friends and relatives, it is somewhat of a relief to them, although the loss of assets means a struggle to restore their old lifestyle during the period of transition. If they are resettled at a place far away from the original village, the problem of coping with the new environment becomes traumatic for them. If they have to resettle themselves in clusters in some other villages, they have to face the problem of adjusting to the wishes and convenience of the host villagers and face their prejudices and suspiciousness. In addition to the long gap between the first announcement of the notification of land acquisition and the displacement/loss of land/common property resources, there was a further gap between displacement and resettlement. The decision-making process about the future resettlement sites also varies with the composition of castes in a particular project as expected, the decision of selecting sites tended to be hijacked in most cases by the upper caste and influential people in such a way as to suit their own interests.

There was a feeling of insecurity and a sense of helplessness. They had no precise information whatsoever about the nature of the resettlement site, and the officials were not specific about the types of facilities that the displaced persons would receive. They did not even know where to go for information, as the offices were located in distant places, and even if they managed to reach their offices, the security guards kept them out. When they succeeded in meeting them, the latter made a lot of promises, but nothing seemed to happen. They felt lonely since they had to leave their ancestral homes and friends and relatives and had to adjust to the new surrounding and live with new people. The problems faced by the displaced persons at the resettlement sites stemmed not only from the changed environment, but also from the sea change in the economic and social lifestyle at these sites that they had to adjust to and which created severe psychological stress for them. The mode of transporting the displaced persons to the new site, and the organisation of the resettlement colony itself, are important indicators of the nature of the arrangements provided by the project authorities, as well as the capabilities of the outstees themselves. The type of transport facilities available to the majority of the oustees when they were transferred to the resettlement sites reflects to a great extent, their economic status and the way they were treated because of it. Hence, the oustees undergo a tremendous psychological trauma and face an identity crisis.

Social disintegration

Relationships among the family members were normally cordial and intimate before land acquisition and displacement. The loss of agricultural land in the projects, disruption in the sources of earning, decrease in family income and increase in the number of family members have created conflicts and bitterness among them. Sometimes they take violent forms.

Development-induced displacement has resulted in widespread traumatic psychological and socio-cultural consequences. These include dismantling of production systems, desecration of ancestral sacred systems, desecration of ancestral sacred zones or graves and temples, scattering of kinship groups and family systems, disorganisation of informal social networks that offer mutual support, disruption of trade and market links etc. There is also a loss of social relationships which provide avenues of representation, mediation and conflict resolution. The very cultural identity of the community, as well as the individual within it, is disrupted causing immense psychological and physiological stress. Thus, it is clear that mining and subsequent displacement brings in its turn certain adverse effect and risks which have economic, social and communal implications. The impoverishments are not only the economic but also the social and cultural dimensions of impoverishment. The people have to lose all forms of capital i.e. natural, man-made or Physical, human and Social Capital, they had acquired for generations. Jobs were provided only to those who possess permanent patta for their land. The landless, labourers, artisans, share croppers, and petty businessmen among others had lost to the mining projects but did not have registered patta in their name which they could produce for their lands and, hence, were left out from the scope of job opportunities in the mining company and remained jobless. The traditional social values like mutual help arrangements, labour exchange relationships and production oriented informal organisations and other socio-economic relationships have been totally disrupted leading to social disintegration. Common property resources (CPRs) such as grazing grounds, forests etc., Community assets like temples, ponds, gathering hall which people usually call as mandap, which villagers have traditional rights and play a vital role for their livelihood have never been compensated under the R & R Act. Loss of usual rights accentuates further social disintegration process.

Loss of Sustenance and Sources of Livelihood

Forest land, where scheduled tribes and indigenous communities settled for generations are acquired for mining projects. There was great loss of sources of livelihood and sustenance for these people, primarily Adivasis who were not justly compensated for the loss of their lands, including community assets and property resources for which no compensation was paid at all. The forest, upon which people fell back for non-wood forest produce that sustained them especially during lean months, has dwindled. Loss of sustenance, sources of livelihood and subsequent impoverishment of those displaced people was, therefore, an obvious consequence. Resettling displaced populations through an equity compass would include land based resettlement, re-employment, house reconstruction, social inclusion, improved healthcare, adequate nutrition, restoration of common property resources, and community rebuilding. There is no dearth of empirical evidence to measure and prove such impoverishment risks among the displaced. Empirical studies support the fact that not only displaced families but also project affected families (PAFs) are subjected to the risks of getting impoverished. Because the project affected families also, similar to the displaced ones, lose their productive assets such as agricultural lands and their livelihood, they too face the risk of economic impoverishment. But many R&R policies ignore to address their needs, except to compensate for the loss of immovable properties. Hence, the R&R policies should consider economic rehabilitation benefits to the PAPs who were reduced to wage-earners after their agricultural lands were acquired for the projects.

Gender Dimensions

Mining activities and its process including displacement and rehabilitation brings problems and plight, woes and adversities for women. The whole family suffers the impact of joblessness, reduction in income and many of them move below the poverty line. But women and children feel its ill effects more than men do. For example, in the absence of alternatives, 56 percent of the displaced families in Assam and 49 percent in West Bengal pulled their children out of school to turn them into child labourers. Because of the trauma of forced displacement, others take to crime. Once the project alienates the land that was the foundation of their relatively high status, tribal women have lower access to work than men do. So, many of them spend their time gossiping or drinking liquor as are noticed in rehabilitation colonies in Odisha. When men waste money on alcohol, women try to maintain their families by selling their bodies since that is the only asset they own. Domestic violence increases, so do malnutrition based diseases such as tuberculosis.[45] Domestic violence and drinking existed in the past too but has increased. Alienation of land and forests deprives women of the resources that met the family's food, water and other needs but their role of providing for these needs remains unchanged. They have to attend to them with it with reduced resources[46] particularly because after displacement, joblessness is higher among women than among men. For example, a study in the coal mines of Singhbhum district of Jharkhand shows that in the past too, displaced women who wanted to work were forced to take up low paid unskilled employment. But they got at least food for their survival. Even that possibility has disappeared with mechanisation. Besides, the coming of mines changes the economy of a village drastically. The villagers were used to a barter economy in which women played an important role. They have now to compete with the salaried class to buy food in the market with no control over its price. This combination of landlessness, joblessness and lack of exposure to the market economy reduces women's access to food. Malnutrition is one of its consequences.[47] If jobs are given in a project, they go almost exclusively to men. It reduces women to being housewives alone dependent on the husband's single salary. But for exceptions, women who want to work have to be satisfied with unskilled daily wage labour. Tribal women who are deprived of the CPRs, on which their work and relatively high status depended, experience downward economic and social mobility. Many displaced men and women cope with this reality by internalising the dominant ideology. For example, because of the influence of consumerist values that accompany the project, men spend much of their income on clothes and entertainment and reduce the share of the family. Hence, even when they earn a higher monetary income than in the past, the real income accruing to the family declines. When less food is available, many follow the dominant custom of the woman eating last after feeding the elders, men, boys and girls in that order. In case of shortages women and girls live on gruel as studies of the DP/PAP in Odisha and other areas show.[48] Another facet of internalisation of the gender ideology is the attitude towards unskilled work. Some men whom the project employs as maintenance staff do not want their wives to do domestic or other unskilled work because "it is against an office worker's dignity to have his wife doing menial work". Even women internalise the ideology of their place being in the kitchen and of not being intelligent enough for skilled work.[49] In exceptional cases as at Bhilai, some women had permanent jobs while men were daily or temporary workers. When mechanisation reduced jobs one witnessed frequent cases of the project luring the man away with the offer of a permanent job far away. The woman was thus compelled to give up her permanent job "voluntarily," and accompany her husband.[50]

The problems and the plight of the women and children are usually sidelined in the resettlement and rehabilitation policies adopted by the various development project authorities. Women and children who were the most adversely affected by the displacement caused by the projects but were not given any special attention. The development projects have not adopted any special measures to look after the women and children who have been ousted by their activities. The nature and type of problems, the marginalised and deprived women and children will confront in their day-to-day life, as also in their socio-economic and cultural environment in the resettlement colonies, need serious attention. Their problems are quite different from those faced by the male oustees, and have often become psychologically traumatic for many among them. They vary from region to region. Studies by various authors and researchers have identified different adverse impacts of mining and serious repercussions of displacement on women and their daily existence.[51] There is; (i) increasing drudgery in collecting fuel, fodder and water for the family, (ii) loss of earnings due to the absence of non-farm activities of their past rural existence, (iii) devaluation of the women's status in the family because of the loss of their traditional sources of earnings, (iv) Even the marriageable girls among the oustees were not considered socially valuable partners by the host community in the areas where they resettled. DFs were away from the network of their kinsmen, the women, of a marriageable age, faced difficulties in negotiating marital alliances, (v) The compensation payments are usually made only to the male heads of the displaced households, the female members of these households feel deprived of the rights that they used to enjoy in their homes in the pre-displacement period, (vi) Women and children are more vulnerable to Health hazards, and malnutrition, from which both tend to suffer, further aggravates their health condition. (vii) A rising trend in prostitution is often associated with displacement. In the case of families ousted by development projects, displacement leads to disruption of legitimate sources of earning by women and out of despair, they ultimately resort to prostitution to earn their living. In some cases, they are forced to do so in the course of their working as domestic helps in houses of other people. The reasons for the increasing trend of prostitution are mainly due to disruption of the traditional sources of earning and decline in opportunities for self employment, (viii) Even working women face various problems. Women affected by the projects make a living by working as domestic servants in the houses of the officials. But they are subjected to immense exploitation by their male partners in the colonies where they live. Besides, they are required to carry out hazardous tasks which are detrimental to their physical safety. (ix) The absence of sanitation and hygiene facilities in resettlement areas also caused problems mainly for women, who require greater privacy than their male counterpart.

Problems of Children

Development-induced-displacement affects various categories of population, the children among them being the most vulnerable. Children in mining areas and children of displaced families face various health hazards. Malnutrition among children is a regular feature. As a result of displacement, the school going children are pushed into the labour-market, and drop-out rates in schools increase. Due to the deterioration in living standards, the people are unable to spend on educating their children.

Increasing Expenditure and Indebtedness

The PAP's expenditure has increased considerably, particularly in the case of Scheduled Tribes. This is due to change in the lifestyle of the people engaged in mining activities which exposed them to the forces of urbanisation and market economy. Previously, they were more or less self-sufficient and most of the things were obtained either through barter or from the forest and surrounding environment. Now they have to depend on the market for everything, which has made them severely indebted. Similarly, industrialisation and subsequent urbanisation has made these people, who were earlier not linked to any type of consumerism, come under the demonstration effect. They are now buying things from the market, such as two-wheelers, electronic gadgets, and so on.[52]

Resistance Movements

There were agitations/ resistances by displaced villagers in Talcher Coal Mining area for proper rehabilitation. One such agitation by Kalamachhuin during March, 2011 had affected Coal transportation. It was reported[53] that Coal transportation and production in two mines in Talcher were disrupted for the ninth consecutive days following an indefinite agitation by displaced villagers at Kalamachhuin. The villagers are demanding jobs and proper rehabilitation. Purna Chandra Pradhan, the president of Kalamachhuin village committee, alleged: "Though Mahanadi Coalfields Limited (MCL) authorities had promised to absorb 631 displaced persons (DPs), they did not keep their word. We have lost all our agricultural and homestead land for the Balarama and Hingula coal mines, but we are yet to be properly rehabilitated and resettled. We would continue our agitation till our demands are met". As many as 700 families were displaced and 1,100 acres of land were acquired by MCL. So far, 44 displaced villagers have got jobs. Resentment was brewing among the villagers following a notice by MCL, asking the villagers to vacate their homes before getting jobs, However, MCL authorities maintained that the villagers were refusing to honour the terms of the rehabilitation package. "We have provided compensation to the affected villagers in lieu of their land and houses. We can only give them jobs only after the villagers dismantled their houses and vacate the village," an MCL officer said. But villagers refuted this. The reason for this is that some of the villagers do not want to vacate their land even after getting employment and compensation packages from the company. Eviction notice was issued by the State Officials. But even after the best of efforts by the company officials, they have not vacated their dwellings. Sources from MCL revealed that Upendra Bhukta, a resident of Padmabatipur near Bharatpur open cast mines, refuses to vacate his dwelling located near the coal mines demanding employment in the company.[54] Advisory Committee has been constituted for rehabilitation, employment and peripheral development activities of the displaced persons in Talcher coal mining area under the chairmanship of Revenue Divisional Commissioner of Northern Division. It was reported that[55] in the Advisory Committee meeting held on 08.11.2006, two local MPs (one from Lok Sabha and another from Rajya Sabha) and one Local MLA walked out of the meeting on the issue on employment of displaced persons. However, there was discussion of Peripheral development activities in the meeting. There were cases of sufferings of displaced persons due to mistakes of MCL authorities. Such case was reported in a local daily.[56] One Sri Prasanna Kumar Nayak, a displaced person who lost 2.80 acres of land running from Pillar to post for compensation and job due to mistakes by MCL authorities in recording his title as 'Naik' instead of 'Nayak'. He also produced an affidavit, as directed by MCL authorities but of no avail. He put forth his grievance before District Collector and Secretary, Revenue Department, Government of Odisha. They directed the MCL authorities to fulfil the genuine claim of Sri Nayak. But still, this was not done and Sri Nayak was suffering.

A Lok Sabha Petition Committee under the chairmanship of Sri Prabhunath Singh reviewed the position of 'Rehabilitation of displaced families displaced by MCL Company' on dt. 24.12.2006.[57] It transpired in the review meeting that out of 9000 families displaced till that date, 7000 families were rehabilitated. Balance 2000 displaced families were yet to be rehabilitated. These displaced families have time and again agitated and took recourse to strike and Road blockade. However, it was decided by the committee that MCL authorities should adopt the 'Rehabilitation policy of Government of Odisha and decided the rehabilitation of displaced people jointly with the local MPs. The issues on rehabilitation of families displaced by MCL were also discussed in the State Legislative Assembly on 29.11.2006. Several MLAs across the Party lines demanded to rehabilitate the displaced families on war footing basis. However, the then Revenue Minister assured the house that a House Committee will be formed and the committee will study the rehabilitation aspect and recommend the Government for proper rehabilitation of displaced persons. Government will thereafter take a decision on the matter.[58] There were incidents of strikes of Road blocks by the families displaced due to MCL. One such case was observed in Siarmal village under Hemgir block of Sundargarh district.[59] Land acquisition for the coal mines of Mahanadi Coal Fields Ltd (MCL) at Siarmal village under Hemgir block of Sundergarh district has hit a road block. Sources said a section of the prospective displaced families under the banner of Siarmal Displaced Persons Committee stalled a resurvey work carried out by a team over a host of issues. The survey team was deputed by the Claim Commission appointed by the Supreme Court. According to sources, a survey in February 2012 had identified about 400 houses coming under the mining area for acquisition. Subsequently, 116 houses were excluded citing them as 'new constructions' as the cut-off date was fixed at September, 2010. Committee president and Zilla Parishad member of Hemgir (A) Rabiratna Patel said after appeal, 50 of the total 116 houses, had been taken up for inquiry. However, he demanded inclusion of all the houses in the resurvey. Patel claimed that no one of the affected persons was earlier informed about the provision of the cut-off date and all should be adequately compensated. He further demanded that those who would lose farm land and homestead land alongwith other poor families who got four decimal of homestead land from the government should be treated as Category A. He threatened that the committee would intensify the agitation if their demands go unheard. MCL's General Manager for Basundhara-Garjanbahal (B-G) area, AK Singh said since the Claim Commission is looking after the issue, the MCL has no direct role in it. Sources at the Sundargarh-based camp office of the Commission said if necessary, the matter would be referred to the Commission's headquarters for further action. As per the norms, the Claim Commission receives records of the land rights and compensation roles from MCL. Compensation packages are prepared as per the provisions of the R&R policy of 2006 for final approval. The Odisha High Court has asked both the Centre and the State, besides Coal India Limited (CIL) and Mahanadi Coalfields Limited (MCL), to file an affidavit each on the "unscientific mining" in Talcher. The thickly populated business district at Hat-tota in Talcher town faces the danger of sinking under the surface, the court was told earlier. This is because of the abandoned coal mines, which haven't been filled with sand, Purna Chandra Sahu, a councilor of Talcher municipality, said in his petition. With water seeping into the tunnels that run some 200 ft. under the surface, a few places are already subsiding, said his counsel. "Methane, a highly combustible gas, is coming out in a few areas. It is only anticipated that things will get worse, if steps are not taken immediately" he added. Hattota is a prime business location. A number of banks offices and several educational institutions, a primary health care centre, and even civil courts, are situated in the area. The councilor pleaded with the high court for setting up of an expert's committee to assess the danger facing Talcher town and its neighbouring areas as a result of the "unscientific" mining of coal.[60]

The case study of rehabilitation and impoverishment effects of some mining projects reveals that there is wide scale discontentment among the displaced families. There is lack of necessary infrastructure for their sustenance. Instances of apathetic attitude of implementing agencies as well as project authorities have come to the fore. Corporate Social Responsibility need be strongly monitored and implemented. In order to have a sustainable mining, it is imperative that responsibilities and accountability have to be clearly determined and lapses should be dealt with seriously. Mining has, however, some favourable social impacts. Mining projects create good infrastructure like roads, electricity, school buildings, drinking water facilities, housing etc. which are of immense help to the local people. Most importantly, mining creates job opportunities not only for the locals but also outside the mining region. Besides, mining provides education, training and skill development opportunities in numerous fields. Now reputed mining companies and corporations have developed Eco-tourism centres in mine sites. Influx of tourists to these areas helps in enhancing the socio-economic standards of the people in those areas both directly and indirectly.

Conclusion

In almost every country including India, mining projects have adverse social impacts. The most affected persons in mining projects are tribals and other indigenous people who have no land rights to support their claim for any compensation. Financial and other benefits are paid for loss of land and assets and not for any rights whatsoever, therefore, in every possibility, may not come under the purview of enumeration. However, the available information is a pointer to the fact that the displacement is huge in areas of rich mineral deposits and the trend will continue in the future. Greater the displacement, higher will be the risk of impoverishment of the displaced. Besides, there is adverse environmental impact. Though the mining affected persons experience similar problems like persons displaced by other types of development projects but its negative socio-environmental impact must be greater than that of other projects. State-wise situation of major mining states reveals that mining area differs from state to state but its social impact and impoverishment risks on the displaced families are more or less similar. Government of Odisha have taken some effective steps to alleviate negative impacts of mining and put in place an elaborate administrative structure and monitoring mechanism in mining sector. Inspite of all these arrangements and measures, there is, however, no significant improvement in the economic conditions of people in the mining areas. Rehabilitation of PAFs/DFs of mining projects is far from satisfactory.

End Notes & References

1. Fernandes and Bharali, 2006.
2. Ramanathan; 1999.
3. Fernandes and Asif, 1997.
4. IBM Report, 1998.
5. Report of Government of India, '*Mining and Rural Development*', 1993.
6. Reort of Govt. of India, 1985.
7. Pattanaik and Panda, 1992.
8. The Times of India, Bhubaneswar, Saturday, August-18, 2012 captioned "*SC stipulates norms for Karnataka Mines*".
9. Arrparmpil, Mathew, "*Displacement due to mining in Jharkhand*", Economic and Political Weekly, Vol. XXXI, No. 24, June 15, 1996.
10. Times of India, September, 2010, "*Whose land is it anyway – It's more says farmer, we'll take it if we need it. Says the Government*".
11. Fernandes, Dr. Walter & Mohmd. Asif, '*Basachyutanka Durdasha*', Indian Social Foundation, Gaunli Bichar, Bhubaneswar, p. 6.
12. World Bank Operational Policy, 4.12, approved 23 October 2001.
13. The first National Release of the 6th State of India's Environment Report titled, "*Rich land, Poor People-Is Sustainable Mining Possible?*", December, 2007.
14. Ministry of Rural Development, Government of India, 1925.
15. Civil appeal Nos. 4602 and 4602/97, Supreme Court of India.
16. Letter No. 16/48/97-MV.VI dated 10th July 2000 of Ministry of Mines, Government of India.
17. (i) Indian Social Institute, *Development, Displacement and Rehabilitation in all Tribal areas of Odisha*-1992- New Delhi. (ii) Fernandes. 1991:250.
18. Dalua, Dr. A. K. "Development Projects and Induced Displacement",– Mass Media ltd. Cuttack 1993.

19. Council Professional Social Workers (CPSW), 'St*ate of Odisha's Environment*', Bhubaneswar, 1994.
20. Fernandes, 2007.
21. *Ibid.*
22. Fernandes and Raj, 1992.
23. Downing, 2002: P.5; http://www.worldbank.org.csesmp
24. Mathur- HM and Marsaden Leds, David, *Op.cit,* pp. 22-23.
25. Das, Veena, *"Dislocation and Rehabilitation, Defining Field"*, Economic & Political Weekly, Op.cit.
26. Mathur, HM and Marsdenleds, David, *Op.cit,* pp. 24-25.
27. Background Paper - Mines, minerals and people for the Indian women and mining seminar, New Delhi, April, 2003.
28. Mohapatra, 1960.
29. Pandey, Balaji, *'Depriving the under privileged for development',* Institute for socio-economic development, Bhubaneswar, p. 36.
30. *Ibid,* p. 38.
31. Mishra, Banikanta, *"Agriculture, industry and Mining in Odisha in the post Liberalisation Era: An inter-District and inter-State Panel Analysis".*
32. Nayak, Birendra in odia, *"Bauxite Alumina and Aluminium in Odisha"* Sameekshya, 98,odia Odisha Bikalpa Sandhani Manch (e.d) 1998, pp. 60-65.
33. *Ibid,* p. 66.
34. Mishra, Banikantha.
35. Financial Express, 16/12/05.
36. CEC Report, Sept. 2005.
37. *Ibid.*
38. *Ibid.*
39. Blum, 2003, p. 263.
40. Kalam, A.
41. Graham, 1982.
42. Pandey, Balaji, *'Impoverishing effects of coal mining projects: A case study of five villages in Odisha'*, in H.M. Mathur and Marscen (Ed) "*Development Projects and impoverishment Risks",* New Delhi, Oxford, 1998.
43. Fernandes, Walter, Mines, Mining and Displacement in India, pp. 2-9.
44. Cernea, Michael, 2000.
45. Fernandes and Raj, 1992.
46. Ganguly Thukral and Singh, 1995.
47. George, 2002, p. 17.
48. Fernandes and Raj, 1992; pp. 153-155.
49. Menon, 1995, p. 101.
50. Sen, 1992, pp. 392-394.

51. Pandey Balaji, 'Impoverishing effects of coal mining projects: A case study of five villages in Odisha', in H.M. Mathur and Marsden (Ed) Development Projects and impoverishment Risks, New Delhi, Oxford, 1998.
52. *Ibid.*
53. Times of India, "Agitation by displaced villagers hits coal production in Talcher", March, 13, 2011.
54. The Times of India, "Land stir hits coal mines" dated 12.06.2011.
55. The Samaj, "Parbatara Mushika Prasaba, Khyatigrasta Prajanka Nijukti O Thaithana hela Nohin".
56. The Samaj, "MCL rabhal Pain (Misery displaced due to mistakes MCL dated 16.11.2006.
57. The Samaj, "Koila Mantralayara Talatula Niti, 2000 Basachyta Paribara Thaithanaru Banehita, (policy of coal ministry, Govt. of India, 2000, displaced families are deprived of rehabilitation) dated 25.12.2006.
58. The Samaj "Kampani dwara bisthapita Lokanka thaithan na hebaru Vidhan Sabhare udbegu" (concern in State Assembly for rehabilitation of displaced families) dated, 30.11.2006.
59. The New Indian Express, "MCL's Land Acquisition Hit" Bhubaneswar, dt. 08.08.2013.There are reports of unscientific mining by Coal India Limited (CIL) and MCL.
60. The Times of India "HC notice to centre, State on 'unscientific mining', dated 05.11.2006.

Chapter- 9

Policies and Guidelines

Hundreds of thousands of people around the globe are forcibly evicted for development projects leaving them impoverished and subject to social exclusion. The practice of involuntary displacement without alternatives and adequate compensation is illegal, unjust and unlawful in terms of international law and fundamental human rights principles. It is disquieting, disconcerting with devastating consequences for those affected. Besides, mining operation itself, from the beginning to closure, leaves a trail of adverse impacts which needs to be mitigated through proper protection and preservation measures. Hence, for smooth Management of Land acquisition for public purposes/development projects, management of displacement, resettlement action plans, creation of public utilities and regulate mining operations, the international bodies, Central as well as State Governments have enacted laws, framed rules and regulations and issue guidelines from time to time.

This chapter highlights the stipulations/guidelines on displacement and policy prescriptions on rehabilitation/ relocation by the international law enforcing organisations besides Acts, Laws and Rules guiding resettlement at the national and state level. There are discussions on the policies and provisions of R&R by different State Governments and Government of Odisha in particular. At the end, there is a concluding note.

International Laws and Guiding Principles

Internationally, a majority of the countries have in place Acts, Laws, rules and regulations for the mining sector. Over the last decade, most of the countries have modified/ revised these Acts and regulations and included elaborate provisions. However, their process of implementation and enforcement differ from country to country. These regulations pertain to, inter alia, the control, governance and monitoring mechanisms, implementation procedure, ensuring ethical business behaviour and non-corrupt practices, protection of environment, ecology and natural resources, social needs, needs of local communities and community development, safety at workplaces, social programmes, welfare supports and economic compensation for individuals and communities who are more likely to face the marginalisation and vulnerabilities associated with indigeneity, lower socio-economic status or rural living and necessary measures for closure and the post-closure of mines.

As regards human rights, there are international human rights laws, global United Nations (UN) human rights treaties[1] such as the Convention on the Elimination of all Forms of Discriminations, the International Covenant on Economic, Social and Cultural Rights, the Convention on the Rights of Persons with Disabilities, the International Covenant on Civil and Political Rights, and the Convention on the Elimination of Racial Discrimination,[2] to mention a few. Over a period of time, an elaborate human rights has emerged,[3] in the form of human rights based approaches (HRBAs) to development involving both State and non-State actors.[4] In 2008, the then UN High Commissioner for Human Rights observed that "the international human rights legal framework, to which all States have subscribed, must be seen as part of the solution and the baseline commitment on development".[5] These HRBAs are meant to address many outstanding issues viz. inequality relating to gender,[6] age[7], ethnicity[8], disability[9], poverty[10], sexual orientation,[11] geography[12] etc., discrimination, social exclusion, marginalisation, needs and rights of especially vulnerable or marginalised groups[13] such as indigenous peoples[14], persons with disabilities[15] and people living in remote rural areas.[16] International human rights law has established that human rights are universal, inalienable and inherent to being human. This implies that human rights apply to all people, everywhere and always. The human rights are indivisible and interdependent i.e. all human rights are equally important and that often the realisation of one right is dependent on and or reinforced by the realisation of another right[17]. This was emphasized in the 1986 UN Declaration on the Right to Development which described the right to development as "an inalienable human right by virtue of which every human person and all peoples are entitled to participate in, contribute to, and enjoy economic, social, cultural and political development".[18] Environmental protection concerns were added to this list by the 1992 Rio Declaration on Environment and Development.[19] Human rights dimensions have also to address environmental contents and concerns such as the air quality, water, waste disposal, environmentally sound management of chemicals and wastes so as to reduce their negative effects on both human health and the environment, combating and adapting to climate change, conserving and sustainably using the marine environment, forest management and land degradation[20].

The Commission on human rights of United Nations developed a framework i.e. “Guiding principles” for the internally displaced people / families. These Guiding Principles address the specific needs of internally displaced persons worldwide. They identify rights and guarantees relevant protection of persons facing involuntary and forced displacement and assistance during displacement as well as during rehabilitation, resettlement and reintegration.[21] Paragraph 1 of the Guiding Principles approach displacement from the perspective of the needs of internally displaced persons. Paragraph 2 is a descriptive identification of the category of persons whose needs are the concern of the Guiding Principles. It highlights two elements: (1) the coercive or otherwise involuntary character of movement; and, (2) the fact that such movement takes place within national borders. The second requirement is to be understood in a broad sense. It refers to the place where the displaced persons find refuge. Paragraph 3 of the Guiding Principles is not a UN declaration on the rights of internally displaced persons, nor do they constitute, as such, a binding instrument. However, they reflect and are consistent with international human rights law and international humanitarian law. The Guiding Principles restate in more detail those legal provisions that respond to the specific needs of internally displaced persons and spell them out in order to facilitate their application in situations of internal displacement.[22] The five sections of these Guiding Principles are: Section I – General Principles, Section II – Principles relating to protection from displacement, Section III – Principles relating to protection during displacement, Section IV – Principles relating to humanitarian assistance, Section V – Principles relating to return, Resettlement and Reintegration. These five sections contain 30 Guiding Principles on Internal Displacement (ID). International human rights and humanitarian law is remarkably developed in the case of forced evictions. The United Nations Commission on Human Rights has described forced evictions as a ‘gross violation of human rights’ and intergovernmental bodies have issued numerous and detailed standards. International and regional courts and quasi judicial expert bodies have frequently declared forced evictions to violate a plethora of human rights while certain forced evictions fall foul of international humanitarian, criminal and refugee law. A dense web of international law seeks to ensure that human rights are protected in cases of proposed or actual evictions. They commonly require that evictions can only proceed in accordance with national law, in certain prescribed circumstances and with due process. The details of the various international standards and jurisprudence have emerged in the context of various human rights and international humanitarian, criminal, labour and refugee law. The *right to adequate housing* was recognised as long ago as 1948, in the Universal Declaration on Human Rights, as well as in 1966 in the International Covenant on Economic, Social and Cultural Rights. The General Comment set out a number of positive obligations which require that states take steps to reduce the risk of forced evictions. The guidelines on development–based displacement, developed by a group of experts and submitted by the UN Secretary-General to the Commission on Human Rights in 1997[23], are notable for their detailed prescriptions on adequate resettlement and compensation.

United Nations have also issued Comprehensive Guidelines on Development Based Displacement. Apart from preamble, the comprehensive guidelines have several sections. Section-I outlines background issues, Section-II and III deals with general obligations and specific preventive obligations respectively. The sights of all persons like integrity of the Home, Assurance of security of tenure are reflected in Section IV. Section V focuses on legal remedies including compensation, restitution and return and resettlements. Monitoring issues are highlighted in Section VI and Savings clause is in Section-VII. The comprehensive guidelines stipulates that 'States should secure by all appropriate means, including the provision of security of tenure, the maximum degree of effective protection against the practice of forced evictions for all persons under their jurisdiction. In this regard, special consideration should be given to the rights of indigenous peoples, children and women, particularly female-headed households and other vulnerable groups. These obligations are of an immediate nature and are not qualified by resource-related considerations. States should ensure that no persons, groups or communities are rendered homeless or are exposed to the violation of any other human rights as a consequence of a forced eviction. All persons have the right to adequate housing which includes, inter alia, the integrity of the home and access to and protection of common property resources. The home and its occupants shall be protected against any acts of violence, threats of violence or other forms of harassment, in particular as they relate to women and children. The home and its occupants shall further be protected against any arbitrary or unlawful interference with privacy or respect of the home. In full cognizance of the contents of the guidelines, there may be instances in which, in the public interest, or where the safety, health or enjoyment of human rights so demands, particular persons, groups and communities may be subject to resettlement. Such resettlement must occur in a just and equitable manner and in full accordance with law of general application. All persons, groups and communities have the right to suitable resettlement which includes the right to alternative land or housing, which is safe, secure, accessible, affordable and habitable'.

Policies and Procedures of the World Bank

The World Bank was the first international development agency to respond to the complexity of the problems endangered by displacement. The World Bank Guidelines on Involuntary Settlement play an important role in regulating World Bank funded development projects and have been partially reflected in the policies and procedures of the regional development banks, for example, the Asian Development Bank and Inter-American Development Bank. Together with the World Bank Inspection Panel, these guidelines provide the form of recourse for affected communities.[24] The World Bank adopted a policy resolution in 1990, which laid down a formal policy and institutional procedures to address issues relating to the displacement process and subsequent resettlement needs.[25] The negative consequences of displacement are typically aggravated by (a) the absence of a strategy for redressing the problems resettlement creates and (b) by inadequate planning and execution.[26]

The Salient features and important points of World Bank R&R Policy are as follows.

(i) Displacement of human settlement should be avoided as far as possible, if unavoidable, displacement should be reduced to the minimum.

(ii) All resettlement programmes must be development programmes so that the productive base and income earning ability of resettlers are improved and they share the benefits of the project which displaced them.

(iii) Socio-economic survey of the area and population to be affected should be conducted well in advance.

(iv) It is the responsibility of concerned Government to relocate the affected persons and for which an appropriate rehabilitation action plan with time schedule is prepared.

(v) While planning resettlement programmes, as well as in executing them, the affected people should be consulted either directly or through their representatives.

(vi) To avoid misunderstanding, friction among the resettlers and the host population, it is necessary to meet equitable treatment in the matter of development activities, civic amenities etc. (similar to the people of adjoining areas).

(vii) The approach to resettlement programme should be to ensure that the oustees regain at least their previous standard of living and as far as practicable, they should be economically and socially integrated into the host communities. To achieve this, a realistic action plan should be prepared in a manner that would give the settlers the opportunity to become physically established and economically self-sustaining in the shortest possible time.

The Bank also published in 1982, an operational manual statement entitled "Tribal People in Bank finance Projects" which deals with tribal displacement exclusively. The manual inter alia, speaks; (a) special measures should be adopted for tribal displaced persons as they are likely to be more harmed than helped by the project, (b) the project report and estimate should include measures necessary to safeguard the interests of tribal oustees and to enhance their wellbeing as far as practicable, (c) there should not be any encroachment on the traditional territories being used or occupied by the tribal people and (d) efforts should be underlined to protect tribal integrity etc.

The Bank went a step further by introducing IPDP (Integrated People's Development Plan) to safeguard the interest of indigenous people who are not only displaced but also who indirectly affected by the project within a given radius. It is commonly understood that the R&R Package aims at payment of compensation for the lost assets and adequate rehabilitation opportunities so that the affected person is not marginalised economically after displacement. Tribals being extremely vulnerable, World Bank aims at assessing the social and cultural impact of the development project on the tribals and their development needs. For this limited purpose, the affected (tribal) people are those who live either within villages or making up whole villages, or hamlets which are affected by the project e.g. by losing community assets essential to the well being of that community. The assessment will include two kinds of village communities: (i) villages surrounding relocated villages within a radius of five kms. which contain members of the same tribal community as those relocate, and (ii) those villages where some families will be relocated to be left with unsustainable means of living, and others who will not be affected but are left behind.

Economic, Social, Cultural, Civil and Political Rights on Forcible Eviction

Forcible eviction of people from their homes commonly violates a wide range of other economic, social and cultural rights. In India, the Supreme Court first derived the prohibition on forced evictions from the right to a livelihood[27]. Other commonly affected rights include education, health and social security. However, what is missing from much legal analysis of forced evictions is the manner in which forcible evictions from the home are accompanied by direct violations of economic, social and cultural rights, in particularly the right to work, food, health, water and education. Forced evictions in rural areas often result in removal of people from land they use for agriculture, while other large-scale forced evictions result in the destruction of property used to maintain a livelihood. In order to partly address this deficiency, the opening section of the proposed guidelines addresses the need for States to ensure sufficient protection against evictions from productive land and confiscation and destruction of property used to sustain livelihoods or protect other economic, social and cultural rights.

The interdependency of human rights is clearly demonstrated in the case of forced evictions. The most voluminous jurisprudence on forced evictions has emerged from courts and quasi-judicial bodies concerned with civil and political rights, namely the right to respect for the home, right to privacy, right to life and right to freedom of movement and right to property. What is clear from recent judgments is that there is a discernible trend towards strengthening protections against forced evictions through the enforcement of the human rights.

Guidelines for Indigenous people

While international law has long concerned itself with regulating the confiscation of private property, redressing centuries of brutal evictions and dispossession of indigenous people has received attention of international body, sadly large-scale forced evictions and systematic dispossession of indigenous peoples continues in many countries. Much attention has focused on the development of the Draft Declaration of Indigenous Peoples by the former Working Group on Indigenous Peoples under the UN Commission on Human Rights. The document is clear on the subject of redress: “Indigenous peoples have the right to restitution of the lands, territories and resources which they have traditionally owned or otherwise occupied or used; and which have been confiscated, occupied, used or damaged without their free and informed consent. Where this is not possible, they have the right to just and fair compensation. Unless otherwise freely agreed upon by the peoples concerned, compensation shall take the form of lands, territories and resources equal in quality, size and legal status."[28] The draft declaration also provides for continuing protection of indigenous lands. However, 'despite resting in the bowels of the Commission for two decades, the declaration is yet to shed its status of draft'.[29] The ILO Convention No. 169 (Indigenous and Tribal Peoples Convention, 1989) likewise provides similar textual protections, namely three separate articles on recognition of ownership (Article 14), safeguarding of natural resources (Article 15) and protection from removal (Article 16). Relocation is forbidden except in exceptional circumstances and where there is free and informed consent of the peoples concerned. Where possible indigenous peoples shall have the right to return and compensation shall be provided in the form determined by the victims. Perhaps the strongest protections can be found in the widely ratified International Bill of Rights (ICCPR and ICESCR). Malcolm Longford and Jean du Plesis[30] have developed guidelines for the prevention and remedy of forced evictions. The guidelines seek to comprehensively reflect the existing state of international law and in a number of instances provide additional detail. A section on the obligations of non-state actors has been added which reflects emerging international law, as encapsulated in the OECD Guidelines on Multinational Enterprises, the UN Sub- Commission's draft Resolution on Responsibilities of Transnational Corporations and Other Business Enterprises with Regard to Human Rights and the World Bank guidelines on resettlement. But the challenge is to not only incorporate international law in legislation, policies and procedures but to establish a political and social culture that does not tolerate forced evictions.

Governance of Mineral Resources in India

Institutional, Regulatory and Legal Framework and Institutional Governance of Mineral Resources in India are mainly divided into five categories: fuel, metallic, non-metallic, atomic and minor. Government of India has set up dedicated ministries for undertaking the work of survey, exploration, mining, production, demand-supply management, auctioning, marketing and pricing of minerals in correspondence to the different mines and mineral categories. Ministry of Mines is the apex institutional organisation responsible for formulation and administration of policies, rules, regulations and laws relating to mines in India with regards to survey, exploration, mining, production, auctioning, pricing and revenue collection of all minerals (including minor minerals), other than natural gas, petroleum, atomic minerals and coal. Ministry of Coal solely deals with survey, exploration, mining, production, auctioning and pricing of coal and lignite reserves while the Ministry of Petroleum & Natural Gas is responsible for exploration, production, supply distribution, marketing and pricing of petroleum including natural gas and petroleum products. Department of Atomic Energy is responsible to carry out geological exploration and discovery of mineral deposits (mainly uranium reserves) required for the Atomic Energy power programme of India.

Acts, Rules and Guidelines of Government of India

The Indian mining sector is governed by several policies, acts, rules and regulations that are jointly managed by the Central and state governments. The main policies and acts are deliberated, formulated and administered by specific ministries through a central legislation under the leadership of cabinet ministers, while the state governments are required to enact these regulatory policies. The main acts that govern the entire mineral sector of India are the Mines Act, 1952 and Mines and Minerals Development & Regulation Act (MMRD) of 1957 under the administration of MoM while the National Mineral Policy of 1993 is the key reference document of rules, regulations, principles and procedures for regulating, developing and controlling the mineral sector (excluding coal, petroleum and atomic minerals). "Significant rules are in force under the MMDR-1957 are the Mineral Concession Rules (MCR), 1960 and the Mineral Conservation and Development Rules (MCDR), 1988. This act was amended in 1972, 1986, 1994, 1999 and 2015 and after each amendment corresponding changes were carried out in MCR and MCDR".

Mines and Minerals Development and Regulation Act, 1957 (MMDR-1957)

MMDR-1957 is the principal legislation that lays down the regulatory and governance framework of mineral and mining sector at the central level in India. Over the years of economic developments and corresponding changes in the mineral sector, this Act was amended multiple times. This Act is divided into eight chapters with 33 specific sections. Administration and regulatory governance of major minerals is controlled by the Central Government while that of minor minerals are entirely delegated to respective state governments. Section 3 (e) of MMDR 1957 gives the Central Government the power to notify 'minor minerals' (by adding or deleting the minerals in the list) while Section 15 provide state governments complete powers for making rules regarding minor mineral mining, levy, collection of royalty and so on. On the other hand, mining activities for all types of major and minor minerals in offshore areas like territorial waters, continental shelf, exclusive economic zone and other maritime zones of India is the exclusive responsibility of the Central Government. For regulating the mining and development of minerals in the offshore area, the Parliament has enacted the Offshore Areas Minerals (Development and Regulation) Act, 2002 which empowers the Central Government under the administrative authority of the Indian Bureau of Mines (IBM) to grant mineral concessions for offshore areas and collect royalty. Over a period of time certain flaws were experienced in MMDR 1957. Considering the lacunas in various clauses under this Act, the Ministry of Mines (MoM) amended it for the fifth time in 2015. As a result, the MMDR Amendment Act, 2015 replaces the MMDR, 1957 with some additional provisions. More minerals were added in the notified minerals category under an additional Fourth Schedule, a new category of mining licence i.e. the prospecting licence-cum-mining lease was created, provided relaxation for maximum area limits. Now, the Central Government can increase the area limits for mining, instead of providing additional leases. Under this act the mining lease period for all minerals other than coal, lignite and atomic minerals has been increased from 30 years to 50 years and after the licence lapses, the lease will be put for auction and cannot be renewed. Besides, the Act provided the state governments the power for granting mining leases and prospecting license-cum-mining leases for both notified and other minerals with the approval of Central Government. The Act has provided for the creation of a District Mineral Fund (DMF). Under this Act, every mine lease holder is mandated to contribute 30 percent of the royalty amount (for leases granted prior to January 12, 2015) and 10 percent of royalty (for leases granted after January 12, 2015) to the DMF which is required to be utilised by the District Collector/Magistrate for development and socio-economic and environmental activities in and around areas where mining operations are carried by the lease holder. The Act has also laid down rules and regulations for the creation of a National Mineral Exploration Trust (NMET) Licencees and lease holders are mandated to contribute two percent of royalty towards NMET exploration activities.

National Mineral Policy, 2008

The first National Mineral Policy (NMP, 1993) for non-fuel and non-atomic minerals which was the most important legal framework for the mines and mineral sector in India was implemented in the year 1993. The policy lays down provisions with regards to regulation of minerals, role of the state in mineral development, survey and exploration, national inventory of mineral resources, strategy of mineral development, foreign trade, fiscal aspects and research and development. NMP 1993 recognised the need for encouraging private investment including Foreign Direct Investment (FDI). There have been considerable changes in the Indian mining sector since then which led to the revision of this policy and enactment of NMP, 2008 (for non-fuel and non-coal minerals). This policy stresses the importance of practicing sustainable mining in order to preserve and augment the exhaustible minerals reserves and aims for optimal utilisation of natural mineral resources of the country. All mining, according to new policy shall be undertaken within the parameters of a comprehensive Sustainable Development Framework which includes guiding principles for a miner to leave the mining area in a better ecological condition after mining.

National Mineral Exploration Policy, 2016

The policy provides for an action plan to ensure comprehensive exploration of mineral resources (non-fuel and non-coal) in the country. The policy addresses the gaps of data consolidation, calls for the establishment of a National Geo-scientific Data Repository (NGDR) – a repository that will assemble all mineral baseline and exploration data, puts special impetus to explore deep seated mineral deposits, provides for the establishment of a National Centre for Mineral Targeting (NCMT) dedicated to increasing the discovery rate in mineral exploration and the quality of discoveries.

Other Mineral Policies, Acts, Rules and Regulations in India: An overview

Besides the Policies, Acts, Rules and Regulations discussed above, there are various other policies enacted by Government of India pertaining to mining sector.[31]

1. Mines Act, 1952: This Act regulates the working conditions of workers in mines, and their safety measures. It also oversees regulations for carrying out mining operations and management.

2. The Mines Rules, 1955: These rules outlined the framework for the basic health examination of persons employed in mines, and sanitation provisions and welfare facilities for miners and their families.

3. Mines Concession Rules, 1960 (Amended on 2012): As per provisions of Section 13 of MMDR Act 1957 these rules define the process of grant of mineral concessions with regards to reconnaissance permits, prospecting licences and mining leases which are granted by state government. These rules were further amended in 2012.

4. Oilfields (Regulation and Development) Act, 1948: This Act regulates the grant of licences for exploration of oilfields by the Central Government.

5. Petroleum and Natural Gas Rules, 1959 Amendments in 2013: This Act regulates the grant of exploration licences and mining leases in respect of petroleum and natural gas.

6. Coal Bearing Areas (Acquisition and Development) Act, 1957: This Act establishes public control over the coal mining industry and its development by providing for acquisition by the State of unused land containing or likely to contain coal deposits.

7. The Coal Mines (Nationalisation) Act, 1973: This Act regulates the acquisition and transfer of the right, title and interest of the owners in respect of the coal mines specified in the Schedule.

8. Coal Mines (Conservation & Development) Act, 1974: This Act aims to provide for conservation of coal and development of coal mines and for matters connected therewith or incidental thereto.

9. Coal Mining (Nationalisation): This Act was amended to allow private coal mining for captive consumption for generation of power, washing of coal obtained from a mine and other end users to be notified by Government from time to time in addition to the existing provisions.

10. Colliery Control Rules 2004: These rules specify the role of the coal Controller, who will be responsible for overseeing and regulating the compliance of mandatory requirements for the colliery owners.

11. Granite Conservation and Development Rules, 1999: These rules were framed for conservation and systematic development and scientific mining to conserve the granite resources and to prescribe a uniform framework with regard to exploitation of granite throughout the country.

12. Marble Development and Conservation Rules, 2002: These rules were framed for systematic prospecting and quarrying or mining of marble and to provide a uniform framework with regard to scientific exploitation of marble.

13. The Offshore Areas Mineral (Development and Regulation) Act 2002: This Act regulates the development of mineral resources in territorial waters, continental shelf, exclusive economic zone and other maritime zones of India.

14. The Offshore Areas Mineral Concession Rules, 2006: The rules lay down the process for grant and renewal of reconnaissance permits, exploration licences and production leases as per provisions of Section 35 of the Offshore Areas Mineral (Development and Regulation) Act, 2002. The rules prescribe for measures for protecting the marine environment and safety measures to be followed in the leased area.

15. Minerals (Evidence of Mineral Contents) Rules, 2015: This provides the set of rules that set the procedures to be followed for conducting any exploration to determine the mineral content and to take up the mineral blocks for auction and mineral concessions.

16. Mineral (Non-exclusive Reconnaissance Permits) Rules, 2015: It laid down the process to be followed for grant of Non-exclusive Reconnaissance Permit - any operation undertaken for preliminary prospecting of a mineral through regional, aerial, geophysics, or geochemical surveys and geological mapping, but does not include pitting, trenching, drilling, or sub-surface excavation.

17. Mineral Conservation and Development Rules, 1988 Amended on 2011, Mineral Conservation and Development Rules, 2017: As per the provisions of sections 18 of MMDR Act 1957, these rules prescribe guidelines for conservation and development of minerals to ensure scientific management of the mining process.

18. National Mineral Exploration Trust Rules, 2015: Under Section 9C (1) of MMDR 2015, the Central Government established a Trust, called the National Mineral Exploration Trust, which is a non-profit organisation. The object of the Trust is to use funds accrued to the Trust for the purposes of regional and detailed exploration in such manner as may be prescribed by the Central Government.

19. Mineral (Auction) Rules, 2015: The rules specify auctioning mechanisms with regards to value of estimated resources for auction of block; value of mineral dispatched; sale price of mineral (grade-wise and state-wise); Reserve Price; Auction prerequisites; Bidding Process; Upfront Payment; Performance Security; and Mine Development and Production Agreement (MDPA).

20. Atomic Minerals Concession Rules, 2016: The rules prescribe the procedures for regulating the grant of reconnaissance permits, prospecting licences and mining leases in respect of Atomic minerals and for purposes connected therewith.

Mining Sector Environmental Protection Policies, Rules and Regulations in India

There are also some policies, Acts and regulations in India for protection and conservation of ecology and environment by the mining sector which is perceived as a major contributor to environmental degradation throughout the globe. Mining sector has a dubious distinction of a 'polluting industry' with substantial amount of environmental footprint that exacerbates ecological degradation. Extraction and processing of minerals create considerable negative environmental and social effects like destruction of flora and fauna, clearance of large tracts of land, air, water, soil, noise pollution, disrupting local ecological balance, wiping out local biodiversity etc. All the stages and phases of mining are having a specific environmental footprint. Some of these Acts[32] are as follows:

1. EIA under Environment Protection Amendment Act 2006 (EIA, 2006) regulates the construction, expansion and modernisation of developmental projects that have a potential threat to the environment in different parts of India. This law mandates prior environmental clearance to be obtained for a listed project before starting any operation. As per 2006 notification, EIA is required for projects listed in category A and B1 project. In the case of mining sector, more than 50 hectares of mining lease area in respect of non-coal mine lease and 150 hectares of lease area in respect of coalmine lease are included in Category 'A', along with offshore and onshore, oil and gas exploration and development. Category 'B' projects cover mining lease area between 5-150 hectares in respect of coal and 5-50 hectares in respect of other minerals. Mineral prospecting, however, is excluded from prior environment clearance.

2. Sustainable Development Framework (SDF) under MMDR 2015: As per this, it is mandatory for every mining company to follow SDF standards in order to achieve five star ratings which are based on mining operation performance on techno, socio-economic and environmental parameters. The star rating of mines is based on self-certification which is verified by the Apex committee constituted under the aegis of the IBM.

3. The Forest (Conservation) Act, 1980: As per this Act, it is mandated on the state government to have consultation and approval before declaring any forest as 'Non-Forest' zone which has been specifically reserved as 'Forest Zone'. The power to make rules under the Act has been delegated to the Central Government.

4. Coastal Zone Regulation, 2011: This regulation strictly controls the setting up and expansion of any industry, operations or processes and manufacture or handling or storage or disposal of hazardous substances as specified in the Hazardous Substances (Handling, Management and Trans-boundary Movement) Rules, 2009 within a strip of 520 metres from the shore in coastal areas.

5. The Water (Prevention and Control of Pollution) Act, 1974, Amended in 1988: The prevention and control of water pollution and the maintenance or restoration of wholesome quality of water is provided

as per the Act and for this purpose, it vests power in the State Pollution Control Boards (SPCB) to lay down and enforce effluent standards for 'trade effluents' i.e. any liquid, gaseous or solid substance discharged by industrial establishments including mines and processing plants.

6. Water (Prevention and Control of Pollution) Cess Act 1977: This Act stipulates for the levy and collection of a cess on water consumed both by persons carrying on certain industries and by local authorities to augment resources for the pollution control boards. Mining industries are included under this the Act and accordingly a cess is charged per kilolitre water used for different activities. However, industries and operations which have installed mechanisms to keep water pollution under control are given a rebate in the cess.

7. The Air (Prevention and Control of Pollution) Act, 1981 Amended in 1987: This Act stipulates that all industries (including mines) operating within designated air pollution control areas must obtain permits (consent) from SPCBs. The standards of emission of air pollutants in to the atmosphere from operational activities and vehicles are prescribed by state boards in consultation with the central board and this is monitored regularly.

8. Sustainable Sand Mining Guidelines, 2016: With a view to maintaining and restoring the ecology of river and other sand sources, these guidelines specify sustainable mining practices and environment friendly management practices.

Land Acquisition, Displacement and R&R Policies of Government of India

India has a legislation at the national level for the acquisition of land and property for setting up the projects and there is a comprehensive policy for the process of displacement and rehabilitation of persons displaced as a result of various development projects. The National Working Group on Displacement in 1989 notes "in the face of such serious human impact, the efforts made for the resettlement of displaced persons are utterly disappointing, hardly touching the fringe of the problems. The approach has been ad hoc, provisional and unprincipled. It varies from State to State, project to project, and from authority to authority. No single, uniform comprehensive common national policy has been adopted". Till 2007, under the then existing rehabilitation policies, rules and guidelines are formulated by various authorities and agencies of the State government dealing with the resettlement of the oustee families. But these rules are made within the framework and principles of the Land Acquisition Act, under which lands and other properties are acquired for most of the projects (except the coal mining projects, where Coal Bearing Areas-Acquisition and Development, Act 1957 is used for the purpose). But this act is absolutely unsuitable to tackle the problems of the community displacement, compensation for lost resource-base, loss of livelihood, and of rehabilitation.[33] Denied any participation in the planning, or a share in the benefits from the projects, the oustees have moved from a state of property to pauperisation in most of the cases.[34]

For the first time, the Ministry of Home Affairs issued a set of instructions in 1965 for grant of concessions to persons whose lands are acquired for any project, in the matter of employment in the public sector projects. In the initial phase of plan development period, employment in projects was an easy means of rehabilitation. But it created a very unhealthy tendency. Every displaced person claimed employment, for which scope was not there for permanent absorption. Due to legal constraints, there may be little scope for retrenchment. In 1984, the committee of secretaries of different ministries of Govt. of India constituted a small group to take a look at the guidelines issued by different ministries in the matter of R&R of the tribal oustees. The group reported that although the instructions issued by different ministries were very sound in principle, very little was done to implement them and there is need to spell out the mode of implementation of these guidelines. However, in consonance with the deliberations in the committee of secretaries referred to above the ministry of Industry, Dept. of Public Enterprise, Bureau of Public Enterprises, issued a set of comprehensive guidelines broadly touching upon the issue of Land Acquisition and Rehabilitation in their office memorandum No. 15/13/84, B.P.E. (c) Dt. 3.2.86. These guidelines are perhaps the first and foremost in spelling out a definite approach to L.A. and R&R activities. So far as the R&R activities are concerned, the guidelines contained that (i) land based rehabilitation as well as economic rehabilitation through self employment should be provided. (ii) Further, keeping in view that the growing population will not permit land based rehabilitation to all the displaced persons, provision was made to limit land based rehabilitation to those whose land including homesteads are acquired for the project and provided further that the persons concerned were themselves cultivating those lands or residing in those homestead land. (iii) Absentee land lords were not entitled to any rehabilitation assistance. (iv) Priority was given to a person losing land completely, over a person losing it partially. (v) Alternative house sites with all civic amenities were to be provided to those incumbents who lost their homestead land. (vi) Training facilities at project cost were to be offered to eligible affected persons so as to enable them to get employment but there will be no commitment for ultimate employment. (vii) The project authorities will also assist the displaced persons in taking to useful avocations like poultry, goatery, piggery, dairy, fishery and various crafts etc.

The working group on development of scheduled tribes during seventh plan highlighted the need for formulating a national policy on rehabilitation of displaced persons with provision for special treatment of the tribal oustees. The major recommendations of the working group are the following:

(i) There should be a national policy on rehabilitation of displaced persons with provision for special treatment of the tribal oustees in the matter of employment in projects and vocational training for self employment.

(ii) Group rehabilitation should be the rule instead of individual rehabilitation and all civic amenities should be provided.

(iii) Land in command areas should be allotted or else irrigation facility should be provided in areas other than command area.

(iv) Agriculturist tribal oustees should be given preference for land based rehabilitation and landless should be settled in other avocations of their choice.

(v) There should be a full fledged Directorate of Rehabilitation in every State. State level Rehabilitation committees under the Chairmanship of Chief Secretary should be formed to review the programme and progress of rehabilitation activities.[35]

Land Acquisition Act; 1894 (Amended in 1984 and From time to time)

The Land Acquisition (LA) Act ; 1894 of Government of India is a Law of British times by which Government forcibly acquires Private property for "Public Purposes" and is different from market purchase of land. This LA Act of 1894 was amended in 1984 and from time to time to address all issues and concerns related to LA. The coal bearing Act was implemented in 1957 for LA in coal bearing areas. A systematic and comprehensive R&R policy was promulgated by Government of India in 2007. However, The Right to Fair Compensation and Transparency in Land Acquisition, Rehabilitation and Resettlement Act, 2013 was promulgated by Government of India w.e.f. September, 2013. The salient features and pertinent points of the aforesaid Acts/Policies and draft bills are illustrated in foregoing sections.

Salient Features of Land Acquisition Act, 1894 (Amended in 1984 and from time to time) are as follows:

(1) Whenever it appears to Government that land in any locality is needed or likely to be needed for public purpose, notification to that effect shall be published by Government in Revenue Department in the official gazette, and the Collector shall cause public notice or the substance of such notification to be given at convenient places in the said locality {Section 4(1)}.

Thereupon, it shall be lawful for any officer, either generally or specially authorised by Government in this behalf and for his subordinates to enter upon and survey and take levels of land, and other preliminary activities.

(2) Damages caused if any in any course of preliminary work referred to above shall be paid for. (Section 5).

(3) Aggrieved person/party if any may file objection within thirty days after the issue of notification under section 4(1); to the Collector. {Section 5(A)}

(4) Declaration shall be made by Government in Revenue Department to the effect that land of given specifications will be required for the specific purpose. No such declaration shall be made unless the compensation to be awarded for such property is to be paid. (Section 6)

(5) After declaration under section-6, Government (in Revenue Department) shall direct the Collector to take order for acquisition of land. (Section 7).

(6) Collector shall cause the land to be measured and make a plan of the same. (Section 8).

(7) Collector shall then cause public notice to be given at convenient places stating that Government intends to take possession of the land and that claims to compensation for all interests in such land may be made to him. (Section 9).

(8) Collector may require any such person to make delivery to him, at a time and place mentioned about interest of other persons if any. (Section 10).

(9) Collector shall enquire into the objections if any in response to notice under section 9 and into the interests of persons claiming the compensation, and shall make an award indicating the land to be acquired, compensation money to be paid etc. (Section 11).

(10) The award made under section 11 shall be final and the Collector shall give immediate notice of the award to such of the persons interested as are not present personally or by their representatives when the award is made. (Section 12).

(11) In case of emergency, Government if so desires, possession of land can be taken over on the expiry of fifteen days from the publication of notice under section 9. (Section 16).

(12) On making award under section 11, Collector shall tender payment of compensation or may deposit in the Civil Court. (Section 31).

(13) In the event of non-payment of compensation (including deposit in the Court) interest shall have to be paid on the awarded compensation amount from the time of taking possession. (Section 34).

(14) Determination of market value of land:

(i) The market value of land is to be determined taking into account the sale(s) transactions at the date of publication of 4(1) notification. This is the best type of determination of market value. If the market value of the land at such date is not available, the market value of the land prevailing on or close to the date of submission of the proposal may be taken into consideration.

(ii) If no such sale(s) transactions are available, sales of such quality of land in the vicinity within a fairly recent date can be taken into account.

(iii) If no such sale(s) in the vicinity is available, the net annual produce of the land can be taken into account for valuation.

(iv) Value awarded in previous Land Acquisition proceedings is also relevant evidence on the question of market value.

(v) For determination of market value of agricultural lands, the annual letting value is the best course open. The net profit the tenant derives from the land over a certain number of years should be taken as the letting value.

(15) While determining the market value of the land, the Collector should also include the value of trees, houses, Tanks, Wells, standing crops etc. those are standing on the land.

(16) What would include valuation:

(i) In addition to the market value of the land, additional market value @12 percent of the said original market value shall be paid in every case for a period from publication of notification U/s 4(1) to the date of award of the Collector or the date of taking possession of the land which ever is earlier, Similarly, solatium @30 percent of the said original market value of the land shall also be added in every case.

(ii) Capitalised value of 25 years purchase of the annual revenue is to be provided in the preliminary estimate in all Irrigation, Power and Multipurpose Projects which are treated as Commercial "Projects" except for Irrigation Research Institute and Flood Control Works.

(iii) Besides the above, 20 percent of the estimated compensation amount of land towards establishment cost, news paper publication expenses, law charges and contingencies will be realised from Projects not bearing expenses of Special Land Acquisition Organisation. Where the Project bears expenses of Special Land Acquisition organisation, 10 percent of the estimated compensation amount of land will be realised and added in the estimate.

The National R&R Policy, 2007

The National R&R Policy, 2007 of Government of India, Department of Land Resources, Ministry of Rural Development was adopted to minimise displacement and to ensure adequate rehabilitation package.

Salient Features of the National Rehabilitation and Resettlement policy, 2007 of Government of India are as follows:

(1) Social Impact Assessment (SIA) of Project

Whenever it is desired to undertake a new project or expansion of an existing project, which involves involuntary displacement of four hundred or more families en masse in plain areas, or two hundred or more families en masse in tribal or hilly areas, DDP blocks or areas mentioned in the Schedule V or Schedule VI to the Constitution, the appropriate Government shall ensure that a Social Impact Assessment (SIA) study is carried out in the proposed affected areas.

(2) Rehabilitation and Resettlement Plan

The procedure mentioned in this chapter shall be followed for declaration of the affected area, carrying out survey and census of affected persons, assessment of government land available and land to be arranged for rehabilitation and resettlement, declaration of the resettlement area or areas, preparation of the draft rehabilitation and resettlement scheme or plan and its final publication.

Where the appropriate Government is of the opinion that there is likely to be involuntary displacement of four hundred or more families en masse in plain areas, or two hundred or more families en masse in tribal or hilly areas, DDP blocks or areas mentioned in the Scheduled V or Scheduled VI to the Constitution due to acquisition of land for any project or due to any other reason, it shall, declare, by notification in the official Gazette, area of villages or localities as an affected area.

(3) Rehabilitation and Resettlement Benefits for the Affected Families

(i) The rehabilitation and resettlement benefits shall be extended to all the affected families who are eligible as affected families on the date of publication of the declaration under paragraph 6.1, and any division of assets in the family after the said date may not be taken into account.

(ii) Any affected family owning house and whose house has been acquired or lost, may be allotted free of cost house site to the extent of actual loss of area of the acquired house but not more than two hundred and fifty square metre of land in rural areas, or one hundred and fifty square metre of land in urban areas, as the case may be, for each nuclear family.

(iii) Each affected family owning agricultural land in the affected area and whose entire land has been acquired or lost, may be allotted in the name of the khatedar(s) in the affected family, agricultural land or cultivable wasteland to the extent of actual land loss by the khatedar(s) in the affected family subject to a maximum of one hectare of irrigated land or two hectares of un-irrigated land or cultivable wasteland, if Government land is available in the resettlement area. This benefit shall also be available to the affected families who have, as a consequence of the acquisition or loss of land, been reduced to the status of marginal farmers.

(iv) The land or house allotted to the affected families under this policy shall be free from encumbrances.

(v) The land or house allotted to the affected families under this policy may be in the joint names of wife and husband of the affected family.

(vi) In case of allotment of wasteland or degraded land in lieu of the acquired land, each khatedar in the affected family shall get a one-time financial assistance of such amount as the appropriate Government may decide but not less than fifteen thousand rupees per hectare for land development.

(vii) In case of a project involving land acquisition on behalf of a requiring body which involves involuntary displacement of two hundred or more Scheduled Tribes families, a Tribal Development Plan shall be prepared, in such form as may be prescribed, laying down the detailed procedure for settling land rights due but not settled and restoring titles of tribals on alienated land by undertaking a special drive together with land acquisition. The plan shall also contain a programme for development of alternate fuel, fodder and non-timber forest produce (NTFP) resources on non-forest lands within a period of five years sufficient to meet requirements of tribal communities who are denied access to forests.

The Right to Fair Compensation and Transparency in Land Acquisition, Rehabilitation and Resettlement Act, 2013

Salient Features of The Right to Fair Compensation and Transparency in Land Acquisition, Rehabilitation and Resettlement Act, 2013[36] are as follows:

The Act has started with the following words with concerns for displaced persons.

'To ensure, in consultation with institutions of local self-government and Gram Sabhas established under the constitution, a humane, participative, informed and transparent process for land acquisition for industrialisation, development of essential infrastructural facilities and urbanisation with the least disturbance to the owners of the land and other affected families and provide just and fair compensation to the affected families whose land has been acquired or proposed to be acquired or are affected by such acquisition and make adequate provisions for such affected persons for their rehabilitation and resettlement and for ensuring that the cumulative outcome of compulsory acquisition should be that affected persons become partners in development leading to an improvement in their post acquisition social and economic status and for matters connected therewith or incidental thereto'.

The Act has 13 chapters and four schedules. The broad provisions of the Act are: determination of social impact and public purpose, Preparation of Social Impact Assessment Study, Public hearing for Social Impact Assessment, Publication of Social Impact Assessment Study, Appraisal of Social Impact Assessment Report by an Expert Group, examination of proposal by a committee constituted, by the appropriate government, Constitution of committee to examine proposals for land acquisition and the Social Impact Assessment Report, Exemption from Social Impact Assessment, Special provisions to safeguard food security, notification and acquisition, Publication of preliminary notification and power of officers thereupon, Preliminary survey of land and power of officers to carry out survey, Payment for damage, Lapse of Social Impact Assessment Report, Rescission of preliminary notification, Hearing of objections, Preparation of Rehabilitation and Resettlement Scheme by the Administrator, Review of the Rehabilitation and Resettlement Scheme, Publication of declaration and summary of Rehabilitation and Resettlement, Land to be marked out, measured and planned including marking of specific areas, Notice to persons interested, Power to require and enforce the making of statements as to names and interests, Enquiry and land acquisition award by Collector, Land acquisition process under Land Acquisition Act, 1894 shall be deemed to have lapsed in certain cases, Period within which an award shall be made, Determination of market value of land by Collector, Determination of amount of compensation, Determination of value of things attached to land or building, Award of solatium, rehabilitation and resettlement award, Rehabilitation and Resettlement Award for affected families by Collector, Provision of infrastructural amenities in resettlement area, Corrections to awards by Collector, Adjournment of enquiry, Power to summon and enforce attendance of witnesses and production of documents, Power to call for records, etc, Awards of Collector when to be final, Power to take possession of land to be acquired, Special powers in case of urgency to acquire land in certain cases, procedure and manner of rehabilitation and resettlement, Appointment of Administrator, Commissioner for rehabilitation and resettlement, Rehabilitation and resettlement committee at project level, Provisions relating to rehabilitation and resettlement to apply in case of certain persons other than specified persons, national monitoring committee for rehabilitation and resettlement, Establishment of National monitoring committee for rehabilitation and resettlement, Reporting requirements, establishment of land acquisition, rehabilitation and resettlement authority, Establishment of Land Acquisition, Rehabilitation and Resettlement Authority, Composition of Authority, Qualifications for appointment as Presiding Officer, Terms of office of

Presiding Officer, Staff of Authority, Salary and allowances and other terms and conditions of service of Presiding Officers, Filling up of vacancies, Resignation and removal, Orders constituting Authority to be final and not to invalidate its proceedings, Powers of Authority and procedure before it, Proceedings before Authority to be judicial proceedings, Members and officers of Authority to be public servants, Jurisdiction of civil courts barred, Reference to Authority, Collector's statement to Authority, Service of notice by Authority, Restriction on scope of proceedings, Proceedings to be in public, Determination of award by the Authority, Form of award, Costs, Collector may be directed to pay interest on excess compensation, Re-determination of the amount of compensation on the basis of the award of the Authority, Appeal to High Court, apportionment of compensation, Particulars of apportionment to be specified, Dispute as to apportionment, payment, Payment of compensation or deposit of same in Authority, Investment of money deposited in respect of lands belonging to person incompetent to alienate, Investment of money deposited in other cases, Payment of interest, temporary occupation of land, Temporary occupation of waste or arable land, procedure when difference as to compensation exists, Power to enter and take possession and compensation on restoration, Difference as to condition of land, offences and penalties, Punishment for false information, *malafide* action, etc., Penalty for contravention of provisions of Act, Offences by Companies, Offences by Government departments, Cognizance of offences by court, Offences to be non-cognizable, Offences to be cognizable only on complaint filed by certain persons, miscellaneous, Magistrate to enforce surrender, Service of notice, Completion of acquisition not compulsory but compensation to be awarded when not completed, Acquisition of part of house or building, Acquisition of land at cost of a local authority or Requiring Body, Exemption from stamp duty and fees, Acceptance of certified copy as evidence, Notice in case of suits for anything done in pursuance of Act, No change of purpose to be allowed, No change of ownership without permission to be allowed, Return of unutilised land, Difference in price of land when transferred for higher consideration to be shared, Provisions to be in addition to existing laws, Provisions of this Act not to apply in certain cases or to apply with certain modifications, Power to amend Schedule, Power of State Legislatures to enact any law more beneficial to affected families, Option to affected families to avail better compensation and rehabilitation and resettlement, Power of appropriate Government to make rules, Rules made by Central Government to be laid before Parliament, Rules made by State Government to be laid before State Legislature, Previous publication of rules made by Central and State Government, Power to remove difficulties and Repeal and saving.

Rules and Guidelines of Government of Odisha

Since there was no comprehensive displacement and R&R Policy of the centre at all India level to be adopted by all the States till 2007, few states initially developed R&R policies for their own states. Maharashtra was the first state to formulate such an Act as early as in 1976. The Maharashtra Resettlement and project Displaced Persons Act, 1976 has in fact been the only legislation in this field for many years, and a trend setting one in India at that point of time. In 1977, Odisha for the first time conceded the right of a displaced landless person to the same land based rehabilitation as that for one who lost the legally owned land. This greatly reduced impoverishment risks of the Scheduled Tribes, who were mostly landless.[37]

Displacement, Resettlement and Rehabilitation (R&R) Policies of Government of Odisha

Our case study region, Odisha, did not have a detailed and comprehensive policy on displacement and R&R till 2006, due to the lack of which, the process of resettlement and rehabilitation of uprooted people has been minimal and not very successful.[38] Although Odisha has gained the recognition of being one of the early states to have come-out with R&R policies and a set of guidelines for the irrigation and hydroelectric projects, there are no uniform guidelines or policies for the mining projects. There is a need for a comprehensive R&R policy that is applicable uniformly to the mining, industrial sector and other sectors as well. The R& R policy should provide as and when needed, a specific focus and treatment to the mining and industrial projects because these projects are different from infrastructure projects such as dams or roads because they are separate in nature in terms of its impact and operation. The former has a "business" or "commercial" connotation and allows for more private sector participation, change of ownership, etc. The Industrial Policy Resolution of 2001 for the first time stresses the need for bringing about "a new relief and rehabilitation mechanism" that would take into account, the interests of all the stakeholders.[39] Till 2006, different projects in different sectors have adopted different compensation measures and rehabilitation procedures. A Chronology of Development induced Displacement Policies and rules for different projects in Odisha are as follows:

1948- Promulgation of Odisha Development of industries, irrigation, Agriculture, Capital construction and Resettlement of Displaced Persons Act: 1948.

1950- Hirakud Dam Policy

1960- Rourkela Steel Plant Policy

1973- Rengali Dam Project Policy (For the first time a Rehabilitation Policy was declared by State Government)

1977- Provision for compensation to landless displaced persons was made. (R&R Policy: 1977)

1984- Policy for NALCO displaced persons was declared.

1989- South Eastern Coal Fields under Coal India Ltd. Policy (Now Mahanadi Coalfield Ltd.)

1993- Chandipur (Interim Test Range displaced persons) Policy

1994- A comprehensive State Policy for all displaced persons displaced by irrigation projects (R&R Policy: 1994)

1996- Gopalpur displaced Policy person for proposed TATA Steel Plant

1997- Kalinga Nagar Industrial Complex at Jajpur-Policy for the displaced persons

1998- Policy for displaced persons as a result of extensive mining

1999- Policy for displaced people for Aditya Aluminum Project

2003- Utkal Alumina and Vedanta Alumina: Policy for the displaced

2005- (18th September 2005) Rehabilitation policy for the displaced families of Kalinga Nagar.

2006- Odisha Resettlement and Rehabilitation Policy (R&R Policy: 2006)

There was no single comprehensive policy till 2005 applicable for all projects including mining projects. Salient features of some of the important policies are given below.

Odisha Development of Industries, Irrigation, Agriculture, Capital Construction and Resettlement of Displaced Persons Act: 1948

Soon after Independence, in 1948, the Government of Odisha adopted the Odisha Development of Industries, Irrigation, Agriculture, Capital Construction and Resettlement of Displaced Persons (Land Acquisition) Act, 1948 for payment of compensation/ award to be paid by the competent authority to the persons entitled thereto as against the acquired land, standing crops if any and buildings thereon. However, nothing was mentioned regarding the resettlement of those ousted under the Land Acquisition Act, on alternative sites.

Resettlement and Rehabilitation Policy, 1977

It was the agitation of the oustees who were evicted by the Rengali

Dam Project in the 1970s which compelled the Odisha State Government to seriously think of formulating a systematic policy for rehabilitation of development project displaced people.[40] A revised resettlement and rehabilitation policy was announced on April 20, 1997 vide Irrigation Department Resolution No. 13169. The policy was an improvement on the earlier one. The two main features that marked out this policy were:

(a) Lands for resettlement in Government sponsored colonies were to be allotted to the oustees, irrespective of the fact whether the oustees were landless or not;

(b) Displaced persons who wanted to make their own rehabilitation arrangements in lieu of the land that was acquired from them, were to be given the cash that they might require to buy the land.[41]

However, the revised rehabilitation and resettlement policy suffered from a number of shortcomings. The policy did not provide for any special measures for the tribals.

Uniform Policy: 1990

A number of amendments were made later to the 1977 resettlement and rehabilitation policy in view of the multi-faceted problems faced by the oustees to rebuild their homes, restore their livelihoods and cope with the new environment, among other things. Hence, a uniform resettlement and rehabilitation policy for the displaced persons in various projects was announced during August, 1990. Special Features of the Uniform Policy were : A son above 18 years of age irrespective marital status, Physically handicapped and mentally retarded persons irrespective of age and sex, Divorcees with dependent child/children, and divorcees without children but no source of livelihood were to be treated as a separate family unit and Minor orphans, who had lost both parents and no body to fall back upon, were also recognised as oustee families and thus eligible for benefits.

The Odisha Resettlement and Rehabilitation (R&R) of Projects Affected Persons (PAPs) Policy, 1994

In 1994, the Government of Odisha, in consultation with civil societies, NGOs and the World Bank came up with Odisha Resettlement and Rehabilitation of Projects Affected Persons policy vide Resolution No. 25296 dated August 27, 1994. Among the significant points in this new policy, the concepts of 'affected person', 'affected zone' and 'affected village' proved to be an improvement on the conceptual frameworks of previous policies.

Odisha Resettlement and Rehabilitation (R&R) Policy, 2006 (as amended up to June-2007)

In order to ensure a sustainable development through a participatory and transparent process, Government of Odisha have framed R&R Policy vide Revenue and Disaster Management (R&DM) Department Resolution No. 18040 14th May, 2006 (as amended up to June, 2007) with the objectives to avoid displacement where possible and minimise it, exercising available options otherwise, to facilitate resettlement/rehabilitation process, recognising voices of the displaced communities (emphasising the needs of the indigenous communities and vulnerable section) and ensuring environmental sustainability through participatory and transparent process. The Policy has provisions for Survey and Identification of Displaced Families, Land Acquisition and Payment of Compensation/Award, Resettlement and Rehabilitation plan, Rehabilitation Assistance for Industrial Projects, Mining Projects, Irrigation Projects/National Parks and Sanctuary, Urban Projects and Linear Projects and any other Projects along with Special benefits to displaced Indigenous Families and Primitive Tribal groups.

Mining Policies of Government of Odisha

Government of Odisha had brought out a policy for rehabilitation of displaced persons/ families in case of Mining projects in the State during 1998.

Rehabilitation Policy:1998

The policy for rehabilitation of displaced persons/ families of Mining projects; 1998 was circulated among all concerned including Revenue Divisional Commissioners and all Collectors vide erstwhile Revenue and Excise Department letter No. 58349 dated 09.11.1998. The policy outlines the eligibility criteria for rehabilitation benefits, Definition of family for the purpose of benefits, Definition of family for the purpose of rehabilitation, survey and identification of displaced families, Employment, Land for homestead purpose, facilities to be extended to homeland less and landless encroachers, incentive for timely vacation, indexation of Rehabilitation grant and formulation and monitoring the rehabilitation plan. The Salient points of the policy are illustrated as under.

1. A family person shall be termed 'displaced' and, hence, eligible for rehabilitation benefits if such family/person has been a permanent resident of Odisha and ordinarily residing in the project area for at least three years prior to the date of publication of notification.

2. The family in relation to displaced person would mean husband and wife and their unmarried children, except unmarried daughters above 30 years, on or before the date of notification.
3. For the purpose of treating a woman as divorcee, legal document if any, in support of the claim will be sufficient The claimant can also make an affidavit and file her claim along with a copy of such affidavit for eventual decision after due enquiry by the R&RO.
4. For the purpose of treating divorcees and widows having no source of livelihood as separate family, financial benefits like old age pension, etc. availed of by them should not be taken into account.
5. Physically handicapped and/or mentally retarded person irrespective of age and sex to the extent and in the manner as indicated below:
   (a) The blind, the deaf, the orthopedically handicapped and mentally retarded persons suffering from more than 40 percent permanent disability may be enumerated as separate family for the purpose of rehabilitation.
   (b) Certificate issued by a registered and duly authorised medical practitioner in the particular discipline will form the basis for acknowledging the claims of physically and mentally retarded persons.
   (c) The certificate furnished by the medical practitioner should indicate the nature and percentage of disability.
6. All unmarried brothers and sisters whose parents are not alive shall constitute a separate family.
7. Ordinarily within two months of publication of notice U/S 4(I) of the Land Acquisition Act or U/S 7(I) of CBA (A.D) Act of 1957 for acquisition of land for the project, a survey would be undertaken in the manner to be decided by the Collector for identification of displaced families and for preparing their socio-economic profile.
8. An identity card to each displaced family shall be issued under the signature of the Collector concerned.
9. The company setting up the Mining Project will have to give preference to the nominated members of the displaced families in providing employment either in the Project or in its ancillary units in the following order of preference.
   (a) Displaced families losing all lands including homestead land.
   (b) Displaced families losing more than 2/3rd of land and homestead land.

(c) Displaced families owning more than $1/3^{rd}$ of land and homestead land

(d) Other displaced families

10. For nominated members of displaced families not provided with employment in the manner indicated above, the company will assist in their self employment through allotment of shops, service premises and other activities in the company's township or otherwise.

11. The Company shall provide appropriate training facilities to the nominated persons of the displaced families to acquire skill as required for such employment.

12. With a view to encouraging the displaced families for self employment or in useful avocations, the company will provide necessary guidance and support and assistance to the families.

13. The families who are not provided employment or self employment as indicated above or who refuse to avail of such employment/self employment would be entitled to a onetime cash grant as follows.

1.	Families coming under category a and b of para-4	Rs 1.00 lakh.
2.	Families coming under category c and d of para-4	Rs 50,000/-

14. Each displaced family will be given $1/10^{th}$ of an acre of land for homestead purpose free of cost in a resettlement colony with basic infrastructure facilities.

15. Each of the displaced families who opts for self relocation elsewhere other than in the resettlement colony will be given a onetime cash grant of Rs 50,000/-.

16. Encroachers who are landless (as defined in the OPLE Act, 1972) will get ex-gratia equal to compensation admissible under L.A. Act, 1894 for a similar category of Rayati land to the extent of land under their physical possession up to a maximum of one standard acre, If the encroachment is unobjectionable.

17. The encroachers who are homestead less (as defined in the OPLE Act, 1972) will get ex-gratia equal to compensation for similar category of Rayati land against the encroached homestead land up to a maximum of $1/10^{th}$ of an acre if the encroachment is unobjectionable and in addition the actual cost of structure thereon. If the encroachment is objectionable, they will be entitled to the cost of the structure only.

18. In order to ensure timely vacation an allowance of Rs 500/- per month per displaced family shall be provided for a period of one year on vacation by the date fixed by the Collector.
19. Rehabilitation Grant in each will be indexed to the official wholesale price Index (WPI) with 1.4.99 as the reference date and will accordingly be revised once in two years.
20. The Rehabilitation Advisory Committee (RAC) as constituted by the Government from time to time shall prepare a comprehensive rehabilitation plan on the basis of the policy parameters outlined above and also monitor its implementation.
    (a) However, the inclusion of major son as separate family among oustees under the rehabilitation scheme is a welcome measure. But it is a shortcoming of the policy not to include the physically handicapped, divorcees, old people and women among the oustees as separate families in whose case the chances of pauperisation due to displacement are the highest.
    (b) The project authority has to resettle the oustees in rehabilitation colonies and provide employment in the project according to the policy. As a result the principle of land-for-land, the main criterion in irrigation projects is neglected. Agriculture, which is the main occupation of the people, cannot be practiced on the pre-displacement scale, as the compensation money is usually far below the market price of land.
    (c) The project provides employment and other self-employment benefits which may not benefit all oustees. The additional sources of occupation are not restored. Since the policy provides employment for one person in the family, other eligible members of the family remain largely unemployed.

Odisha Mineral Exploration Policy, 2015

Government of Odisha, vide resolution No. 6065 dated 04.07.2015 have promulgated Odisha Mineral Exploration Policy, 2015 with the objective to strengthen the institutional set up for the mineral exploration and to upscale the mineral exploration activities in the State with an aim of upgrading the exploration level to G2 level in keeping with the UNFC norms within a reasonable time frame and eventually to G1 level so that the potential of mining sector is achieved through scientific exploitation of mineral resources. The strategies like institutional mechanism, updation of status report, preparation of master plan, role of permit holders, licencees and lessees, financial support for mineral exploration and prioritisation for achieving those objectives has been outlined in the policy resolution.

R&R Policy Coal India Ltd.

In absence of any national policy, the Coal India Ltd. formulated its R&R policy in 1993 for the oustees of coal mining projects. Its main features are:-

(a) The project affected people included land owners including tribals who cultivate land under traditional rights, people who lose homestead land because of acquisition, share croppers, landless labourers, tribals depending on forest produce etc.

(b) Compensation is paid for acquisition of lost immovable assets.

(c) For the purpose of resettlement and rehabilitation, adult members of a Displaced Family are taken in to consideration.

(d) DPs desirous of land based rehabilitation will be assisted by the project.

(e) Shifting allowance will be given.

(f) Project will help in absorption in self-employment avenues like dairy, poultry, shop keeping etc.

(g) Those DPs who opt to be on their own will get one time cash for subsistence.

(h) Women DPs will get particular attention to have access to income generating opportunities.

(i) Resettlement colonies for the DPs will have all minimum basic civic facilities.

(j) The land loser will get a job in lieu of land acquired.

(k) Landless PAPs will also get opportunity for jobs when outside recruitment is made.

Government of Odisha had issued an Uniform Guidelines for rehabilitation of families/persons displaced due to acquisition of land for Coal mining operations in the districts of undivided Sambalpur and Dhenkanal during 1988 which was modified during 1989. The initial guidelines of 1988 was notified vide the erstwhile Revenue and Excise Department letter No. 35318-GE-(SBI)-96/88/R dated 2nd June, 1988, and the modified guidelines of 1989 came into force vide the erstwhile Revenue and excise Department letter No. GE (SPB) 96/88 711, dated 02.01.1989 addressed to the collectors of Sambalpur and Dhenkanal (undivided). Sailent features of the guideline of 1988 are as follows:

1. The policy is for rehabilitation of families/persons displaced due to acquisition of land for Coal Mining Operation in the districts of Sambalpur and Dhenkanal.
2. A family person shall be eligible for rehabilitation benefit if:
   (a) All the land of the family have been acquired
   (b) More than 1/3rd agricultural land along with homestead land has been acquired
   (c) Only homestead land has been acquired
   (d) Homestead less persons residing in the village

   This is to be ascertained through joint enquiry by a committee consisting of representative of the South Eastern Coal Fields Ltd. and the Government (Collectors of the Districts).
3. A family in the aforesaid context shall include the husband and wife as the case may be and their un-married children but does not include a major son who has been separated from the family on or before the date of notification under Section 4(1) of the Land Acquisition Act, 1894 in respect of area village. He shall be treated as a separate family.
4. (i) one member from each family sustaining loss of dwelling houses, homestead land and agricultural land not less than 1/3rd of the total holding shall be provided with employment on a priority basis.

   (ii) One member of each family having sustained loss of three acres of non-irrigated land or two acres of irrigated land shall be provided with employment on second priority.

   (iii) In case of families having lost only homestead land, one member from each family shall be provided with employment according to availability.

   (iv) Besides, contractors working under the S.E.C.F shall provide employment according to their requirement to members of displaced families according to eligibility. S.E.C.F shall prepare a list of such person to be provided with employment and supply the same to the contractors working under them with copy to Collectors of the Districts concerned.

   (v) Rehabilitation of Other displaced families shall be made through self employment schemes.
5. Each displaced family shall be allotted/reclaimed Government land up to 10 dec. for homestead purpose free of cost.

6. In the rehabilitation colonies, facilities like road, street lights, school building, health centre, provisions of drinking water, recreation centres, ponds, tanks, play ground, veterinary centres, tree plantation, shopping centre shall be provided according to necessity.
7. Each displaced family shall be given assistance at the rate of Rs. 2000/- for construction of temporary shed.

The modified guidelines of 1989 specifies for rehabilitation of families/ persons displaced due to acquisition of land for Coal Mining Operation in the districts of Sambalpur and Dhenkanal.

1. A family/person shall be eligible for rehabilitation benefit, if;
   (a) All the lands of the family have been acquired
   (b) More than 1/3rd agricultural lands along with homestead land have been acquired.
   (c) Only homestead land has been acquired.
   (d) Homestead less persons residing in the village.

   This is to be ascertained through joint enquiry by a committee consisting of representative of the South Eastern Coal Fields Ltd. and the Government (Collector of the District).
2. A family in the aforesaid context shall include the husband and wife as the case may be and their unmarried children but does not include a major son who has been separated from the family on or before the date of notification under section 4(1) of the land acquisition Act, 1894 in respect of area/ village. He shall be treated as a separate family.

   Regarding Employment;
   (a) One member of each family sustaining loss of dwelling house or homestead land and agricultural land not less than 1/3rd of the total holding shall be provided with employment on a priority basic.
   (b) One member of each family sustaining loss of three acres of non-irrigated land or two acres of irrigated land shall be provided with employment on second priority.
   (c) In case of families having loss only homestead land or the total agricultural holding, one member from each family shall be provided with employment according to availability.
   (d) In case of families who have lost 1/3rd of the total agricultural holding, one member from each families shall be provided with employment according to availability.

(e) Rehabilitation of other displaced families shall be made through self employment schemes.

(f) Each displaced families shall be allotted reclaimed Government land up to 10 dec. for homestead purpose free of litigation.

(g) Each displaced family shall be given maintenance of Rs 1,000/- for construction of temporary shed.

(h) Expenditure in respect of paras e, f and g shall be borne by the Government.

(i) Any rehabilitation package finalised by the Government of India, Department of Coal, towards compensation in lieu of loss of earnings, it would automatically apply to the rehabilitation schemes.

Conclusion

Some Policies, Acts and Rules enacted by the Government of India and different states as for example our case study of Odisha State as described above specifically relate to mining sector and others are general in nature applicable to all sectors. However, these Policies, Acts and Rules give a bird's eye view on the regulations in place pertaining to mining sector. Land acquisition by the Government and subsequent involuntary displacement is beset with several issues. These issues include resettlement and rehabilitation, compensation for the loss of livelihood, accentuation of conflicts etc. which are to be addressed on a priority basis. Hence, legal rules and Government policies for the process by which land is to be acquired and how current owners are compensated will play a critical role. There is, thus, a need to integrate relief measures with land acquisition initiatives, protect the interest of the displaced people, secure land rights and livelihood of the natural resource-based communities, take an optimal decision to confer legal rights on the losers of cultivable land belonging to loss of habitats, environment and bio-diversity consequent upon land acquisition and resettlement. Besides, R&R need to take into cognizance the needs of women affected by displacement. In our case study state of Odisha, among forest gatherers and farming societies, women are equal contributors to the family income, sustenance and its dynamics. However, rehabilitation packages have always ignored this aspect. In fact, gender disparities in rehabilitation policies are glaring, according to a 2005 study by the sociologist Dr Rita Ray and Vikramaditya of UNDP. The compensation money goes to the head of the family - usually a male. The Land Acquisition Act 1894, amended in 1984, which is the core law regarding acquisition and compensation rules, recognises only male ownership of land. Since land is rarely registered in a woman's name, she is out of the ambit from the beginning. The lack of formal legal title to land should not be a bar to entitlements under the new draft policy. This can make every woman a potential beneficiary. Moreover, currently, jobs-for-land types of benefits go to the married and unmarried sons; rarely to the unmarried daughters, never to the married ones. Families themselves want their sons to get the jobs. Families with only one or more daughters lose out; women-headed families with only daughters are completely invisible. For women, life after displacement is full of endless hurdles, woes and miseries. Something as crucial but often-ignored in policy parameters, are facilities like toilets. If no toilets are provided or toilets are there but there is no water, women are left with the option of using the forestland and common grazing land. However, they are not allowed by the usually hostile host community to use such places. Besides, loss of access to forest and cash-for-land compensation results in complete loss of income for women. This, combined with reduced access to water and fuel wood, is the most common problems faced by women who are relocated. That apart, post-displacement, women and minor girls are often forced to work as daily wage labourers, domestic servants or even as sex workers. The women also have to cope with alcoholism and domestic violence.[42] R&R policy should encompass to address all these issues of gender disparity in R&R activities and ensure women empowerment. Besides, Ecology and Environment, their sustenance, preservation and maintenance is an important issue particularly in the mining sector. Despite the fact that mining contributes significantly to the economic development and is a building block of the economy, the planet, our earth must also be protected from environmental degradation. Stipulations and strict regulations must be imposed by the government to control the environmental pollution.

End notes and References

1. Karin Arts, *"A Human Rights Perspective, Responsible Behaviour"* International Institute of Social Studies, The Hague, Netherlands, 2017.
2. UN Treaty Collection, for the status of multilateral human rights treaties.
3. Nelson, PJ, Dorsey, E: "*At the nexus of human rights and development: new methods and strategies of global NGOs*", World Dev. 2003, 31:2013-2026.
4 (i) Nelson, PJ: "*Human rights, the millennium development goals and the future of development cooperation*", World Dev 2007, 35:2041-2055.

   (ii) Gready, P, Vandenhole, W: "*Human Rights and Development in the New Millennium: Towards a Theory of Change*", Routledge; 2014.
5. UN Office of the High Commissioner for Human Rights: "*Claiming the Millennium Development Goals: A Human Rights Approach*", New York and Geneva: United Nations (HR/PUB/08/3); 2008.

6. Cornwall A, Rivas A: "*From gender equality and 'women's empowerment' to global justice: reclaiming a transformative agenda for gender and development*", Third World Q 2015, 36:396-415.
7. Dornan, P, Woodhead, M: "*How Inequalities Develop Through Childhood: Life Course Evidence from the Young Lives Cohort Study*". UNICEF Office of Research—Innocenti; 2015.
8. UNICEF: "*For Every Child A Fair Chance: The Promise of Equity*", UNICEF; 2015.
9 (i) Lord, JE, Raja, DS, Black, P: *Beyond the orthodoxy of rule of law and justice sector reform: a framework for legal empowerment and innovation through the convention on the rights of persons with disabilities*. World Bank Leg Rev 2012, 4:5-65.

   (ii) Bukola, RA: "*Poverty and the realisation of the millennium development goals in Nigeria: disability rights the missing link*". East Afr J Peace Hum Rights 2011, 17:532-550.
10. UN Human Rights Council: *Final Draft of the Guiding Principles on Extreme Poverty and Human Rights*, Submitted by the Special Rapporteur on Extreme Poverty and Human rights, Magdalena Sepu´lveda Carmona; 18 July: UN Doc. A/HRC/21/39; 2012.
11. Ibhawoh, B: "*Inclusion versus exclusion*", in The Sage Handbook of Human Rights. Edited by Mihr, A, Gibney, M. Sage; 2014:324- 339 (at 335–337).
12. UNICEF: "*For Every Child A Fair Chance: The Promise of Equity*", UNICEF; 2015.
13. Kabeer, L: "*Can the MDGs Provide a Pathway to Social Justice? The Challenge of Intersecting Inequalities",* United Nations Development Programme; 2010.
14. Enns, C, Bersaglio, B, Kepe, T: "*Indigenous voices and the making of the post-2015 development agenda: the recurring tyranny of participation*", Third World Q 2014, 35:358-375.
15. Stein, MA, McClain-Nhlapo, C, Lord, JE: "*Education and HIV/AIDS: disability rights and inclusive development. In The Millennium Development Goals and Human Rights: Past, Present and Future*", Edited by Langford, M, Sumner, A, Yamin, AE. Cambridge University Press; 2013:274-294.
16. Kuruvilla, S: "*The millennium development goals and human rights: realizing shared commitments*", Hum Rights Q 2012, 34:141-177 (at 153).
17 (i) Carey, SC, Gibney, M, Poe, SC: "*The Politics of Human Rights: The Quest for Dignity"*. Cambridge University Press; 2010:. especially Chapter 1.

   (ii) Bantekas, I, Oette, L: "*International Human Rights Law and Practice*", Cambridge University Press; 2013.

   (iii) Office of the United Nations High Commissioner for Human Rights: *Frequently Asked Questions on a Human Rights-Based Approach to Development Cooperation*. United Nations (HR/PUB/06/8); 2006.
18. UN General Assembly: *Declaration on the Right to Development*, UN Doc. A/RES/41/128; 1986: adopted on 4 December.
19. UN General Assembly: *Rio Declaration on Environment and Development,* UN Doc. A/CONF.151/26; 2017: Annex 1 adopted on 1 August 1992.
20. Karin Arts, "*A Human Rights Perspective, Responsible Behaviour*", *Op.Cit.*

21. Walter Kalin, "Guiding Principles on internal Displacement; The Americal Society of International Law and the Brookings Institution, 2008
22. *Ibid.*
23. 'The Practice Of Forced Evictions: Comprehensive Human Rights Guidelines on Development Based Displacement' E/CN.4/Sub.2/1997/7, 2 July 1997.
24. Dana Clark, *'The World Bank and Human Rights: The Need for Greater Accountability'*, Harvard Human Rights Journal, Vol. 15, Spring (2002), 205. Clark notes that resettlement programmes have rarely provided adequate compensation or livelihoods. The inability of the Panel to supervise its recommendations means it has little control over the remedying of violations.
25. World Bank report on funding developmental programmes.
26. World Bank (1990): Operational Directives 4.30, Involuntary Resettlement, Washington, DC (1994).
27. Olga Tellis v Bombay Municipality Corporation [1985] 2 Supp SCR 51 (India); (1987) LRC (Const) 351.
28. Draft Declaration on the Rights of Indigenous Peoples, E/CN.4/Sub.2/1994/2/ Add.1 (1994).
29. 'UN Commission Debates Draft Declaration on Indigenous Peoples', Oral Statement of Sanchay Chakma of AITPN under Agenda Item 15 of the 61st Session of the United Nations.
30. Longford, Malcom and Jean du Plessis, *"Dignity in the Rubble? Forced Evictions and Human Rights Law"*, COHRE, 2005.
31. Information extracted from official government documents of Ministry of Mines, Government of India, 2016.
32. (i) ISID, 2012

    (ii) Ministry of Mines, Government of India, 2016.
33. Thukral, E.G. *"Dams for whose Development"* in fernades, walter and E.G. Thukral (eds) *"Development Displacement and rehabilitation"* Indian Social Institute, New Delhi, 1989.
34. *Ibid.*
35. Dalua, A.K., *Op.Cit.*
36. The Gazette of India, Act No.30 of 2013, The Right To Fair Compensation And Transparency In Land Acquisition, Rehabilitation And Resettlement Act, 2013, 26th Sept. 2013.
37. Mathur, H.M. *Op.Cit.*
38. Morse, Bradford and Berger, T (1992): 'Sardar Sarovar: Report of the Independent Review' Resource Futures International, Ottawa.
39. Ravindran, Latha- *Op.Cit.*
40. Pandey, Balaji- *Op.Cit.*
41. Dalua, A.K.- *Op.Cit.*
42. Jena, Mani Padma, 'India together', 2006.

Chapter - 10

Towards Inclusive, Sustainable, Responsible and Ethical Mining

This is the concluding section that presents a conspectus of the whole discussion. There is a summary of preceding chapters in capsulation, steps 'Towards Inclusive, Sustainable, Responsible and Ethical Mining' with a prescriptive note and 'The Way Forward'. The chapter ends with a Conclusion.

Summary of Discussions

Emergence of 'Inclusiveness in Mining' is a recent approach but concern for 'Mining Sustainability' was there long before evolution of SDGs in development framework. Responsible Mining practices and Ethical Mining are not oft-discussed issues. Initially, the approach and practices were limited to efforts for remission of carbon and other airborne emissions and building some community infrastructure. Over a period of time, when mining gathered momentum as an important revenue generating activity in sectoral development framework, other complex elements appeared and mitigating measures, management equations were conceptualised. A multipronged approach on environmental, economical and social issues such as; land, water, air, income, employment, safety, quality of life, health, education and communication need to be adopted to address the adverse impact, make positive contribution and achieve inclusive and sustainable of mining through responsible and ethical mining practices. These economic, social, and environmental initiatives along with profit generation for mining companies/corporations and all its stakeholders will have a positive contribution to society and the communities as well as long term lasting impact on environment and ecosystem.

The approach to Inclusiveness aims at reckoning and including everything concerned whereas principles of Sustainability is that the resources used for meeting current needs should also be available to future generations. The approach and principles of non-exclusion and uncompromising future availability are keys to development of critical sectors. Inclusive and Sustainable mining holds the promise of economic development for both developing and under developed countries. Mining companies as well as government need to adhere to the principles of Responsible and Ethical Mining practices so that negative impacts can be avoided or reduced to a maximum extent possible. Three elements, e.g., environmental protection, economic growth, and social equity constitute Responsible Mining. Principles and practices of Responsible Mining and Ethical Mining respectively are that; the mining industry should respect human rights and the aspirations of affected communities; provide safe, healthy as well as respectful workplaces and avoid or minimises harm to the environment; and the mining companies should adopt and maintain business ethics in their operations right from sourcing to closing including ethical mining in supply chain management of various products and ethical sourcing. In responsible mining companies bear the responsibilities and commitments on environmental, economic, social and governance issues. Ethical mining is concerned with human rights to a great extent and business and profit are not the sole motive of the companies/corporations. Refraining from illegal mining practices, shouldering socio-economic and environmental responsibilities, respecting human rights, absence of bribery and corruption in operation, maintaining transparency and good governance are some of the key principles of ethical mining. However, responsible and ethical mining are integral parts of inclusive and sustainable mining.

Mining should meet the development across environment, economic and social needs or the triple bottom line. Mining must be ecologically, economically and socially sustainable by adopting inclusive approach, responsible and ethical practices. Triple Bottom Line refers to 'three Ps' i.e. the planet, the people and the profit; people being the stakeholders, planet being our environment and profit has economic connotations as the main driver of any business. Along with economic or profit considerations, mining must take in to account two other bottom lines i.e. social, and ecological effects when making decisions. With an objective to maximise profit, mining companies must not do away their responsibilities for mitigating the social and ecological impacts. A balanced development in mining sector can be maintained when all the three needs i.e. economy, ecology and social needs are fulfilled. A number of rules and laws have been promulgated, Acts enacted, regulations framed and guidelines issued from time to time by the international authorities, Central as well as State Governments for regulating mining activities/ operations and controlling adverse socio-economic and environmental impacts. There are international laws, national guidelines and regional / state government rules, policies and provisions pertaining to mining leases/auctions, Management of Land acquisition for public purposes/development projects, MIDR, management of displacement, resettlement action plans and R&R activities, creation of public utilities, management of mining operations, protection and preservation measures. Many countries have, time and again, modified and revised these Acts and rules to include/delete some clauses and add elaborate provisions. However, these policies, principles and guidelines are only regulatory measures. A lot many actions need to be initiated towards Inclusive, Sustainable, Responsible and Ethical Mining. Mining's contribution to the economy of developed countries needs no reiteration. With globalisation, mining has become an important source of investment and profit for the private sector. The mining scenario, like the rest of the present economy, is linked to globalisation. It is, therefore, a crucial factor in the development dynamics of international economy. Mining activities and operations are potentially disruptive to the natural environment, have inherent deleterious impacts on ecology, detrimental to the socio-economic structure of affected indigenous people and deteriorate their cultural heritage. Despite these adverse impacts, there is no denying the fact that mining is essential due to its significant contribution towards revenue generation and economic development at large. The role of mining to the economic growth and prosperity need no overemphasis. Mining and Minerals are necessary for all countries; under developed, developing and developed, due to its fiscal impact and economic development. Mining is always an important and vital source of foreign exchange revenue and fiscal receipts for all governments. Besides being a catalytic agent for establishment of mineral based industries thereby providing large scale direct and indirect employment, it is a perennial source of income generation. All these usher faster economic growth and development at local, regional and macro level. Hence, it is not a pertinent question whether mining should be taken up or not, rather it is about the ways and manner how mining operations should be undertaken or in other words, how to ensure inclusive, sustainable, responsible and ethical mining. Minerals, like many other natural resources, are not infinite and cannot be explored indefinitely. Hence there is always a question mark about its Sustainability.

The forest land diversion for mining has been going up. So has water use and air pollution in the mining hotspots.

Mining of major minerals generated enormous amount of waste, most of which has not been disposed of properly. Inclusive growth will require ways to value local resources - be it water, minerals or energy - so that it gives back more than it takes. The mineral industry degrades the land, uses local water, but does little to return back wealth. Worse, the royalty on minerals goes to state exchequers, not to local communities. This needs to be changed. With sustainable public and private management, inclusive approach, responsible and ethical practices the mining and metals industry can contribute to poverty alleviation across the world, maintaining ecosystem integrity while ensuring sustainable development of the economy. Mining's long time horizon, its need for both skilled and unskilled labour, its links to regional infrastructure and service development as well as the importance of the products that it produces, means that a Sustainable Mining can make a unique and powerful contribution to sustainable development. Challenges in sustainable and inclusive, responsible and ethical mining are many. But there are some crucial challenges and outstanding issues which need to be tackled and subverted. Economic development and faster economic growth is one of the major challenges vexing the minds of political leaders, development practitioners, planners and economists alike. Economic development and growth can ensure sustainable mining. Without economic development and progress, sustainable mining is not possible. At the same time, a sustainable mining has a vital role to play in economic development. Both sustainable mining and economic development are, therefore, interrelated and interconnected. Sustainable Mining contributes significantly to economic development. Mining is a vital source of income generation through foreign exchange earnings and revenue receipts for governments worldwide, particularly, in countries like Bolivia, Brazil, Colombia, Tanzania, Burkina Faso, Ghana, Congo, India, Indonesia, and China. It is an important source of energy through coal resources. It helps in development of local economy in many ways. Mining operations take place in more than 100 countries worldwide. Mining sector helps in reduction of poverty and improvement in living standards in low and middle income countries. Mining plays an important role in promotion of employments and creation of Jobs both directly and indirectly. It is reported that there were 1.5 million people employed in the mining sector in developed countries, and 2.2 million in developing nations, besides, employment of an estimated 25 million people in small scale mining operations and indirect support to more than 150 million people worldwide. Mining promotes mining machinery and drives tourism demand. In spite of all these, there is a flipside also. Mining has a trail of negative social and environmental impact. Loss of land, loss of livelihood, mining-induced displacement and resettlement (MIDR) has become a major risk from the point of view of social sustainability. State-wise situation of major mining states reveals that mining contribution to the state's economy differs depending on its mining area but the status of rehabilitation, impoverishment risks of the displaced families and impact on environment are more or less similar.

Towards Inclusive, Sustainable, Responsible and Ethical Mining

The states and the Government of India should develop effective implementing machinery and a robust monitoring mechanism in order to have an inclusive and sustainable mining. Mining activities should not be a potential threat to the health and livelihood of the poor and vulnerable groups who have little mobility or means of alleviating negative impacts. That apart, the harmful impacts of mining should be addressed through development of new technology that uses fewer chemicals during extraction and processing. Moreover, adequate and sufficient rehabilitation provisions should have been guaranteed for the rehabilitation and relocation of the indigenous tribes. However, it may be noted that, in spite of experiencing its share of environmental and health related problems that adversely impact human quality-of-life, small-scale mining plays a pivotal role in alleviating poverty in the developing world and contributes significantly to national revenues and foreign exchange earnings. These important socio-economic contributions make small-scale mining an indispensable economic activity. Countries should, therefore, take some steps to obtain the greatest benefits from mining. They need to monitor the social and environmental impacts in communities and regions affected by mining and in particular, by mine closure. Countries can establish a sound mineral regulation and licencing system for commercial-scale mining. This involves creating "a level playing field," with ease of entry and responsible exit, sound fiscal policy and avoidance of subsidised state-owned mining enterprises (SOE), or, if they already exist, the privatisation of SOEs. They can ensure sound macro-economic policies so that mineral-rich countries benefit from the developmental impact that mining can have, rather than seeing non-mining sectors obstructed and opportunities wasted and attract private-sector investment through appropriate laws and regulations. Questions of ownership, land and water use, socio-environmental standards, and responsibilities need to be reliably clarified and appropriately implemented and monitored. Moreover, it should encourage planning for mine closure, already upon the first investment by a mining company, by supporting the build up of local administrative and management capacity and by designing and implementing appropriate regulation and oversight for mine closure. Besides, countries can alleviate health, environmental and socio-cultural risks for the population engaged in mining through information and education, as well as through agreements with the project authorities. Some of the target oriented actions for mitigating outstanding social issues and environmental challenges are presented below which are prescriptive and suggestive in nature.

Promulgation and Strict Enforcement of Mining and Minerals Laws and Policies should be ensured. Mining and Minerals Laws with provisions of control, regulatory measures and legal framework for all stakeholders should be framed, promulgated and strictly enforced by the government with zero tolerance for the breach of law. To address the emerging issues, outstanding challenges and new problems, the law should be amended from time to time whenever required as per the need and necessity. The law should spell out in clear terms the modus operandi of government, other regulatory authorities and agencies, mining companies and all other stakeholders as well as enforcement modalities. Mining rules and policies must also address other important issues viz. procedure for fixation of royalties/taxes and payment to the government, determination/demarcation of mining areas, quantum and period of extraction, land acquisition and eviction, displacement, rehabilitation, resettlement and restoration, development of mining areas and communities, ecology and environmental protection, mine closure regulations, water conservation measures in mining areas, setting up of appropriate departments/commissions/corporations/boards for implementation and monitoring of Mines and Minerals activities etc.. Apart from empowering the regulatory authorities and agencies, the law promotes as much as possible, peaceful co-existence between mining companies, their host communities and the government. There should be strong and strict provisions in the law for avoiding unbridled and rampant mining, illegal mining of minor and major minerals and corruption at any level. The Laws/Acts/Regulations/Notifications should be very transparent, clear, intelligible and without any scope of ambiguity, confusion and misinterpretation. Provisions, stipulations and sections should cover all stages and all aspects of mining and include all stakeholders encompassing whole gamut of issues that may crop up at different phases of implementation and monitoring.

Transparency and openness, accountability and responsibility, efficiency and effectiveness, precautions and safeguards are the key indicators for gaining the trust and confidence of people and ensuring smooth and unhindered implementation of the programme.

Issues on land acquisition by the Government and subsequent involuntary displacement include resettlement and rehabilitation, compensation for the loss of livelihood, accentuation of conflicts etc. which are to be addressed on a priority basis. Hence legal rules and Government policies for the process by which land is to be acquired, and how current owners are compensated, will play a critical role. There is, thus, a need to integrate relief measures with land acquisition initiatives, protect the interest of the displaced people, secure land rights and livelihood of the natural resource-based communities, take an optimal decision to confer legal rights on the losers of cultivable land belonging to loss of habitats, environment and bio-diversity consequent upon land acquisition and resettlement. Besides, R&R need to take into cognizance the needs of women affected by displacement. In many states, among forest gatherers and farming societies, women are equal contributors to the family income, sustenance and its dynamics. However, rehabilitation packages have always ignored this aspect. In fact, gender disparities in rehabilitation policies are glaring, according to a 2005 study of UNDP. The compensation money goes to the head of the family, usually a male Land Acquisition Act 1894 as amended in 1984 was the base law on land acquisition and compensation. Over the years, many Acts and Rules were promulgated on displacement, rehabilitation and resettlement. Mostly male ownership of land is recognised in these rules. This bar on formal legal title to land should be avoided as far as practicable in entitlements to enable every woman a potential beneficiary. Moreover, benefits of jobs-for-land should also go to the unmarried daughters and families with only one or more daughters. R&R policy should encompass to address various other issues[1] of gender disparity in R&R activities and ensure women empowerment.

Achieving inclusive sustainable development through effective community relation/engagement and socio-economic development would lead to the company achieving its social licence to operate. This is because all issues and concerns of the community members are addressed in a timely manner. They are, therefore, respected and see themselves as part of the company. Livelihood Restoration Programme and Community Training and Employment were the strategies being adopted by the Southern African Institute of Mining and Metallurgy Surface Mining in 2008. Only provision of socio-economic development in mining communities will lead to the company securing a social licence to operate, however, is not true rather effective community relations should be combined with socio-economic development. The strategies to ensure socio-economic development involves effective community relations, community development programme, livelihood restoration programme and community training and employment.

Besides these social commitments, mining companies and corporations should take effective initiatives to turn high-level environmental action on the ground. All these will ultimately lead to inclusive and sustainable mining. The mineral industries have to play an effective role for fostering the inclusiveness and long-term sustainability. In order to foster socially inclusive and environmentally sustainable resource development, mining companies and corporations need to evolve a set of guiding principles and performance indicators as well as comprehensive, integrated and inclusive approach in various key areas like; energy use, greenhouse gas (GHG), emissions management, external outreach, crisis management, tailings management, Aboriginal People, Aboriginal Relations, Displacement, Rehabilitation of DPs, Health and Safety, Water, Biodiversity, Conservation of Environment, finance, taxation, land access, workplace, workforce, Mine Closure, community, Multi stakeholder engagement, various institutional actors, monitoring and reporting requirements etc. This will enhance and improve the social, environmental and economic performance of mining industries.[2] These should be initiated suo-moto without waiting for any direction from regulating agencies or government as voluntary initiatives are a convenient way for industry to avoid government regulations that were seen as cumbersome, expensive and time consuming.[3]

The Way Forward

In order to ensure inclusive, sustainable, responsible and ethical mining, the initiatives, actions and responsibilities of both the implementing agencies and regulatory authorities like mining companies, government and all other stakeholders (primary, secondary and tertiary) are many. They have to adopt an inclusive approach and ensure sustainable mining practices by effectively delivering their responsibilities through targeted actions. Some of the important principles of responsibilities, prescriptions and policy initiatives for targeted actions to be adopted by the appropriate agencies which will ultimately lead to inclusive and sustainable mining are discussed below.

Continuous Peoples' (stakeholders') Participation and Effective Community Engagement should be a basic principle for ensuring inclusive and sustainable mining. Participatory decision-making is vital for success and sustainability of any development programme. Rights of people should be respected and recognised. Mining should not take place without the consent of the people or the local communities. Upholding people's rights in turn motivates and reminds them of their duties. Community members should be involved in the decision-making process at all stages. Continuous Peoples' Participation and their Effective Involvement are very important in achieving sustainable mining. The mining company may select or nominate a member of each of the affected community to its core committee or the standing committee as the company's primary member. The mining company may form 'Community Committees' for each community. These committees may comprise the community leader, an elderly person, a youth representative, women's representative and any other educated member from that community with a representative from the company. Periodical meetings of these committees may be held regularly in order to discuss their problems, identify stumbling blocks and find out solutions. The primary member may put up the issues before the core committee or the standing committee for effective final decision and resolution of conflicts.

Crisis Management and Conflict Resolution can be made through Continuous Peoples' Participation and Effective community Involvement. The community members may monitor and track the complaints/conflicts/problems/crises and held continuous dialogue with the authorities to resolve a particular conflict or address the problem. The company may maintain a grievance register and track it regularly to ensure that all grievances are addressed in the shortest possible time. This responsibility should be assigned to a specific designated officer with fixing of accountability. He can institute a management plan and follow it regularly. Continuous dialogue with stakeholders, peoples' participation, community involvement and timely resolution of conflicts is paramount to effective community relations.

This is one suggested relative model and can be modified depending on the place of operation and concerned stakeholders. For continuous community engagement, the mining companies may devise and adopt any similar model depending on persons in charge of implementation, place and time period of implementation. For sustainability in mining, therefore, effective community relations should be established and Community members need be continuously engaged in every decision-making process and Conflict Resolution.

There should be proper Resettlement, Restoration and Rehabilitation of displaced families involuntarily displaced due to mining. It has been suggested from some quarters that mining won't help displaced and hence solution lies in proper and adequate compensation to the displaced families. Different proposals were mooted in India and other parts of the globe on this issue. In the midst of a debate over development vis-à-vis displacement and environment protection, the then Finance Minister of India Pranab Mukherjee said the solution does not lie in stopping mining projects but in properly compensating people who are displaced. "The answer does not lie in companies stopping mining activities. The answer lies in providing alternatives to those displaced ...... in what form we can compensate them and make them beneficiaries of economic development," he remarked in a coal summit in New Delhi. These comments came within a month of the environment ministry rejecting earlier clearances to Vedanta's Rs 7,000 crore bauxite mining project in Odisha citing violation of green and tribal norms. During 2010, the then, Steel Minister of India, Virbhadra Singh mooted a proposal to give concessions to PSUs in the proposed law for 26 percent profit-sharing with people displaced by mining projects. However, PSUs Coal India and SAIL, differ on the issue. A group of ministers (GoM), headed by the then, Finance Minister also suggested a new mining legislation which proposes that companies share 26 percent of their profit from mining with those displaced. "That (the issue of compensating those displaced) is the challenge and we are addressing that issue," the minister said. The profit sharing proposal was questioned by the industry on the ground that 26 percent was too high and not practical. However, it was agreed that a part of the profit from mining companies should be earmarked for providing adequate compensation and other facilities to the DFs. Proper rehabilitation and resettlement of mining induced DFs is a daunting task. Proposed plans/models suggested in different forums is pertinent and worth mentioning. The CSE's 6th State of India's Environment Report highlighted proper rehabilitation, compensation to displaced families and mining affected communities in mining areas along with arrangement for Jobs at least one per displaced family and others. The London Declaration specifies that[4] (1) A moratorium on new large-scale mining projects in Greenfield areas of Asia, Africa and Latin America; (2) Companies must clean up the terrible damage caused by their past and current mines, without drawing on public funds, and be held morally, legally and financially responsible for their misdeeds; (3) The World Bank/IMF cease funding of industry-initiated mining codes which are imposed on the governments of Africa, Asia-Pacific and Latin America; (4) Mandatory higher standards in all mining; (5) That surface and subsurface rights of indigenous peoples and all mining-affected communities be unequivocally respected and enforced. The London Declaration said "We seek solidarity from civil society and specifically from development and environment NGOs, in response to the global outcry from communities affected by mining. We ask these organisations: (1) To ensure that mining affected communities are fully informed in advance on all aspects of mining and minerals projects and empowered to speak for themselves in response; (2) To refuse to participate in initiatives, such as MMSD, which are primarily spearheaded by the industry to serve its own purposes; (3) To advocate for politically and legally enforceable measures that will hold the mining industry accountable, above all, to mining and exploration-affected communities".

Be that as it may, following pertinent issues must be taken care of in a way to address the problems of Resettlement/relocation, Restoration & Rehabilitation.

Allocation of Benefits: A fraction of royalty from mining and licence fees of mines must automatically accrue to mining-affected communities and to communities whose land is being used in the mining project.

Resettlement plans: Rehabilitation and clean-up plans and budget should be a mandatory requirement for permission to begin a project. In addition, restoration of existing damage caused by previous projects by the company seeking a new permit should be completed as a condition of being awarded the new permit. Companies must routinely be held legally and financially accountable for remediation of any damage caused by their past and current mines.

Restoration of Livelihood: Along with the compensation to Displaced Families (DFs) in the form of cash or land, programmes to restore their livelihoods should also be launched. The mining company should make all negotiations with the displaced families under the active involvement of the community leaders who own land in the area. In all compensation issues, the displaced families should be given cash compensation and also alternative land so that they can continue with their farming or traditional activities. The traditional farmers should be given raw materials for cultivation. The families who are not interested in arable farming, should be offered the option of livestock like cows, Sheep, Goats, Pigs rearing, cloth weaving, Bakery, Soap Making or petty business as per their choice. Self Help Groups may be formed for involvement of community members and quality inputs as well as appropriate technology should be provided where ever required. Marketing arrangements and credit linkage with the banks for these activities must be ensured. The stakeholder communities must be assisted compassionately in all possible manners. All these will go a long way in helping the community members to earn their livelihood and maintain the pre-displacement economic status.

Provisions for the Indigenous Tribes: Displacement mostly of indigenous tribes and other communities is the natural outcome of mining projects may be with varying degrees and dimensions. Hence, their proper rehabilitation is the foremost criteria of any Sustainable mining. Since mining projects affect the indigenous communities at the most, besides rehabilitating/ relocating the displaced Scheduled Tribes in colonies or other suitable locations, their food security should be guaranteed with provisions for agricultural land and water supply. Mining activities should not be a potential threat to the health and livelihood of these poor and vulnerable groups who have little mobility or means of alleviating negative impacts. Moreover, adequate and sufficient rehabilitation provisions should have been guaranteed for the rehabilitation and relocation of the indigenous tribes.

Farm Productivity: Besides the compensations and provision of local infrastructure, it should also be the responsibility of the government to enhance farm productivity – such as improving soil quality or developing local irrigation facilities of the remaining land of the farmers in the locality. The displaced land owners/evacuees should be settled in well built up colonies if they lose their home with essential infrastructures. Additional land, if available, should also be provided to them for cultivation since they have developed special skills in farming and other allied activities.

Delivery of R & R packages: There is a strong need of change of present process of Land acquisition and delivery of compensation packages. It is the Top-Down approach which government follows all through. There is need for consultative process with local communities. Local residents should be consulted time and again and involved in selecting areas to be acquired for setting up Industries or development projects. Bottom-up approach should be followed strongly and local community should be given the responsibility for designing and implementation of compensations. This will go a long way to check snowballing the public outrage and protests.

Employment to members of DFs and local community in mining companies/corporations is an important challenge. Mining companies, when established, have provisions of employment to local people. In most of the MOUs signed between the government and the mining companies, there is a clause of reservation of unskilled jobs for local people and skilled jobs as per availability. But sometimes that do not happen. According to CSE's 'State of India's Environment' reports, "If people were carried away by the tall talks of politicians justifying mining as a means of employment, then think twice. This is a chimera. Employment is a promise not kept in many cases. All governments justify mining arguing that the sector will provide employment, but this is a myth". The report using government data shows how employment has fallen in the mining sector as a whole. The fact is that the modern mining industry does not require people. Between 1999 and 2005, in Odisha state, the value of mineral production increased three-fold - at the same time, employment reduced by 20 per cent. Chandra Bhushan, CSE's associate director and the coordinator and co-author of the aforesaid report remarked, "In fact, modern industrial growth requires resources of the region - minerals, water or energy. It does not require people. In other words, it does not necessarily provide local benefits. If it provides employment benefits, it is outside the poor region in which it is based". The report suggested that there should be arrangements for Jobs at least one per displaced family and others by the mining project. It should be mandatory for the company to earmark all the unskilled jobs for the displaced families and local people and skilled or semi-skilled jobs as per qualification for the personnel in the surrounding area. There may be disputes in local employment; in that case, an Employment Committee may be constituted comprising of community leaders, selected community members and others. The Committee will be responsible for resolution of conflicts and disputes and their recommendation for employment should be worked out.

Vocational Training and educational facilities for the children of the displaced families and local people up to the secondary level must be provided by the company. Adequate educational infrastructure in the communities should be created in order to make education accessible to all. For higher education, the company should introduce a scholarship scheme for students to study outside the local area. Most of the people in the working age group in rural areas and stakeholder communities, lack employable skills due to illiteracy and less education over the years. There is a necessity to make arrangements for providing Vocational Training for them in different trades. Establishment of Industrial Training Institutes (ITIs) including mining-related courses in mining areas can also be considered to facilitate the Vacation employment.

Development of Community Infrastructure is a basic requirement to involve and evoke the participation of local people in mining activities. Along with development of agriculture and allied sectors, Community Infrastructure Development Programmes (CIDP) in various other sectors such as; communication, electricity, health and sanitation, drinking water, education etc. should also be implemented for the socio-economic development of the DFs and the Project Affected Families (PAFs) who are the primary stakeholder communities. It is necessary that Primary Health

Care Facilities should be guaranteed to the stakeholder communities since when the company start its operations. Health Facilitators with required training can be deployed by the company to deal with minor cases and assist in giving first aid to the community members. Arrangements should be made to refer the serious cases to the nearby government hospital. As regards water supply and sanitation, provisions of bore wells/tube wells as per the norm must be made. In large settlements, provision of piped water supply with proper water treatment can be taken up. Sanitation facilities in the settlement colonies need be ensured. Garbage dumping bins or refuse containers should be supplied to the communities. Arrangements for maintenance and repair of water supply facilities and garbage disposal and waste/ refuse clearance should be ensured. A trained community member or dedicated personnel from the company should be responsible to monitor the water supply and sanitation facilities. Roads and communication and electricity are harbingers of economic development. In order to facilitate the regular movement, marketing and other activities of the community members, road communication at least to Gram Panchayat connecting block headquarters should be ensured. Electricity connections to the houses of the community is a necessary requirement which should be provided in order to make them self sufficient in gainful economic activities.

Since the communities facing displacement due to mining projects have always high expectations from the Government as regards compensation and rehabilitation which the government is never able to meet, mining companies should come forward to ensure socio-economic development of the affected communities by providing them with development projects in order to gain their confidence and avert any chances of resistance movements. Community Development Projects should be grounded not just in an ad-hoc manner or only for meeting the requests of communities but the mining company should be committed for development of its primary stakeholder communities on sustainable basis. Both mining company and Government should be more humane in dealing with their problems and adopt humane practices.

Environmental degradation is the immediate outcome of any mining operation and therefore, Protection and preservation of environment and ecology of the surrounding areas must be the prime responsibility of the mining companies. Adequate protection measures against all adverse and negative impacts of mining must initially be put in place by the mining companies. They should bear the responsibility for safety of the people and the environment. According to the CSE's 'State of India's Environment' report, "It is about ensuring that mining is conducted in an environmentally and socially acceptable manner. Mining must not be allowed if the environmental and social costs outweigh its economic growth". Mining operations cause noise, blast, ash and dust. It pollutes air, water and environment. There are standard operating procedures (SOPs) for mining companies at international level with guidelines prescribed by International Standards Organisation's Occupational Health and Safety Management System (OHSAS-18001) and Environmental Management System (EMS- 140001). In India, there is a system of Environment Impact Assessment (EIA) for granting clearance to any mining project. A signatory to the Stockholm Declaration (1972) on environment, India enacted laws to control water (1974) and air (1981) pollution soon after. It was only after the Bhopal gas leak disaster in 1984 that the country legislated an umbrella Act for Environmental Protection in 1986. Under the Environment (Protection) Act, 1986, India notified its first EIA norm in 1994, setting in place a legal framework for regulating the activities that access, utilise, and affect (pollute) natural resources. Every development project has been required to go through the EIA process for obtaining prior environmental clearance. Thus, the EIA came in to force in 1994, after the regulatory shortcomings in the Environment (Protection) Act, 1986 were brought to light. The 1994 notification was replaced with a modified draft in 2006. Although the subsequent 2006 act did encompass gaps in the procedure for grant of Environmental Clearance (EC) for an industrial/development project, it was the first of its kind to lay down a set of guidelines before the establishment of such a project. The draft EIA Bill 2020 of India was notified in March, 2020 incorporating the amendments and relevant court orders issued since 2006, and to make the EIA "process more transparent and expedient". EIA is the process or study which predicts the effect of proposed industrial/infrastructural projects on the environment. EIA system is vital to confirm development projects to environment safety and thereby ensure sustainable mining development. Environment should, however, be protected at any cost. No exemptions should be granted to the mining companies in meeting standard environment safety and protection standards. There should be strict environmental regulations in both physical and biological environment.

Environmental cost–benefit analysis of projects must invariably be taken up at initial stage and if benefit outweighs cost, such projects can be cleared. Initial Environmental Examination (IEE) should be conducted at pre-feasibility level to assess and identify possible environmental impacts. Only ecologically sound projects should be cleared. No exemptions should be granted to the mining companies in meeting standard environment safety and protection standard. A comprehensive environmental monitoring programme should be in place and carried out by the Regulatory agencies such as the Forest and Environment/Mining Department or Environmental Protection Agency/Pollution Control Board/Commission as the case may be, to check noise, blast, ash, dust, water and air quality for each mining project and ensure that the company is within statutory and permissible limits. Any figures above the minimum requirement need be investigated and resolved as soon as possible. Third party independent surveys/monitoring exercises and concurrent evaluation studies can be undertaken to ensure that the company is within acceptable limits and suggest feedbacks for mid-course corrections. A robust Monitoring Mechanism is required to deal with violations. Ambit of scrutiny should be broad based encompassing far-reaching impacts like ecological crisis, climate change etc. In order to ensure that a mining project will not be detrimental to environment, there must be an independent and impartial preparation of Environment Impact Assessment (EIA) report as well as a stringent environmental clearance system. People have procedural environmental rights; they require clean air and drinking water. Mining operations should not be a potential threat to their health and hence public consultation process is a must before grounding of a project. There has to be transparency and sufficient scope for public participation in environmental law making. That apart, the harmful impacts of mining should be addressed through development of new technology that uses fewer chemicals during extraction and processing. Advanced and modern technology must be adopted in managing the water, air, land, ecology and the environment in mining operations. The companies have to preserve the environment and mitigate any negative impact on the community and the company. The three strategic issues pertaining to environment and ecology are land and its biodiversity, water and air. One most important aspect in preserving land and biodiversity, water and air is to conduct Environmental impact assessments (EIA) studies regularly before mining operation till closure and initiating appropriate steps accordingly. This should be a basic prerequisite and the minimum requirement of mining companies/corporations irrespective of its size and scale.

The potential negative impact of a mine on land and biodiversity should be managed efficiently by mining companies without disrupting landscapes and ecosystems. They should protect critical species, make land restoration and create offsetting areas. Water is vital for life and living of local communities. It is used not for drinking water alone with a direct bearing on their health status, but also for agriculture, horticulture and fisheries, and therefore, any disruption both on water quality and flow will be fatal for local people. Another most important challenge is acid mine water management. The effect of acid mine water remains even for centuries after a mine's closure and hence considered as a mine's biggest contingent liability. Leakages of acid mine water pond and its flow to natural water sources must immediately be checked on detection. Mining corporations should make adequate provisions for water treatment and provide clean and potable water for drinking, domestic, irrigation and other uses. Such community-oriented comprehensive solutions (COCS) by the company will go a longway in establishing a good relationship with affected community. Huge emissions of carbon dioxide (CO_2) and chlorofluorocarbons by mining companies cause air pollution. Mining activities are normally carbon dioxide intensive operations through the consumption of fossil fuels and, therefore, prone the threats of climate change. Use of green energy, alternative sources of energy/renewable energy may be some of the alternative initiatives which mining companies can explore. For establishing a good community relation, a mine authority can provide power to the surrounding villages by involving people through assisting them in construction and maintenance of biomass plants or solar and wind plants.

Acid Mine Drainage (AMD) has been considered as one of the greatest environmental debt to posterity by a mine. Prevention, mitigation, controls and after all stopping the release of AMD is an important challenge of all mines, large and small, irrespective of its scale of operation. Acid Mines, left unmanaged will be dangerous not only to pristine environments and ecology but also to economically precious natural resources. The long term objective and strategy should, therefore, be complete prevention of AMD rather than temporary mitigation through stopping or remediation. Current best practices for mitigating AMD involve a thorough site analysis as well as extensive geochemical, biological and hydrologic analysis during the evaluation phase. The analysis includes: 'location, access, climate, ecology, history of previous mining, waste materials, geology, hydrology, mineralogy, descriptions of all materials that will be excavated or exposed, soils, reclamation objectives, end land uses, data tables, relevant figures, other pertinent information and most importantly, improving the knowledge and understanding of watersheds, water patterns, water conveyance systems, rainfall intensities and distributions, and the probabilities of natural disasters'.

Another vital issue is to reduce the mine waste. Most miners fail in reducing the waste because they have to cope with growing demand with increasing pace of population. Since they cannot reduce or limit the extraction, exploration or production of minerals, they should constantly endeavour to reduce waste through reuse and recycling of minerals and metals. Most minerals/metals can be recycled infinitely though recycling process but it is difficult for alloys and complex technological products than for pure metals. However, to reduce extraction and waste, national Governments can incentivise for reusing and recycling of metals besides providing modern and innovative facilities for recycling. Moreover, better practice for conservation of metals must be adopted through strict mining regulations by the Government.

The most important step for avoiding lasting negative impact on ecology is post-closure reclamation. This requires a long term detailed closure planning during the production and exploration phase. Before closure, the mine site should complete the wind-up operations such as; removal of stored fuels and chemicals, waste detoxification, removal of roadways, removal of erosion, sediment control measures and removal as well as capping of tailings. This will facilitate the reclamation, rehabilitation and long term maintenance of the mine area. Just like in the social practices, there is a distinction of practices and requirements for active mining as well as mine closure or material stewardship approaches. The former is mainly focused on the reduction of emissions, effluents, spillages, waste, and water usage'.[5] There are mostly general calls to "minimize" emissions and treat water resources "responsibly"[6] or to "maximize efficiency of water-use".[7] Beyond this, management systems and emergency response plans are required by the Multi Stakeholder Initiatives (MSIs) to prevent the often long-lasting damages of spillages, dam bursts and other contaminations. Such pollution may happen in an abrupt event, but it may as well occur unnoticed over a longer time period. The latter case is addressed by some MSIs that require the implementation of environmental monitoring systems to detect long-term degradations of wildlife or water quality.[8] More complex issues such as the management of impacts on biodiversity[9] or the negative effect on the quality of life in neighbouring communities[10] are addressed, but only at their surface with calls to minimise them during operations. The operation time for individual mines is limited. After that, the mining areas has to be restored to make them safe and to allow other kinds of land use. Planning the controlled ending of mining operations and proper measures for achieving this are to be started well before commencing mining operations and be developed throughout the project's lifecycle with the broad-based participation of local residents and other stakeholders. Post-mining land use is a way towards sustainability, and hence, closure of a mine also requires functional and tested technical and scientific methods so that the quarries, waste areas and other infrastructure can be restored in a way that allows further sustainable use of the area. These approaches and measures can be considered as canons of Inclusive, Sustainable, Responsible and Ethical Mining.

According to Katarzyna Pactwa and others, "both the government and industry need to take a comprehensive view of sustainable development that should cover dimensions other than environment, such as stakeholder engagement and consultations, local area socio-economic development and transparency in communication and accountability. Preparation of a socio-economic assessment report for a mining project should be made a part of the permitting process for the grant and administration of mineral concession to a mining enterprise. Local socio-economic development works should preferably be executed by mining enterprises rather than government and semi-government agencies to avoid the problems of inadequate capacity, political manipulation and corruption. In order to alleviate the limitations of small mines in carrying out sustainable development activities, consortia of small mining enterprises in a region should be promoted. Technical advisory services should be made available to them in the relevant areas. Mineral development in a region should be carried out within its available social and environmental "carrying capacity" and infrastructural facilities at a given point of time".[11] In order to ensure sustainable mining, the responsibilities of both the implementing agencies and regulatory authorities like mining companies, government etc. are many. They have to ensure sustainable mining by effectively delivering their responsibilities through ethical principles and display responsible practices and evolve policy initiatives for targeted actions to be adopted by the appropriate agencies.

One of the outstanding challenges and preponderant problems that appeared in the forefront is rampant, unabated, unbridled as well as illegal mining and corruption at different levels. Since mineral resources are finite and there is no scope of their renewal, rampant and unscientific mining must be checked. Expression of interest basis or tender process or competitive bidding process in mining leases as per the provisions of law should be meticulously followed to prevent illegal mining and corruption. There are number of instances of rampant, unabated and illegal mining and corruption in mining sector in major mining states of India. An analysis of the case studies of mining in the states boil down to the fact that unless rampant and illegal mining put to a grinding halt and corruption is checked, states' economic progress will be farfetched. The mineral activities are mostly concentrated in forest lands/ areas. Though there is no massive displacement in the mining sector, indigenous communities are mostly affected and worst sufferers. Adequate and sufficient rehabilitation provisions should have been guaranteed for their rehabilitation and relocation. Comprehensive mining policy of the respective states and the Mines and Minerals Development and Regulations (MMDR) Act of Government of India should have been scrupulously followed. The same is the problem with our case study region of Odisha state. In Odisha Legislative Assembly, to a question whether Odisha has the stock of iron ore to feed the long line of steel plants proposed in the state, the Government replied in the assembly way back August 26, 2008 that the deposits would last for only 25 years. "The total iron ore reserve in Odisha assessed till date is 5,300 million tons. Iron ore can be available for these industries for 25 years," the Government had said. It made a similar prediction for the proposed alumina industries gulping bauxite deposits assessed at 1,793 million tons. A good number of MoUs were signed by the Government and mining companies for private medium and big steel plants would consume approximately 200 million tons of iron ore per annum. Aspiring to be the steel hub of the country, Odisha's dream might face an untimely death unless it puts brakes on unbridled mining. Rampant and unscientific mining has led to apprehensions that the mineral-rich state could face a shortage of iron ore if all 49 plants, envisaging an investment of Rs Two lakh crore for production of 75 million tons per annum (MTPA), for which private companies have signed pacts (MoUs) with the State Government, go on steam. Going by latest estimates made by the State Directorate of Geology, Odisha has around 5230 million tons (MT) of iron ore reserves. The Indian Bureau of Mines (IBM) had a few years ago reported there is 3780 MT iron ore in the State. Keeping these estimates in view, upcoming Steel Plants, both inside and outside the State, dependent on souring iron ore from Odisha, may face scarcity of raw material in future.

According to the Society of Geo-scientists and Allied Technologists, ore production in Odisha stood at 1.85 million tonnes per annum (MTPA). It increased to 7.5 MPTA in 1975 and 9.25 MTPA in 1995. But it jumped to 50 MTPA in 2005. In 2007-08, it shot up to 74 MTPA and is projected to touch 200 MTPA. It is pointed out, "Considering the ore availability and rate of consumption, Odisha has resources for not more than 20 years," expressing the need for reassessment of ore in leasehold and freehold areas. "Unless we plan for the future, there can be no planning for industrialisation". Steel and Mines Department too, sounded a cautionary note and stressed on curbing wasteful use of ore and reduced per capita consumption of steel worldwide. They pointed out that the stress on mineral resources was "tremendous" and said, "Danger is knocking at our door steps. We should be absolutely economical in use of ore".[12] If people believe mining as a means of economic development, studies by the Centre for Science and Environment (CSE) have debunked it as a myth.

Hence, Government of Odisha should be more cautious while allotting mining leases. A perspective plan for next 25 years should be prepared regarding the deposit of minerals in the State, Grant of leases to Private Companies in respect of all minerals and future plan of action. This will spell out a clear road map for mineral sector in Odisha. The Auditor General Odisha (AGO) made a performance audit on mining in Odisha pointed out some irregularities. The December 11, 2012 and January 2, 2013 queries from the Accountant General (AG) to the Government also bring home the points as to how most of the private industrial ventures have been illegally entertained setting the State's future virtually in a ruinous path. The AG's performance audit revealed that Government signed 93 MoUs with promoters from 2001 to 2012, which included 49 for steel and mining industries without 'assessing the need as well as other inputs like land, water and minerals'. "MoUs were signed based on suo-moto offers made by the promoters rather than on expression of interest basis or tender process. Due to such suo-moto offers made by the promoters, interest of the state could not be safeguarded as the state was deprived of best investment proposals", the audit observed. Quoting the September 27, 2012 Supreme Court opinion, the AG said "when precious and scarce natural resources are alienated for commercial pursuits of profit maximising private entrepreneurs, adoption of means other than those that are competitive and maximise revenue may be arbitrary and face the wrath of Article 14 of the Constitution". It questioned public bids for steel industries not being invited and said "there' was no reasonable and exceptional grounds that could warrant the department to reach a conclusion, that there may not be other interested parties specially when land, water, coal and minerals were provided at concessional rates'. "Decision has to be fair, reasonable, non-discriminatory, transparent, non-capricious, unbiased, without favouritism or nepotism, in pursuit of promotion of healthy competition and equitable treatment," the AG said quoting the SC Verdict.[13] Since Minerals are available natural resources of the State, it should be handled with care and caution. A competitive bidding process should invariably be adopted while granting mining leases.

Illegal mining of minor minerals is very rampant in many States. State Governments are empowered to make rules for fixing collection and enhancement of rate of royalty and dead rent of minor minerals. These rules are not uniform and differ from State to State. The power to prevent illegal mining, transportation and storage of minor minerals is also under the jurisdiction of the state government. As a result there is no centrally maintained database of complaints against illegal mining of these minerals. The ongoing construction boom has accelerated the demand for these minerals and the confusion caused by the non-uniformity of laws as well as the lack of a central list of proclaimed offenders has helped the mining mafia in illegally extracting these minerals. Obviously, the cost differential between legal mining and illegal mining also makes the latter lucrative while the nexus often found between politicians, bureaucrats and illegal miners often makes it unlikely any action to be taken against them.

In 2009, the ministry of environment and forests (MoEF), Government of India, constituted a group under the chairmanship of the secretary, Environment and Forests to look into the environmental aspects associated with mining of minor minerals. In 2010, the group submitted its report and made several recommendations relating to definition of minor minerals, size of mine lease, period of mine lease, cluster of mine approach for small sized mines, requirement of mine plan for minor minerals, river bed mining etc. MoEF has asked the states to examine the report and to issue necessary instructions for incorporating the recommendations made in the report. The Supreme Court has also ordered that state and UT governments should take immediate steps to frame necessary rules considering the recommendation of the group. In the Odisha Legislative Assembly Session on 27th November, 2006, Treasury bench MLAs joined forces with their opposition counterparts in the Assembly, grilling the government especially over its failure to impose tax on mineral – bearing land leading to loss of hundreds of crores of revenue every year. The legislators pressed for the cancellation of leases to private companies and handing over of all mines to state owned Odisha Mining Corporation (OMC). MLAs suggested that all private companies requiring ore should buy from OMC at the market price. The government should consider setting up separate corporations for major minerals like iron ore and bauxite to handle the State's resources and stop giving mines to private companies for a "Pittance". Several MLAs, during a debate, also attacked the government for not being able to collect cess from mineral bearing land and sought the enactment of a law for the purpose. The then Mines' minister defended the government, pointing out that the Odisha Rural Infrastructure and Socio-Economic Development Act, 2004, was enacted with the aim of collecting mining cess and there by boost revenue earnings. "We had even collected Rs 14.23 crore through it. However, the Odisha High Court on December 5, 2005, rejected the law, saying the state does not have powers to enact such legislation as mining is a central subject. It is a constitutional matter. The Supreme Court had in the State of West Bengal versus Kesoram Industries case said different states can enact law to collect mining cess. Hence, we have moved SC against the HC decision. We will wait for the apex court verdict".[14]

More than a year after imposing huge penalty on miners for illegal extraction of iron and manganese ores, the State Government is yet to realise a rupee. The mine owners instead have successfully thrown a spanner in the decision by challenging the State Government's decision before the mining tribunal, a revision authority functioning under the Union mines ministry. It is pending for orders. Official sources said the State government imposed around Rs 67,900 crore as fine on 103 mines for extraction of minerals in excess to permissible limits between 2000 and 2010.[15]

Ulliburu Mining Scam which surfaced in the State, is a case in the point. Anti-corruption sleuths have filed charge sheet against 23 people, including 16 Government officers and two firms for their alleged involvement in the multi crore Ulliburu mining scam in Keonjhar district. The charge sheet was filed at the special vigilance court in Keonjhar. According to vigilance sleuths, over 65 lakh metric tons. of iron ore worth Rs 1523 crore was illegally excavated from the Ulliburu mines between 2004 and 2010. The Charge sheet named Deepak Gupta, prime accused in the scam, his wife Nitu, father Harichanda, mother Kantadevi and brother Champak. The 16 Government officials include three Deputy Directors of mines, four mining officers, three senior inspectors of mines, two range officers and four forest officers, Two firms, Deepak Steel and Power Limited and Snehapuspa Marketing Private Limited, through which the Guptas excavated and sold iron ore illegally, have also been named in the charge sheet. Both the firms are owned by the Gupta family. They have been charged for breach of trust (SECTION-409), theft (379), forgery (468), Cheating (420), Stealing property (411), criminal conspiracy and trespassing. Inquiry reports of vigilance, local tehsildar and mines Department have also been attached to the chargesheet.[16] A Mega Mining Scam in Odisha was exploded in 2009 and serious corruption charges were levelled against miners and State Government officials. In November, 2010, the Centre instituted the Justice M B Shah commission for inquiring into the scam. The Shah Commission submitted an interim report to the Centre that pointed out certain irregularities and blatant violation of laws. The Commission submitted the final report in Jan, 2014. It was placed before the Union cabinet. The Union mines ministry after getting the report had sought an action taken report (ATR) from the State Government. It is reported that the Union Cabinet referred the M. B. Shah Commission of inquiry report on illegal mining in Odisha to the committee of Secretaries (CoS) to suggest further action which may include recovery of nearly Rs 60,000 crore from the miners and a CBI probe into roles of public servants in allocating mines to private companies. The Commission has severely indicted both the Centre and the Odisha Government on illegal mining of iron and manganese ores in its 5-volume report. The Committee of Secretaries (CoS), comprising secretaries of mines, environment and law, will submit its suggestions to the Government which is supposed to table the Commission's findings along with 'action taken report' in Parliament. Besides indicting both the Centre and the Odisha Government, the Commission has also asked the State Government to recover over Rs 59,203 crore from the miners and use the recovered amount for the development of two districts, Keonjhar and Sundargarh, "which are badly affected by illegal excess mining". It also recommended granting all future mining leases via auction route only and asked the Government to revisit environment clearances given to all 55 mining leases which are located along the Baitarani River. Till the completion of the exercise, "the mines may be closed", it said. [17]

The Centre may issue show cause notices to companies, including SAIL, Tata Steel, Essel Mining, JSPL and Odisha Mining Corporation (OMC) that hold 75 leases where illegal mining of iron and manganese was found by the Shah Commission. The step would be in addition to legal action being initiated by the Odisha Government against firms that carried out mining operations without environmental and other clearances. As many as 75 firms, including Steel Authority of India Ltd (SAIL), Tata Steel, Additya Birla Group's Essel Mining and industries, Jindal Steel and Power Limited (JSPL); and OMC have been accused of indulging in "illegal/without any lawful authority" for production of ores. According to the ministry of environment and forests (MoEF), Government of India, "It is proposed to issue directions to these 75 mining leases for restricting the production of ore to the level specified in the Environment Clearances letters and issue show cause notice for violations. It is also proposed to write to Odisha State Government to take credible action against these mine projects under the Environment (Protection) Act, 1986, as per the facts of the case". An inter-ministerial panel has recommended de-allocation of 29 coal blocks of companies such as Essar Power, Hindalco, Tata Power and Jindal Steel and Power Ltd, and has suggested "no action at present" for 30 other blocks. The Panel proposed de-allocation of Chakla and Ashok Karkatta coal blocks in Jhakhand allocated to Essar Power; Tubed coal block in Jharkhand jointly allocated to Hindalco industries and Tata Power and Gare Palma IV/6 in Chhatisgarh jointly allotted to firms like JSPL and Nalwa Songe Iron Ltd. The Shah Commission has observed that there are 96 leases which have obtained much-delayed environment clearances under EIA (Environment Impact Assessment) notification and in 75 leases, extraction of ore has been done after January- 27, 1994. "The 22,56,74,380 MT of iron ore and 30, 71,973 MT of manganese have been extracted illegally without having lawful authority," the Shah Commission observed in its report tabled in Parliament.[18] Thus, Illegal mining and corruption practices in mining at different levels are widely prevalent across the states. Hence, utmost care, meaningful measures and exemplary actions should be taken to prevent illegal mining and corruption. Both mining companies and the Government have, therefore, to bear the responsibility of adopting Ethical Practices in mining operations.

Government, for the interest and benefit of persons and areas affected by mining should adopt a 'Mines and Minerals Development Fund'. This may be in the form of a Rolling or revolving Fund/Corpus Fund/ community Investment Fund which can be utilised for the purpose of community development. Government receives Taxes and Royalties from mining which constitute a major source of its revenue. A portion or a fixed percentage of this huge amount should be earmarked for development of mining areas over and above the present CSR Fund. Mines and Minerals Development Fund so constituted should be utilised for the developmental and welfare activities in directly affected areas where mining operations like excavation, blasting, mining, beneficiation, waste disposal etc. are carried on as well as indirectly affected areas where local population are affected due to economic, social and negative environmental consequences such as deterioration of water, soil and air quality, reduction in flows of streams, depletion of ground water, noise pollution, congestion etc. and also for displaced/affected families who have either legal and occupational rights over the land being mined or those with usufruct and traditional rights. The welfare activities and developmental projects which are to be taken up in these areas should be apart from the normal government schemes/ programmes/projects. Besides, special provisions should be made for development works for the affected indigenous people, scheduled tribes, traditional forest dwellers and for scheduled areas. The implementation process to be adopted in these cases should be different from those of normal areas.

In India, central government framed a scheme, namely, 'Pradhan Mantri Khanij Kshetra Kalyan Yojana' to be implemented by the District Mineral Foundations (DMF) of the respective mineral districts of the country. The funds comprise of payment to the trust made by the lessees of major and minor mineral leases at the rates specified by the government. But, DMF should not always be utilised in the development of infrastructure. For example, qualified and trained teachers are more important than constructing more school buildings. Likewise, service delivery and quality improvement is equally or more important than only building infrastructure. Food security and Social welfare measures, providing employment, developing livelihood opportunities with available local resources for vulnerable groups are also important for holistic development of districts. Comprehensive District Plans and or long term perspective plans on DMF should be prepared by the district authorities taking in to account the whole gamut of issues of the district and utilise the DMF accordingly.

Innovations and technologies are key factors for making mining more inclusive and sustainable. Innovations and technologies can be applied in the fields of biodiversity protection, resource saving, robust risk management, clean production, decontaminations and many other fields as per local conditions. Preserving environments in ice-covered areas and ocean floor and achieving clean production are important areas in the development of new mining activity. Besides trading off with other economic sectors, major attention in mining should be on productivity and cost savings, new economic uses, limit health risks, employment and local infrastructure, investment in long term development, reducing energy use and increasing the use of renewable resources.

It is imperative, in order to be Inclusive, Sustainable, Responsible and Ethical, that the mining companies and corporations should establish robust monitoring mechanism, reporting standards, reclamation procedures, put in place a system of concurrent evaluation on socio-economic and environmental performance and protection measures, ensure sustainable land use following mine closure and above all, strictly, invariably control illegal, unethical mining and corrupt practices. They should adhere to the principles of Triple Bottom Line, adopt responsible Practices, make constant endeavours to mitigate and eliminate environmental damage; protect public health and safety; restore the land to a productive condition or to its original condition after closure and provide social/economic benefits for the community. They must find out an inclusive, comprehensive and sustainable solution to air, water, land pollutions, mitigate socio-economic adverse impacts, preserve and maintain environment and ecology. The targeted actions in this direction will go a long way in mitigating the adversities and negative impacts of mining operations and related activities and prove effective in ensuring Inclusiveness, Sustainability, Responsibilities and Ethics in Mining.

Conclusion

All mining irrespective its size, scale of operation, type, location, area, profit, other economic ramifications, surrounding social structures and ecological settings, must consistently and invariably adopt and apply inclusive approach and sustainability principles, responsible and ethical practices encompassing the full array of activities for mitigating the outstanding issues, and deleterious impacts in toto. The objectives, strategy and plan of actions should be predefined preferably from the initial stages of operation when there is scope for avoiding cost and time overrun and initiation of mid-course corrections. The remediation strategy should be a combination of modern technologies, innovative and proven practices. For effective implementation and mitigation of adverse impacts, a standard operating procedure should be devised by the company itself and entrusted to an efficient and committed team of professionals. Some actionable performances can also suitably be delegated to reputed, high ranking NGOs and civil societies. All in all, inclusiveness and sustainability should form the core and essence of mining and responsible and ethical practices should be embedded in to all stages of mining operation right from the policy planning to closure.

End Notes & References

1. Fernandes and Raj have discussed about domestic violence of displaced families; 1992.
2. (i) Gardiner, E., Gelfand, J., TSM in Action: The Mining Association of Canada's *"Towards Sustainable Mining"* Initiative. Paper presented at the Securing the Future: Mining, Metals & the Environment in a Sustainable Society/8th ICARD - International Conference on Acid Rock Drainage, 2009.

   (ii) MAC, Towards Sustainable Mining: Performance Indicators, 2010.
3. Gratton, P., *"Mining, Sustainable Development and First Nations, Our New Frontier"* [speech was given to the Canadian Institute of Mining, Metallurgy and Petroleum (CIM) - North Central Branch, Prince George, British Columbia on June 26, 2008 by Pierre Gratton, President and CEO of the Mining Association of British Columbia].
4. The London Declaration was made in 18- 23rd May, 2001.
5. https://www.globalreporting.org/ resourcelibrary/GRI-G4-Mining-and-Metals-Sector-Disclosures. *"GRI. Mining and Metals Sector Disclosures"*.
6. https://aluminium-stewardship.org/asi-standards/path-asistandardslaunch /#C2017, *"ASI. Performance Standard"*.
7. http://www.responsiblemining.net/images/uploads/IRMA_Standard_Draft_ v2.0_with_MOV. *"IRMA Standard for Responsible Mining"*.
8. http://www.cyanidecode.org/sites/default/files/pdf/18_ Cyanide Code12-2016. *"ICMI. Cyanide Management Code"*.
9. http://www.responsiblejewellery.com/files/RJC_Code_of_Practices_ 2013_eng. "RJC. Code of Practices",2013.
10. (i) DDI. Maendeleo Diamond Standards.

(ii) Fair Stone. International Standard for the Natural Stone Industry. http://fairstone.org/wp-content/uploads/sites/2/2016/06/160520_Fair-Stone-Standard_6th_Edition_MQ.

11. Katarzyna Pactwa, Justyna Woźniak and Andrzej Strempski, *"Sustainable mining – Challenge of Polish mines - The way ahead"*.
12. The Times of India, "Unchecked Mining Plays havoc with ore Reserves; Experts fear shortage when 49 Plants become operational" dated 03.02.2009.
13. The Times of India "Our Mineral Reserves to last only 25 years", dated 06.03.2013.
14. The Times of India 'Govt. does flak on Mining matters', Tuesday, November, 28, 2006.
15. The Times of India, "Govt. fails to extract fine from Miners", dated January 3, 2014.
16. The Times of India, "Vigilance files charges against 23", dated January 3, 2014.
17. The Times of India "Union Govt. sends Shah Panel Report to committee of Secretaries" dated January 3, 2014.
18. The Times of India- "Notices likely to 75 farms for illegal Mining in Odisha", dated 17.02.2014.

Bibliography

Agarwal, Anil & Narain, Sunita, Sharma, Anju and Imchen, Achila (2001). *Poles Apart Global Environment Negotiations,* New Delhi: Centre for Science and Environment.

Agarwal, S.K. (1996). *Environmental Pollution from Mining Industry in Industrial Environment Assessment Strategy*, New Delhi: APH Publishing Corporation.

Areeparampil, Mathew. (1996). "Displacement Due to Mining in Jharkhand," *Economic and Political Weekly,* 31(9):1524-1528.

Behura, N. K. (1990). "Socio-Economic Problems and Social Change," In Sudarshan, V. Kalyan M A. (eds), *The Uprooted*, New Delhi: Gyan Publishing House.

Behura, N. K., and P. K. Nayak. (1993). "Involuntary Displacement and the Changing Frontiers of Kinship: A Study of Resettlement of Orissa, ' in M. M. Cernea and S. E. Guggenheim (eds.), *Anthropological Approaches to Resettlement: Policy, Practice, Theory.* Boulder, Colo.: Westview Press.

Bharti, Indu. (1991). "Bihar's Dams, Tribals' Woes", *Economic and Political Weekly*, June 1-8.

Boon, J.A. (2015). *Corporate Social Responsibility, Relationships and the Course of Events in Mineral Exploration - an Exploratory Study*. Ph D. Thesis in Sociology, Carleton University, Ottawa, Canada.

Canadian International Resources and Development Institute (2017) *Transforming Artisanal and Small Scale Mining.*

Christmann, P., Arvanitidis, N., Gilles Recoché, L.M. and Sclar, S. (2007). *Towards the Sustainable Use of Mineral Resources: A European Geological Surveys Perspective.* Minerals & Energy - Raw Materials Report, Volume 22, 2007 - Issue 3-4: Sustainable Resource Management, No. 2.

C. Michael. (1995), "Social Integration and Population Displacement: The Contribution of Social Sciences," *International Social Science Journal,* XLIII, 1.

C. Michael. (1996), "Bridging the Research Divide: Studying Development Oustees," in Tim Allen (ed.), *In Search of Cool Ground: War, Flight and Homecoming in Northeast Africa,* London: United National Research Institute for Social Development, Africa World Press.

Carnea, Michacl. (1996). "Public Policy Responses to Development induced Population Displacements," *Economic and Political Weekly*, Vol. XXXI, No. 24, June, 15.

Carnea Michael M. (2003). "For a New Economics of Resettlement: A Sociological Critique of the Compensation Principle," *International Social Science Journal.*

Carnea, 1997. "The Risks and Reconstruction Model for Resettling Displaced Populations". *World Development*, 25(10).

Carnea, M. (2000). "Risks, Safeguards and Reconstruction: A Model for Population Displacement and Resettlement," *Economic and Political Weekly*, October, 7.

Carnea, M.M. (ed.) (1994). *The Economics of Involuntary Resettlement: Questions and Challenges,* Washington D.C.: World Bank.

Carnea, Michael M. (1996). "Eight Main Risks: Impoverishment and Social Justice," in *Resettlement,* Washington DC: The World Bank.

Carnea, Michael M. (1998). "Impoverishment or Social Justice?" in H. M. Mathur and D. Marsden, (eds.) *Development Projects and Impoverishment Risks,* New Delhi: Oxford University Press.

Carnea, Michael M. (2000). "Impoverishment Risks and Reconstruction: A Model for Population Displacement and Resettlement," in M.M. Carnea and C. McDowell (eds.) *Risks and Reconstruction: Experiences of Resettlers and Refugees.* Washington, DC: The World Bank.

Carnea, Michael M.(1998). "Involuntary Resettlement in Development Project: Policy Guidelines," in *World Bank Financed Projects*, Washington, D.C.: World Bank.

Cernea, M.M.(1995). "Understanding and Preventing Impoverishment from Displacement: Reflections on the State of Knowledge", *Journal of Refugee Studies*, Vol. 8, Issue 3, pp. 245-264.

Cernea, Michael M. (1988). "Involuntary Resettlement and Development Projects", *Finance and Development,* Vol. 8 No. 3, September 1988.

Colson, E. (1971). *The Social Consequences of Resettlement,* Manchester: Manchester University Press.

Cook, C. C. and Aleki Mukendi. (1993). "Involuntary Resettlement in Bank-Financed Projects Lessons from Experience in Sub-Saharan Africa," In C. Cook (ed.) *Involuntary Resettlement in Africa.* Washington DC: The World Bank.

Council for professional Social Workers (CPSW). (1994). *State of Odisha's Environment*, Bhubaneswar: CPSW.

Dalua, A. K. (1993). *Development Projects and induced Displacement*, Cuttack: Mass Media Ltd.

Dalwai, Ashok. (1998). "Integrating Ongoing Development Programmes of the Government," in *Development Induced Displaced and Rehabilitation* New Delhi: Prachi Prakashan.

Das, C.R. (2003). "Land Acquisition for Development Projects and Rehabilitation of Displaced Persons: A Case Study of NALCO, Odisha," in S.N. Tripathy, (ed.) *Economic Development and Problem of Displacement*, New Delhi: Anmol Publications, pp. 104-105.

Das, J.C., Walter Fernandes and Sam Rao. (1988). "The Extent and Prospects of Displacement," *Social Action*, 38 (3).

Das, J.K. (2001). *Human Rights and Indigenous Peoples*, New Delhi; A.P.H. Publishing House.

Das, V. (1995). "Development or Destruction: New Mining Projects in Orissa," *Economic and Political Weekly*, XXX (22).

Das, Veena (1996). "Dislocation and Rehabilitation Defining Field," *Economic and Political Weekly*, Vol. XXXI, No. 24, June, 15.

Dash. L.N. (2007). "Economics of Mining in Odisha," *Odisha Review,* Nov., pp. 69-76.

Deogharia, Prakash Chandra, (2012). *Development, Displacement and Deprivation*, New Delhi: Shree Publication.

Di Capua, G., Peppoloni, S., and Bobrowsky, P. (2017). *The Cape Town Statement on Geoethics. Annals of Geophysics*, 60, Fast Track 7.

European Commission (2011). *A renewed EU strategy 2011-14 for corporate social responsibility*. (Communication from the Commission to the European Parliament, the Council, the European Economic and Social Committee and the Committee of the Regions No. COM (2011) 681 final). Brussels, European Commission.

Elizabeth, Colson (1971). *The Social Consequences of Resettlement*. Manchester: Manchester University Press.

Femandes, Walter, and Mohammed Asif (1997). *Development-Induced Displacement and Rehabilitation in Orissa, 1951 to 1995: A Database on Its Extent and Nature.* New Delhi: Indian Social Institute.

Femandes, Walter and S. Anthony Raj (1992). *Development, Displacement, and Rehabilitation in the Tribal Areas of Orissa*. New Delhi: Indian Social Institute.

Fernandes, Dr. Walter and Mohmd. Asif. 'Basachyutanka Durdasha', *Gaunli Bichar,* Bhubaneswar: Indian Social Foundation.

Fernandes, Walter (2004). "Rehabilitation Policy for the Displaced", *Economic and Political Weekly*, 20 March, pp.1191-1192

Fernandes, Walter (1991). "Power and Powerlessness: Development Projects and Displacement of Tribals" *Social Action,* 41(3).

Fernandes, Walter (1996). "Land Reforms Ownership Pattern and Alienation of Tribal Livelihood," *Social Action*, 46 (4), Oct-Dec.

Fernandes, Walter (2008). "Sixty Years of Development- Induced Displacement in India: Impacts and the Search for Alternatives," in Hari Mohan Mathur (ed). *India: Social Development Report 2008: Development and Displacement*, New Delhi: Council for Social Development and Oxford University Press.

Fernandes, Walter and Enakshi Ganguly Thukral, (eds.) (1989). *Development, Displacement, and Rehabilitation.* New Delhi: Indian Social Institute.

Fernandes, Walter and S. Anthony Raj (1992). *Development, Displacement and Rehabilitation in the Tribal Areas of Odisha*. New Delhi: Indian Social Institute.

Fernandes, Walter (2006). Developmental Induced Displacement: Impact on Tribals," *Adivasi*, Vol. 46, Dec.

G.C. Kar (1999). "Displacement, Resettlement and Rehabilitation," Odisha *Economic Journal*, Vol. XXXI.

Geisler, Charles (2002). "Endangered Humans: How Global Land Conservation Efforts are Creating a Growing Class of Invisible Refugees", *Foreign Policy*, May-June.

GHoga D'Souza, Nafisa (2002). "Development, Displacement and marginalised Women," in Walter Fernandes and Sanjay Barbora (ed.) *Changing Women's Status in India: Focus on the Northeast,* Guwahati: North Eastern Social Research Centre.

Goodland, Robert (2007). 'India: Odisha, Kashipur, Utkal Bauxite and Alumina Project: Human Rights and Environmental Impacts' March 19.

Government of India, 'Draft national Policy for Rehabilitation of Persons, Displaced as a Consequence of Acquisition of Land', Ministry of Rural Development.

Government of India (2003). National Policy of Resettlement and Rehabilitation for Project Affected Families, New Delhi: Ministry of Rural Development.

Government of Odisha (1991). *Economic Survey 1990-91,* Bhubaneswar: Directorate of Economics and Statistics.

Government of Odisha. *Report on the Activities of the Revenue and Disaster Management Department,* 2013-14.

Government of Odisha (1989). Uniform Guidelines for Rehabilitation of the Displaced Persons/ families due to SECL (Now MCL) Projects of Sambalpur and Dhenkanal Districts, Bhubaneswar: Revenue and Excise Department.

Government of Odisha (1994). Odisha Resettlement and Rehabilitation of Project Affected People's Policy, Bhubaneswar: Water Resources Department, Government of Odisha.

Government of Odisha (2012). *Economy Survey 2011-12*, Government of Odisha: P & C Department.

Government of Odisha (2019). *Economy Survey 2018-19*, Government of Odisha: P & C Department

Government of Odisha (2020). *Economy Survey 2019-20*, Government of Odisha: P & C Department

Government of Odisha (2021). *Economy Survey 2020-21*, Government of Odisha: P & C Department

Government of Odisha (2022). *Economy Survey 2021-22*, Government of Odisha: P & C Department

Govt of India (1996). 'Modified Draft National Policy for Resettlement and Rehabilitation of Persons Affected by Reservoir Project,' New Delhi: Ministry of Water Resources, Central Water Commission.

Govt. of Odisha (2013). *Annual Plan 2012-13, Odisha, Volume-I,* Bhubaneswar: Govt. of Odisha. Planning & Convergence Department.

Govt. of Odisha (2015). *Annual Activity Report, Forest and Environment Department, Govt. of Odisha, 2015-16*, Bhubaneswar.

Govt. of Orissa (2015). *Statistical Abstract of Odisha*, Bhubaneswar: Directorate of Economics & Statistics, Govt. of Orissa.

Goyal, Sangeeta (1996). "Economic Perspectives of Resettlement and Rehabilitation," *Economic and Political Weekly*, Vol. XXXI, No. 24, June 15.

Haimendorf, Christoph von Furer (1982). *Tribes of India: The Struggle for Survival*, Delhi: Oxford University Press.

International Association for Promoting Geo-ethics (2017).

International Council of Mining and Metals (ICMM)-2012. *Community Development Kit.*

International Council of Mining & Metals - ICMM (2017).

Initiative for Responsible Mining Assurance (2017).

International Finance Corporation - IFC (2012). *Performance Standards on Social and Environmental Sustainability.*

International Labour Organisation - ILO (2012). Convention 169.

ISO (2010). ISO 26000 - social responsibility.

Kemp, D. and Owen, J.R. (2013). *Community relations and mining: Core to business but not "core business*". Resources Policy, 38, 523-531.

Kemp, D., Owen, J.R., and van de Graaff, S. (2012). *Corporate social responsibility, mining and "audit culture"*. Journal of Cleaner Production, 24, 1-10.

Kreuzer, O.P. and Etheridge, M.A. (2010). *Risk and uncertainty in mineral exploration: Implications for valuing mineral exploration properties*. AIG News, (100), 20-28.

Kothari, Smitu (1996). "Whose Nation? The Displacement as Victims of Development," *Economic and Political Weekly*, Vol. XXXI, No. 24, June-15.

Li, F. (2015). *Unearthing Conflict – Corporate Mining, Activism and Expertise in Peru.* Durham and London: Duke University Press.

Macintyre, M., Mee, W., and Solomon, F. (2008). *Evaluating social performance in the context of an 'audit culture': A pilot social review of a gold mine in Papua New Guinea.* Corporate Social Responsibility and Environmental Management, 15(2), 100-110.

Mathur, H. M. (2006). "Development Projects and Displacement of Tribal Peoples," *Adivasi,* Vol. 46, No. 2, Dec.

Mathur, H.M. and David Marsden (1999). "Development Projects and Impoverishment Risks Resettling Project–Affected People in India," Oxford: Oxford University Press.

Mathur, Hari Mohan, and David Marsden (eds.) (1998). *Development Projects and Impoverishment Risks*. New Delhi: Oxford University Press.

Mathur, Hari Mohan (1997). *Managing Projects that involve Resettlement: Case Studies from Rajasthan, India.* Washington DC: The World Bank.

Mathur, Harimohan (1998). "Development Displacement and Resettlement: Focus on Asian Experiences," in *Development Induced Displacement and Rehabilitation,* New Delhi: Prachi Prakashan.

Mishra, Banikantha (2010). "Agriculture, Industry and Mining in Odisha in the Post-Liberalisation Era: An Inter-District and Inter-State Panel Analysis," *Economic & Political Weekly*, May 15, Vol. xlv, No. 20.

Mohapatra, L.K. (1991). "Development of whom? Depriving the Dispossessed Tribals", *Social Action,* (4) 1, July-Sept.

Mohapatra, L.K. (1999). *Resettlement Impoverishment and Reconstruction in India: Development for the Deprived*, New Delhi: Vikash Publishing House.

N.K. Behura and P.K. Nayak (1993). "Involuntary Displacement and the Changing Frontiers of Kinship 'A Study of Settlement in Odisha'" in Michael M Carnea and Scot E. Guggenheim (ed.), *Anthropological Approaches to Resettlement.*

N.K. Behura (1996). "Environment, Displacement and Development: Case Study from Odisha," *Journal of Indian Anthropology Society*, Vol. 31.

Nag, Sajal (2001). "Nationhood and Displacement in Indian Subcontinent," *Economic and Political Weakly*, December 22.

Nickless, E. (2017). *Delivering Sustainable Development Goals: The Need for a New International Resource Governance Framework.* Annals of Geophysics, 60, Fast Track 7.

Nurmi, P.A. (2017). *Green Mining - A Holistic Concept for Sustainable and Acceptable Mineral Production.* Annals of Geophysics, 60, Fast Track 7.

OECD (2017). *Guidelines for Multinational Enterprises – MNEs*

Owen, J.R. and Kemp, D. (2013). *Social licence and mining: A critical perspective.* Resources Policy, 38, 29-35.

Oxfam, "*Women, Communities and Mining: The gender impacts of mining and the role of gender impact assessment*", 2009.

O.O. Ibeanu (1999). "Exiles in their Own Home: Conflicts and Internal Population Displacement in Nigeria", *Journal of Refugee Studies*, Vol. 12, No. 2.

Panda, P. and Panigrahi, N. (1989). "The Problem of Displaced People: A Study in the Coal Mines of Brajrajnagar, Sambalpur Odisha," in R.N. Pati and B. Jena (ed.) *Tribal Development in India*, New Delhi: Ashis Publishing House.

Pandey, Balaji, et al. (1996). *Development, Displacement and Rehabilitation in Orissa, 1950-1990.* Bhubaneswar: ISED.

Pandey, Balaji, (1998). "Impoverishing Effects of Coal Mining Projects: A Case Study of Five Villages in Odisha", in H.M. Mathur and Marsden (ed.) *Development Projects and impoverishment Risks*, New Delhi: Oxford.

Pandey, Balaji (1998). *Depriving the Underprivileged for Development*, Bhubaneswar: Institute for Socio-economic Development.

Panigrahi, Nilakantha (2003). "Displacement and Deprivation of Tribal People in Odisha", in S.N. Tripathy (ed.) *Economic Development and Problem of Displacement*, New Delhi: Anmol Publications.

Parasuraman, S. (1994). *Resettlement and Rehabilitation People Displaced by Development Projects: The Evaluation of Indian Experience*, Mumbai: Tata Institute of Social Sciences.

Parasuraman, S. (1996). "Development Projects, Displacement and Outcomes for Displaced," *Economic and Political Weekly,* Vol. XXXI, No. 24, June, 15.

Patnaik, S. M. (1996). *Displacement, Rehabilitation, and Social Change: The Case of Paraja Highlanders*, New Delhi: Inter-India Publications.

Proceedings of the Saintala High Level Meeting For the Rehabilitation of the Displaced Persons.

PUCL (2006). Police Firing at Kalinganagar, A Report by People's Union for Civil Liberty, Bhubaneswar.

PUCL (2006). Maikanch Gulikanda O' Mishra Commission Report: Eka Samikshya, (in Odia) Bhubaneswar.

Porter, M.E. (2011). *Creating shared value: Redefining capitalism and the role of the corporation in society.*

Responsible Mining Index (2017).

Ruggie, J. (2008). *Promotion and protections of all human rights, civil, political, economical, social and cultural rights, including the right to development. Protect, respect, and remedy: A framework for business and human rights.*

Report of the Special Representative of the Secretary-General on the issue of human rights and transnational corporations and other business enterprises, John Ruggie. (No. A/HRC/8/5). New York: United Nations.

Ruggie, J. (2011). Report of the Special Representative of the Secretary-General on the issue of human rights and transnational corporations and other business enterprises, John Ruggie - *Guiding principles on business and human rights: Implementing the United Nations "Protect, respect and remedy" framework.*

Ramanathan, Usha (2008). "The Land Acquisition Act 1894: Displacement and State Power," in Hari Mohan Mathur, (ed) *India: Social Development Report 2008: Development and Displacement*, New Delhi: Council for Social Development and Oxford University Press.

Ramesh, K.S. (1998). "Resettlement and Rehabilitation of Families Displaced By Development Projects: Corporate Social Responsibility," *Journal of Management,* 28(1).

Ravindran, Latha (2007). "Economic Impact of Development Projects on the Displaced and Project Affected Families: An Empirical Study on Industrial Projects in Odisha," *Vilakshan*: XI MB *Journal of Management,* March.

Ravindran, Latha 1996. "Resettlement and Rehabilitation: From Policy Guidelines to Practice-A Case Study". in *Theme Papers of the First Training Seminar on Resettlement and Rehabilitation.* Bhubaneswar: Gopabandhu Academy of Administration.

Reddy, IUB (1993). *Displacement and Rehabilitation,* New Delhi: Mittal Publications.

Reddy, IUB (1993). *Impact of Projects and Social Transformation in Displacement and Rehabilitation*, New Delhi: Mittal Publications.

Report of the Scheduled Areas and Scheduled Tribe Commission VI, 1960-61.

S.M. Patnaik (1996). *Displacement, Rehabilitation and Social Change,* New Delhi: Inter-India Publication.

S.M. Patnaik (2000). "Understanding Involuntary Resettlement" An Anthropological Perspective," *The Eastern Anthropologist,* Vol. 53, No. 1-2, Jan-June.

Sahoo, Sarbeswar (2005). "Tribal Displacement and Human Rights Violations in Odisha", *Social Action,* April-June, Vol.55, No. 2.

Sanjay Khatua and William Stanley (2006). "Integrated Rural Development of Weaker Sections in India, in Ecological Debt: A Case Study from Odisha, India, in Athena K. Peralta, (ed.) *Ecological Debt*, The Ecumenical Centre.

Satya Narayan, G. (1999). *Development, Displacement and Rehabilitation*, Jaipur: Rawat Publications.

Sethi, B.P. (2009). "Indigenous People, Displacement and Development," Punarbas, Vol.1, March, Bhubaneswar.

Sharma, R.N. 2003. "Involuntary Displacement: A Few Encounters", *Economic and Political Weekly*, March 1.

Sagebien, J. and Lindsay, N. (2011). *Systemic causes, systemic solutions - CSR in a social and environmental value governance ecosystems context.* In: Sagebien J. and Lindsay N.M. (Eds.), Governance ecosystems - CSR in the Latin American mining sector (pp. 12-30). Houndmills, Basingstoke, Hampshire, U.K.: Palgrave Macmillan.

Silver, H. (2015). *The contexts of social inclusion.* DESA Working Paper No. 44. UN Department of Economics and Social Affairs.

Serrano, E., Ruiz-Flaño, P. (2007). *Geodiversity. A theoretical and applied concept*, Geographica Helvetica Jg. 62(3), 140-147.

Swaziland National Trust (2017). Iron Mines.

The Asia Foundation (2007). *Responsible Mining Definition.*

The Mining Association of Canada (2014). *Towards sustainable mining.*

Thomson, G. (2016). *Management of Socio-political Risk Arising from Corporate Transitions: the Mt. Milligan Experience*, Master of Applied Arts Thesis, Mining Engineering, University of British Columbia.

Taru (1996). Economic and social Impacts of Bauxite Mining and Alumina Processing in Kashipur and Kalahandi, Orissa (mimeo), Taru, New Delhi.

United Nations (1987). *Our Common Future* - Brundtland Report, Oxford, Oxford University Press.

United Nations (2007). *United Nations declaration of the rights of indigenous peoples.*

United Nations (2011). *Guiding Principles on Business and Human Rights (GPs).*

United Nations (2016). *Development Program, Mapping Mining SDGs – An Atlas.*

United Nations (2017a). *Global Compact.*

United Nations (2017b). *Sustainable Development Goals – SDGs.*

Voluntary Principles on Security and Human Rights (2017).

Viegas, Philip (1991). *Encroached and Enslaved: Alienation of Tribal Lands and its Dynamics*, New Delhi: Indian Social Institute.

Vom Hau, Matthias. *State Capacity and Inclusive Development: New Challenges and Directions* (March 5, 2012). Effective States and Inclusive Development Research Centre Working Paper 2.

World Economic Forum (2016). White Paper: *Voluntary Responsible Mining Initiatives.*

Walker, J., Howard, S., 2002. *Finding the Way Forward: How Could Voluntary Action Move Mining Towards Sustainable Development?* Environmental Resources Management (ERM) and International Institute for Environment and Development (IIED), London.

Webb, K., 1999. *Voluntary initiatives as an alternative to regulation.* In: Gibson, R.B. (Ed.), Voluntary Initiatives and the New Politics of Corporate Greening. Broadview Press, Peterborough, ON, pp. 32-50.

Worrall, R., Neil, D., Brereton, D., Mulligan, D., 2009. *Towards a sustainability criteria and indicators framework for legacy mine land*. Journal of Cleaner Production 17 (16), 1426-1434.

Whitmore, A., 2006. *The emperors new clothes: sustainable mining?* Journal of Cleaner Production 14 (3-4), 309-314. WMI, 1994.

World Bank blogs. *Better Health in Mines and Mining Communities: A Shared Responsibility*, 2015.

World Bank's report. "Poverty and Shared Prosperity 2016: Taking on Inequality", 2016.

World Bank. 'Atlas of sustainable development goals 2017'. World Development Indicators, doi:10.1596/978-1-4648-10. (2017).

World Bank Operational Policy, 4.12, approved 23 October 2001.

World Health Organisation, WHO factsheet.

WBCSD, *"Water and Sustainable Development Program. Water: Facts and trends".*

World Mining Congress, *"Production Energy Optimisation in Mining"*, 2013.

Young, J., Septoff, A., 2002. *Digging for Change: Towards a Responsible Minerals Future, an NGO and Community Perspective*. Mineral Policy Centre.

Young, S.T., 2005. *Leading environmental change: the case of the global mining industry.* Review of Business 26 (1), 34-38.

Zaman, MQ (1990). "Land Acquisition and Compensation in Involuntary Resettlement", *Cultural Survival Quarterly,* Vol. 14, No. 4.

Zoomers, A. (2010). Globalisation and the foreignisation of space: Seven processes driving the current global land grab. *The Journal of Peasant Studies, 37*(2), pp. 429-447.

Zhai, T. T., & Chang, Y. C. "Standing of environmental public-interest litigants in China: Evolution, obstacles and solutions". Journal of Environmental Law, 30, 369–397. doi:10.1093/jel/eqy011, 2019.

Index

F

J

K

S